The Humongous Book of Geometry Problems

Translated for People who don't Speak Math!!

by W. Michael Kelley

ALPHA

A member of Penguin Group (USA) Inc.

ALPHA BOOKS

Published by the Penguin Group

Penguin Group (USA) Inc., 375 Hudson Street, New York, New York 10014, USA

Penguin Group (Canada), 90 Eglinton Avenue East, Suite 700, Toronto, Ontario M4P 2Y3, Canada (a division of Pearson Penguin Canada Inc.)

Penguin Books Ltd., 80 Strand, London WC2R 0RL, England

Penguin Ireland, 25 St. Stephen's Green, Dublin 2, Ireland (a division of Penguin Books Ltd.)

Penguin Group (Australia), 250 Camberwell Road, Camberwell, Victoria 3124, Australia (a division of Pearson Australia Group Pty. Ltd.)

Penguin Books India Pvt. Ltd., 11 Community Centre, Panchsheel Park, New Delhi—110 017, India

Penguin Group (NZ), 67 Apollo Drive, Rosedale, North Shore, Auckland 1311, New Zealand (a division of Pearson New Zealand Ltd.)

Penguin Books (South Africa) (Pty.) Ltd., 24 Sturdee Avenue, Rosebank, Johannesburg 2196, South Africa

Penguin Books Ltd., Registered Offices: 80 Strand, London WC2R 0RL, England

International Standard Book Number: 978-1-59257-864-1
Library of Congress Catalog Card Number: 2008943864

14   8   7

Interpretation of the printing code: The rightmost number of the first series of numbers is the year of the book's printing; the rightmost number of the second series of numbers is the number of the book's printing. For example, a printing code of 09-1 shows that the first printing occurred in 2009.

Printed in the United States of America

Note: This publication contains the opinions and ideas of its author. It is intended to provide helpful and informative material on the subject matter covered. It is sold with the understanding that the author and publisher are not engaged in rendering professional services in the book. If the reader requires personal assistance or advice, a competent professional should be consulted.

The author and publisher specifically disclaim any responsibility for any liability, loss, or risk, personal or otherwise, which is incurred as a consequence, directly or indirectly, of the use and application of any of the contents of this book.

Most Alpha books are available at special quantity discounts for bulk purchases for sales promotions, premiums, fund-raising, or educational use. Special books, or book excerpts, can also be created to fit specific needs.

For details, write: Special Markets, Alpha Books, 375 Hudson Street, New York, NY 10014.

Contents

Introduction

Are you in a geometry class? Yes? Then you NEED this book. Here's why:

- Fact #1: The best way to learn geometry is by working out geometry problems (and doing proofs).

  There's no denying it. If you could figure this class out just by reading the textbook or taking good notes in class, everybody would pass with flying colors. Unfortunately, the harsh truth is that you have to buckle down and work problems out until your fingers are numb.

- Fact #2: Most textbooks only tell you WHAT the answers to their practice problems are but not HOW to do them!

  Sure, your textbook may have 175 problems for every topic, but most of them only give you the answers. That means if you don't get the answer right you're totally screwed! Knowing you're wrong is no help at all if you don't know WHY you're wrong. Math textbooks sit on a huge throne, like the Great and Terrible Oz and say, "Nope, try again," and we do. Over and over. And we keep getting the problem wrong. What a delightful way to learn! (Let's not even get into why they only tell you the answers to the odd problems. Does that mean the book's actual author didn't even feel like working out the even ones?)

- Fact #3: Even when math books try to show you the steps for a problem, they do a lousy job.

  Math people love to skip steps. You'll be following along fine with an explanation and then all of a sudden BAM, you're lost. You'll think to yourself, "How did they do that?" or "Where the heck did that 42 come from? It wasn't there in the last step!" Why do almost all of these books assume that in order to work out a problem on page 200, you'd better know pages 1 through 199 like the back of your hand? You don't want to spend the rest of your life on homework! You just want to know why you keep getting a negative number when you're calculating volume of a right rectangular prism with an oblique cone inside that's full of angry trapezoids. (On second thought, maybe that problem was part of a bad dream.)

- Fact #4: Reading lists of facts is fun for a while, but then it gets old. Let's cut to the chase.

All of my notes are off to the side like this and point to the parts of the book I'm trying to explain.

Just about every single kind of geometry problem or proof you could possibly run into is in here—after all, this book is HUMONGOUS! If a thousand problems aren't enough, then you've got some kind of crazy math hunger, my friend, and I'd seek professional help. This practice book was good at first, but to make it GREAT, I went through and worked out all the problems and took notes in the margins when I thought something was confusing or needed a little more explanation. I also drew little skulls next to the hardest problems, so you'd know not to freak out if they were too challenging. After all, if you're working on a problem and you're totally stumped, isn't it better to know that the problem is SUPPOSED to be hard? It's reassuring, at least for me.

I think you'll be pleasantly surprised by how detailed the answer explanations are, and I hope you'll find my little notes helpful along the way. Call me crazy, but I think that people who want to learn geometry and are willing to spend the time drilling their way through practice problems should actually be able to figure the problems out and learn as they go, but that's just my 2¢.

Good luck and make sure to come visit my website at www.calculus-help.com. If you feel so inclined, drop me an e-mail and give me your 2¢. (Not literally, though—real pennies clog up the Internet pipes.) —Mike Kelley

Dedication

For my extended family, who selflessly change their entire lives whenever I am working on a book, because without them, I'd never be able to even begin writing, let along chug along month after month. Thanks to my Mom, Ron, Wanda, and Fred. They clear the way for me, working harder, babysitting more often, and giving me the occasional (but fairly regular) distraction-free time I need when things get hectic.

A special thanks also to Rob Halstead, who helped me proofread this book, and to the staff of the Calvert Library/Prince Frederick, my office away from home when the home office full of four-year-olds wearing plastic high heels gets a little too noisy.

For my son Nick, the Professor, who makes me proud every day. May we still be battling Galactus and the Joker on some future incarnation of the Xbox when you're old enough to have better things to do.

For my daughters Erin and Sara, who cannot wait the ten seconds it takes me to walk up the stairs when I get home to start telling me absolutely everything that happened at school. You are first in my heart, and I hope I will always have a place on my ballerinas' dance cards.

Finally, for my wife Lisa, who makes me wish every day that I had more to offer her than high-maintenance, stressed-out, sickly me. You're the best there is, and not just because you know where all the warp whistles are in the Super Mario games.

Trademarks

Chapter 1

RATIONAL NUMBERS AND ALGEBRAIC EXPRESSIONS

Fractions and basic algebra rules

The first three chapters of this book are dedicated to the review of the core algebraic and arithmetic concepts necessary for the study of geometry. This chapter investigates rational expressions (fractions) and the fundamental rules governing the simplification and evaluation of mathematical expressions, both rational and nonrational.

Even though the geometry problems don't technically start until Chapter 4, don't skip over these first few chapters. You need some basic algebra skills before you get too far into geometry, so you might as well learn them up front.

This chapter begins with fractions, including creating them from decimals; changing proper fractions into improper fractions; reducing things to lowest terms; and combining fractions by adding, subtracting, multiplying, or dividing. It then shifts out of fractions and moves to working with algebraic expressions, including how to simplify them using the order of operations and how to evaluate them when you know what to plug into the variables.

Expressing Rational Numbers
Converting things into and simplifying fractions

1.1 Express 0.8 as a fraction in lowest terms.

One digit (8) appears right of the decimal point, so divide 8 by 10^1.

$$\frac{8}{10}$$

> If six digits were right of the decimal, you'd divide by 10^6 to create the fraction. Ignore any zeroes that appear at the far right side of the decimal.

Divide the numerator (8) and denominator (10) by 2 to express the fraction in lowest terms.

$$\frac{8 \div 2}{10 \div 2} = \frac{4}{5}$$

> A fraction is in lowest terms when no number (except 1) divides evenly into the numerator and denominator.

Therefore, $0.8 = \frac{4}{5}$.

1.2 Express 0.25 as a fraction in lowest terms.

Two nonzero digits appear to the right of the decimal, so to convert 0.25 into a fraction, divide 25 by $10^2 = 100$.

$$\frac{25}{100}$$

Divide the numerator and denominator by 25, the greatest common factor, to reduce the fraction to lowest terms.

> The greatest common factor is the biggest number that divides into 25 and 100 evenly (leaving no remainder).

$$\frac{25 \div 25}{100 \div 25} = \frac{1}{4}$$

Therefore, $0.25 = \frac{1}{4}$.

1.3 Express 1.659 as a percentage.

To convert a decimal into a percentage, move the decimal point two places to the right. This is the equivalent of multiplying 1.659 by 100.

$$1.659 = 165.9\%$$

1.4 Express 3.61% as a decimal.

To convert a percentage into a decimal, reverse the procedure described in Problem 1.3: move the decimal point two places to the left. Because 3.61% contains only one digit left of the decimal, you must insert a zero placeholder immediately preceding 3 to place the decimal point correctly.

> To check your answer, move the decimal back two places to the right to make sure you get what you started with: 0.0361 = 3.61%.

$$3.61\% = 0.0361$$

1.5   Express $\frac{2}{5}$ as a decimal.

To convert a fraction into a decimal, divide the numerator (2) by the denominator (5) using long division. Because the divisor is greater than the dividend (5 > 2), place a decimal point and a zero to the right of the dividend.

$$5\overline{)2.0}$$

For now, ignore the decimal in the dividend and proceed as though 2.0 were actually 20. Note that 20 is divisible by 5 a total of 4 times (20 ÷ 5 = 4). Place the quotient (4) above the rightmost digit of 20.

$$5\overline{)2.0}^{.4}$$

Multiply the numeral above the division bracket by the divisor (4 · 5 = 20) and write the *opposite* of the result (–20) below the divisor.

$$5\overline{)\,2.0}\\ -20$$

Calculate the difference.

$$5\overline{)\,2.0}\\ -20\\ \overline{0}$$

Because the difference is equal to zero, the division problem is complete.

$$\frac{2}{5}=0.4$$

1.6   Express $\frac{1}{9}$ as a decimal.

Apply the procedure described in Problem 1.5; use long division to calculate 1 ÷ 9.

$$9\overline{)1.0}$$

Note that 10 ÷ 9 = 1 remainder 1. Place the quotient (1) above the division bracket, multiply it by the divisor (1 · 9 = 9), and write the *opposite* of the result (–9) below the dividend. Calculate the difference.

$$9\overline{)\,1.0}^{.1}\\ -9\\ \overline{1}$$

When the number you're dividing by (in this case, 5) is bigger than the number you're dividing into (in this case, 2), the answer is less than 1. Write ".0" after the number you're dividing into and place a decimal above the division bracket as well.

You're still treating 2.0 like 20, so subtract: 20 – 20 = 0. Write a 0 below the horizontal line.

Pretend that 1.0 is actually 10. Because the remainder is not zero, you know you're not done dividing.

The difference (10 − 9 = 1) is less than the divisor, so place an additional 0 right of the dividend (1.0 becomes 1.00) and right of the quotient (1 becomes 10).

$$
\begin{array}{r}
.1 \\
9{\overline{\smash{\big)}\,1.00}} \\
\underline{-9} \\
10
\end{array}
$$

The bottommost number (10) is divisible by the divisor (9) one time, with a remainder of 1 (10 ÷ 9 = 1 remainder 1). As in the preceding step, the quotient (1) is placed above the division symbol and multiplied by 9, its opposite is placed at the bottom of the division problem, and the difference is calculated.

$$
\begin{array}{r}
.11 \\
9{\overline{\smash{\big)}\,1.00}} \\
\underline{-9} \\
10 \\
\underline{-9} \\
1
\end{array}
$$

Each time this process is repeated, the result is the digit 1 in the quotient. Therefore, $\dfrac{1}{9} = 0.11111111\ldots$. Use a horizontal bar to indicate a repeated digit or series of digits in a decimal: $\dfrac{1}{9} = 0.\overline{1}$.

Note that all decimals derived from fractions either repeat infinitely (as in this example) or terminate (as in Problem 1.5).

> Every time you place another zero at the end of the dividend (1.00 becomes 1.000, 1.0000, and so on) and next to the 1 at the bottom of the problem, you get 10 ÷ 9 = 1, the same thing over and over.

> In an improper fraction, the numerator is larger than the denominator.

1.7   Express $2\dfrac{1}{3}$ as an improper fraction.

To convert the mixed number $A\dfrac{B}{C}$ into an improper fraction, multiply the denominator C by the whole number A, add the numerator B, and then divide by the denominator C. In other words, $A\dfrac{B}{C} = \dfrac{CA + B}{C}$. In this example, $A = 2$, $B = 1$, and $C = 3$.

$$
\begin{aligned}
2\frac{1}{3} &= \frac{3 \cdot 2 + 1}{3} \\
&= \frac{6 + 1}{3} \\
&= \frac{7}{3}
\end{aligned}
$$

1.8 Express $5\dfrac{4}{7}$ as an improper fraction.

Apply the formula $A\dfrac{B}{C} = \dfrac{CA+B}{C}$ presented in Problem 1.7.

$$5\frac{4}{7} = \frac{7\cdot 5 + 4}{7}$$
$$= \frac{35+4}{7}$$
$$= \frac{39}{7}$$

1.9 Express the improper fraction $\dfrac{9}{2}$ as a mixed number.

To convert the improper fraction $\dfrac{A}{B}$ into a mixed number, calculate $A \div B$.

If this is an improper fraction, then A > B.

Let the resulting quotient be equal to Q, and let the remainder equal R.

The mixed-number form of the improper fraction is $Q\dfrac{R}{B}$, where B is the denominator of the original improper fraction.

To transform $\dfrac{9}{2}$ into a mixed number, calculate $9 \div 2$ to get 4 remainder 1.

The quotient (4) is the whole part of the mixed number, the remainder (1) is the numerator, and the denominator matches the denominator of the improper fraction (2).

$$\frac{9}{2} = 4\frac{1}{2}$$

1.10 Express $\dfrac{18}{7}$ as a mixed number.

Apply the technique described in Problem 1.9. Because $18 \div 7 = 2$ remainder 4, the whole part of the mixed number is 2 and the numerator is 4. Both fractions share the same denominator, 7.

$$\frac{18}{7} = 2\frac{4}{7}$$

1.11 Reduce the fraction $\dfrac{15}{40}$ to lowest terms.

The numerator and denominator share the common factor 5 (because 5 divides evenly into 15 and 40). Divide both values by 5 to reduce the fraction.

$$\frac{15 \div 5}{40 \div 5} = \frac{3}{8}$$

Because 3 and 8 share no common factors (other than 1), the fraction $\dfrac{3}{8}$ is expressed in lowest terms.

1.12 Reduce the fraction $\dfrac{24}{180}$ to lowest terms.

Note that 24 and 180 are even numbers, so they are each evenly divisible by 2 and thus share a common factor.

$$\frac{24 \div 2}{180 \div 2} = \frac{12}{90}$$

The numerator and denominator of the simplified fraction are once again even, and thus once again divisible by 2.

$$\frac{12 \div 2}{90 \div 2} = \frac{6}{45}$$

Notice that 6 and 45 are both evenly divisible by 3.

$$\frac{6 \div 3}{45 \div 3} = \frac{2}{15}$$

Because 2 and 15 share no common factors (other than 1), the fraction $\dfrac{2}{15}$ is expressed in lowest terms.

You can simplify the fraction all in one step if you divide by the greatest common factor of 24 and 180, which is 12. Actually, this problem divides by 2, then 2 again, and then 3, which is one way of calculating the greatest common factor:
$2 \cdot 2 \cdot 3 = 12$.

Combining Rational Numbers
Add, subtract, multiply, and divide fractions

1.13 Simplify the expression: $\dfrac{5}{6} + \dfrac{11}{12}$.

You cannot add or subtract rational numbers unless they share a common denominator; these fractions do not. Notice that the largest denominator (12) is evenly divisible by the other denominator in the expression (6); therefore, the least common denominator (LCD) is 12.

If the smaller denominator doesn't divide evenly into the larger denominator, don't panic. Check out Problems 1.15 and 1.16.

Multiply the numerator and denominator of the first fraction by 2 to ensure that the rational numbers have common denominators.

$$\frac{5}{6} + \frac{11}{12} = \frac{5 \cdot 2}{6 \cdot 2} + \frac{11}{12}$$
$$= \frac{10}{12} + \frac{11}{12}$$

To add the rational numbers, combine the numerators.

$$= \frac{10 + 11}{12}$$
$$= \frac{21}{12}$$

Reduce the fraction to lowest terms.

$$= \frac{21 \div 3}{12 \div 3}$$

$$= \frac{7}{4}$$

1.14 Simplify the expression: $4 - \frac{6}{5}$.

Although they contain no *explicit* denominators, integers are rational numbers with an *implicit* denominator of 1. In this case, $4 = \frac{4}{1}$.

$$4 - \frac{6}{5} = \frac{4}{1} - \frac{6}{5}$$

The largest denominator in the expression (5) is evenly divisible by the other denominator (1): $5 \div 1 = 5$. Multiply the numerator and denominator of $\frac{4}{1}$ by 5 to express the rational numbers in terms of a common denominator.

$$= \frac{4 \cdot 5}{1 \cdot 5} - \frac{6}{5}$$

$$= \frac{20}{5} - \frac{6}{5}$$

Combine the numerators.

$$= \frac{20 - 6}{5}$$

$$= \frac{14}{5}$$

> Integers are positive or negative numbers that aren't part of a fraction and have no decimal. They're basically whole numbers, like 4 in this problem.

1.15 Simplify the expression: $\frac{1}{8} - \frac{5}{6}$.

The larger denominator in this expression (8) is not evenly divisible by the smaller denominator (6), because $8 \div 6$ produces a remainder. Therefore, 8 is not the LCD. In an attempt to identify the LCD, multiply that value by 2: $8 \cdot 2 = 16$. However, because 16 is not evenly divisible by 6, it is not the least common denominator either.

keep multiplying the larger denominator by increasingly bigger numbers (2, 3, 4, 5, and so on) until the smaller denominator divides in evenly.

Multiplying 8 by 2 did not result in the LCD, so multiply 8 by 3: $8 \cdot 3 = 24$; the LCD is 24 because it is evenly divisible by the other denominator in the expression (6). Divide the LCD by each of the original denominators and multiply the fractions by the corresponding quotients.

$$\frac{1}{8} - \frac{5}{6} = \frac{1 \cdot 3}{8 \cdot 3} - \frac{5 \cdot 4}{6 \cdot 4}$$

$$= \frac{3}{24} - \frac{20}{24}$$

$$= \frac{3 - 20}{24}$$

$$= -\frac{17}{24}$$

Unlike Problems 1.13 and 1.14, BOTH fractions need to be multiplied by something so they have a common denominator. The first fraction gets multiplied by $24 \div 8 = 3$, and the second gets multiplied by $24 \div 6 = 4$.

1.16 Simplify the expression: $\dfrac{1}{3} + \dfrac{3}{10} - \dfrac{17}{15}$.

The largest denominator (15) is not the least common denominator because it is not evenly divisible by the remaining denominators (3 and 10). To continue the search for the LCD, multiply 15 by 2: $15 \cdot 2 = 30$. Because 30 is divisible by all the denominators in the expression, it is the LCD. To rewrite each fraction using the LCD, divide each denominator into 30 and multiply the numerator and denominator of the fraction by the quotient: $30 \div 3 = 10$, $30 \div 10 = 3$, and $30 \div 15 = 2$.

$$\frac{1}{3} + \frac{3}{10} - \frac{17}{15} = \frac{1 \cdot 10}{3 \cdot 10} + \frac{3 \cdot 3}{10 \cdot 3} - \frac{17 \cdot 2}{15 \cdot 2}$$

$$= \frac{10}{30} + \frac{9}{30} - \frac{34}{30}$$

Combine the numerators of the rational numbers.

$$= \frac{10 + 9 - 34}{30}$$

$$= -\frac{15}{30}$$

The expression isn't fully simplified until you reduce the fraction.

Divide the numerator and denominator by 15, the greatest common factor of 15 and 30, to reduce the fraction to lowest terms.

$$-\frac{15 \div 15}{30 \div 15} = -\frac{1}{2}$$

Therefore, $\dfrac{1}{3} + \dfrac{3}{10} - \dfrac{17}{15} = -\dfrac{1}{2}$.

1.17 Simplify the expression: $\dfrac{1}{8} \cdot \dfrac{3}{5}$

To calculate the product of two rational numbers, divide the product of the numerators by the product of the denominators.

$$\frac{1}{8} \cdot \frac{3}{5} = \frac{1 \cdot 3}{8 \cdot 5} = \frac{3}{40}$$

> Multiply the tops of the fractions and multiply the bottoms of the fractions. You don't need a common denominator to multiply or divide fractions.

1.18 Simplify the expression: $\dfrac{7}{9} \cdot \dfrac{3}{5}$.

Divide the product of the numerators by the product of the denominators.

$$\frac{7}{9} \cdot \frac{3}{5} = \frac{7 \cdot 3}{9 \cdot 5}$$
$$= \frac{21}{45}$$

Divide the numerator and denominator by 3, the greatest common factor of 21 and 45, to express the fraction in lowest terms.

$$= \frac{21 \div 3}{45 \div 3}$$
$$= \frac{7}{15}$$

1.19 Simplify the expression: $\dfrac{2}{7} \div \dfrac{4}{11}$.

Dividing by the rational number $\dfrac{4}{11}$ is equivalent to multiplying by its reciprocal: $\dfrac{11}{4}$. Rewrite the quotient as a product.

$$\frac{2}{7} \div \frac{4}{11} = \frac{2}{7} \cdot \frac{11}{4}$$

> Leave the left fraction alone, change the division symbol to multiplication, and flip the right fraction upside down.

Calculate the product.

$$= \frac{2 \cdot 11}{7 \cdot 4}$$
$$= \frac{22}{28}$$

Reduce the fraction to lowest terms.

$$= \frac{22 \div 2}{28 \div 2}$$
$$= \frac{11}{14}$$

1.20 Simplify the expression: $2 \div 4\frac{1}{6}$.

Convert the mixed number into an improper fraction using the technique described in Problems 1.7–1.8.

$$4\frac{1}{6} = \frac{6 \cdot 4 + 1}{6} = \frac{24 + 1}{6} = \frac{25}{6}$$

Substitute the improper fraction into the original expression.

$$2 \div 4\frac{1}{6} = \frac{2}{1} \div \frac{25}{6}$$

> $2 \div 1 = 2$. Any number divided by 1 is equal to the original number.

Rewrite the quotient as a product, as demonstrated by Problem 1.19.

$$\frac{2}{1} \div \frac{25}{6} = \frac{2}{1} \cdot \frac{6}{25}$$

Calculate the product.

$$= \frac{2 \cdot 6}{1 \cdot 25}$$
$$= \frac{12}{25}$$

Therefore, $2 \div 4\frac{1}{6} = \frac{12}{25}$.

Translating Expressions
Create math phrases out of English phrases

1.21 Translate into an algebraic expression: 12 more than a number.

This expression indicates that an unknown value ("a number") should be increased by 12. Use a variable, like x, to represent the unknown value: $x + 12$.

1.22 Translate into an algebraic expression: 5 less than a number.

In Problem 1.21, "12 more than a number" indicates a sum. In this problem, "5 less than a number" indicates a difference. To express 5 less than an unknown value x, subtract 5 from it: $x - 5$.

1.23 Translate into an algebraic expression: 7 less a number.

As in Problem 1.22, the word "less" in this expression indicates subtraction. However, "7 less a number" and "7 less *than* a number" do not describe the same expression. In this problem, "less a number" is synonymous with "minus a number," so the correct expression is $7 - x$.

> The big difference is the word THAN. Ten less THAN a number is $x - 10$, but 10 less a number is $10 - x$. The order is important because those two expressions usually aren't equal.

1.24 Translate into an algebraic expression: one-third of a number.

The word "of," especially when paired with a fractional phrase such as "one-third" or "one-half," typically indicates multiplication. Therefore, one-third of a number is the product $\frac{1}{3} \cdot x$. It is also correct to express x as the rational number $\frac{x}{1}$ and, therefore, write one-third of a number as $\frac{1}{3} \cdot \frac{x}{1} = \frac{x}{3}$.

> Think of it this way: to get one-third of something, you divide it by 3, so one-third of x is $\frac{x}{3}$.

1.25 Translate into an algebraic expression: the quotient of 4 and 36 less than a number.

The final part of this expression is similar to Problem 1.22; 36 less than a number is represented by the expression $x - 36$. The word "quotient" indicates a division problem—the quotient of A and B is $A \div B$.

Therefore, the quotient of 4 and $x - 36$ is $4 \div (x - 36)$. Use parentheses to indicate that 4 is to be divided by the entire expression $x - 36$. Omitting the parentheses results in the incorrect expression $4 \div x - 36$, which translates into the phrase "36 less than the quotient of 4 and x."

You may also correctly express the quotient as a fraction: $\frac{4}{x - 36}$.

1.26 Translate into an algebraic expression: the product of a number and 8 more than half of that number.

> $\frac{1}{2}x^2 + 8x$ is also correct. You need to understand exponential rules and the distributive property to get that answer; you'll develop those skills later in this chapter.

Let x represent the unknown number. Half of x is written $\frac{1}{2}x$, and 8 more than half of x is written $\frac{1}{2}x + 8$. The word "product" indicates multiplication—the product of A and B is AB. Therefore, the product of x and $\frac{1}{2}x + 8$ is $x\left(\frac{1}{2}x + 8\right)$.

1.27 Translate into an algebraic expression: the difference of twice a number and the square of that number.

Twice a number x is written $2x$, whereas the square of a number is that number raised to the second power: x^2. The difference of $2x$ and x^2 is written $2x - x^2$. Note that the order of the terms in a subtraction (and division) problem is important, so when calculating a difference, write the terms in the order that they appear in the problem.

> Because "twice a number" comes first in the phrase, $2x$ comes first in the expression. $2x - x^2 \neq x^2 - 2x$ in most cases.

Simplifying Arithmetic Expressions

Exponent laws and combining expressions

1.28 Simplify the expression: $y^3 \cdot y^2$.

The product of two exponential expressions with the same base (in this case, y) is the common base raised to the sum of the powers.

$$y^3 \cdot y^2 = y^{3+2} = y^5$$

> This y has no exponent written explicitly, which means it has an implied exponent of 1 (just as the coefficient of x in the expression $x + 2y$ is unwritten but also equals 1).

1.29 Simplify the expression: $(2x^2y)(3x^4y^7)$.

Each of the factors in this product contains a coefficient, an x raised to an exponent, and a y raised to an exponent. To calculate the product, multiply the corresponding components, applying the exponential law outlined in Problem 1.28 to the x- and y-expressions.

$$(2x^2y)(3x^4y^7) = (2 \cdot 3)(x^2 \cdot x^4)(y^1 \cdot y^7)$$
$$= 6x^{2+4}y^{1+7}$$
$$= 6x^6y^8$$

1.30 Simplify the expression: $\left(10x^6y^9\right)^2$.

The entire expression $10x^6y^9$ is raised to the second power (or "squared"), so square each of the factors. To raise an exponential expression (like x^6 and y^9) to exponential powers, multiply the powers: $\left(x^a\right)^b = x^{ab}$.

$$\left(10x^6y^9\right)^2 = 10^2 \cdot \left(x^6\right)^2 \cdot \left(y^9\right)^2$$
$$= 100x^{6(2)}y^{9(2)}$$
$$= 100x^{12}y^{18}$$

1.31 Simplify the expression: $\dfrac{9y^4z^4}{12y^2z^5}$.

According to Problem 1.28, the product of exponential expressions with the same base is equal to the common base raised to the sum of the powers: $(x^a)(x^b) = x^{a+b}$. Similarly, the *quotient* of exponential expressions with the same base is equal to the common base raised to the *difference* of the powers: $\dfrac{x^a}{x^b} = x^{a-b}$.

$$\frac{9y^4z^4}{12y^2z^5} = \frac{9}{12} \cdot \frac{y^4}{y^2} \cdot \frac{z^4}{z^5}$$
$$= \frac{9}{12} \cdot y^{4-2} \cdot z^{4-5}$$
$$= \frac{3}{4}y^2z^{-1}$$

> Divide the top and bottom of this fraction by 3 to reduce it.

Notice that z is raised to the -1 power. The negative exponent indicates that z should be placed in the denominator of the fraction. When placed correctly, the exponent is no longer negative.

$$= \frac{3y^2}{4z^1}$$

$$= \frac{3y^2}{4z}$$

1.32 Simplify the expression: $(9x)^2 + 4x^2$.

All expressions must be simplified according to a predetermined order of operations: (1) parentheses and grouped expressions; (2) exponential expressions; (3) multiplication and division, performed left to right; and (4) addition and subtraction, performed left to right. In this expression, the exponent of $(9x)$ must be addressed before addition can occur.

$$(9x)^2 + 4x^2 = 9^2 \cdot (x^2) + 4x^2$$

$$= 81x^2 + 4x^2$$

Combine like terms $81x^2$ and $4x^2$.

$$= 85x^2$$

$81x^2$ and $4x^2$ are like terms because they have the exact same variable, x^2. To combine like terms, add the coefficients.

1.33 Simplify the expression: $2(5y - 3)$.

The distributive property states that $a(b + c) = ab + ac$; if a quantity is multiplied by a value, that value can then be multiplied by each term in the quantity.

$$2(5y - 3) = 2(5y) + 2(-3)$$

$$= 10y - 6$$

In this problem, the quantity $(5y - 3)$ is multiplied by 2, so you can multiply 2 by each term individually: $2(5y)$ and $2(-3)$.

1.34 Simplify the expression: $6x(9y + 1) - 8y(2x - 5)$.

Apply the distributive property, multiplying $6x$ and $-8y$ by the terms in the adjacent quantities.

$$6x(9y + 1) - 8y(2x - 5)$$

$$= \left[6x(9y) + 6x(1) \right] + \left[(-8y)(2x) + (-8y)(-5) \right]$$

$$= 54xy + 6x - 16xy + 40y$$

Combine like terms.

$$= (54xy - 16xy) + 6x + 40y$$

$$= 38xy + 6x + 40y$$

1.35 Simplify the expression: $(x+6)(3x-1)$.

In Problem 1.33, each term of a quantity is multiplied by a single value. In this problem, the quantity on the right is again multiplied by an outside value. However, in this case, each term of $(3x-1)$ is multiplied by two different values, x and 6.

To calculate the product of $(x+6)(3x-1)$, multiply the terms of $(3x-1)$ by x, the first term of the left quantity. Then multiply $(3x-1)$ by 6, the second term of the left quantity. Finally, add all four of the products together.

> You're basically applying the distributive property twice, because you're multiplying $(3x-1)$ by $(x+6)$, which has two terms in it.

$$(x+6)(3x-1)=[x(3x)+x(-1)]+[6(3x)+6(-1)]$$
$$=3x^2-x+18x-6$$

Combine like terms.

$$=3x^2+(-x+18x)-6$$
$$=3x^2+17x-6$$

1.36 Simplify the expression: $(5x+y)^2$.

Squaring an expression is the equivalent of multiplying the expression by itself: $a^2 = a \cdot a$.

$$(5x+y)^2 = (5x+y)(5x+y)$$

Calculate the product using the technique described in Problem 1.35: multiply each term of the right quantity by $5x$ (the first term of the left quantity) and then multiply each term in the right quantity by y (the second term of the left quantity).

$$=\left[5x(5x)+5x(y)\right]+\left[y(5x)+y(y)\right]$$
$$=25x^2+5xy+5xy+y^2$$

Combine like terms.

$$=25x^2+\left(5xy+5xy\right)+y^2$$
$$=25x^2+10xy+y^2$$

1.37 Simplify the expression: $(x-4)(x^2-2x+1)$.

To calculate the product, distribute each term of the left quantity to each term of the right quantity. ◄

> In other words, multiply x by each term of $x^2 - 2x + 1$, and then do the same with –4.

$$(x-4)\left(x^2-2x+1\right)$$
$$=\left[x\left(x^2\right)+x(-2x)+x(1)\right]+\left[(-4)\left(x^2\right)+(-4)(-2x)+(-4)(1)\right]$$
$$=x^3-2x^2+x-4x^2+8x-4$$

Combine like terms.

$$=x^3+\left(-2x^2-4x^2\right)+(x+8x)-4$$
$$=x^3-6x^2+9x-4$$

Note: Problems 1.38–1.40 refer to the expression $(x+4) \div x$.

1.38 Calculate the quotient and remainder using long division.

The long division of polynomials is very similar to the long division of real numbers, as outlined in Problems 1.5–1.6. Rewrite the problem using a long division bracket.

$$x\overline{)x+4}$$

Calculate the quotient of the leftmost term of the dividend and the divisor. ◄

> The dividend is what you're dividing INTO (x + 4), and its leftmost term is x. The divisor is what you're dividing BY (x).

$$\frac{x}{x}=1$$

Write the result (1) above the constant (4) in the dividend, because 1 and 4 are like terms. ◄

> 1 and 4 are like terms because they have the exact same variable: none.

$$x\overline{)\overset{\textstyle 1}{x+4}}$$

Multiply the newly placed 1 by the divisor (x) to get $1 \cdot x = x$. Write the *opposite* of the result $(-x)$ below the dividend and draw a horizontal line beneath it.

$$x\overline{)\,\overset{\textstyle 1}{\underset{\underline{-x}}{x+4}}}$$

Combine the expressions and write the result below the horizontal line: $(x+4)-x=4$.

$$x\overline{)\,\overset{\textstyle 1}{\underset{\underline{-x}}{x+4}}}\\4$$

The degree of x (which is 1, because $x = x^1$) is greater than the degree of 4 (which has no x, so its degree is 0).

The degree of the divisor is greater than the degree of the expression below the horizontal line, so the division problem is complete. The quotient is the expression above the division bracket (1) and the remainder is the expression below the horizontal line (4).

Express the answer as the sum of the quotient and the fraction whose numerator is the remainder and whose denominator is the divisor.

$$(x+4) \div x = 1 + \frac{4}{x}$$

Note: Problems 1.38–1.40 refer to the expression (x + 4) ÷ x.

1.39 Verify the solution generated in Problem 1.38.

Multiply the quotient (1) by the divisor (x), and then add the remainder (4) to ensure that the result is the dividend ($x + 4$).

$$\begin{aligned} (\text{quotient})(\text{divisor}) &+ \text{remainder} = \text{dividend} \\ (1)(x) \quad\quad &+ \quad 4 \quad = \quad x + 4 \\ x \quad\quad &+ \quad 4 \quad = \quad x + 4 \end{aligned}$$

Note: Problems 1.38–1.40 refer to the expression (x + 4) ÷ x.

1.40 Calculate the quotient and remainder by writing the expression as a fraction in lowest terms. Assume that $x \neq 0$.

Write the division problem as a rational expression.

$$\frac{x+4}{x}$$

This does not work for the reciprocal:

$$\frac{x}{x+4} \neq \frac{x}{x} + \frac{x}{4}$$

You can't break addition or subtraction in the denominator into two fractions with a matching numerator.

Write the expression as a sum of rational expressions.

$$\frac{x+4}{x} = \frac{x}{x} + \frac{4}{x}$$

Reduce the left fraction to lowest terms: $\dfrac{x}{x} = 1$.

$$= 1 + \frac{4}{x}$$

Notice that the result is equal to, and therefore verifies, the solution to Problem 1.38.

Note: Problems 1.41–1.42 refer to the expression $(x^2 + 8x - 3) \div (x - 2)$.

1.41 Calculate the quotient and remainder using long division.

Rewrite the problem using a long division bracket.

$$x - 2 \overline{) x^2 + 8x - 3}$$

Begin by dividing the leftmost term of the dividend (x^2) by the leftmost term of the divisor (x): $\dfrac{x^2}{x^1} = x^{2-1} = x$. Write the result above the term $8x$ in the dividend, because x and $8x$ are like terms.

$$x - 2 \overline{) x^2 + 8x - 3} \quad \overset{x}{}$$

Multiply the newly placed x by each term of the divisor and write the *opposites* of the results below the dividend.

$$\begin{array}{r} x \\ x - 2 \overline{) x^2 + 8x - 3} \\ -x^2 + 2x \end{array}$$

> The entire goal of this step is to make the first term of the divisor (x^2) disappear. You do that by writing its opposite ($-x^2$) below it and adding those terms together: $x^2 + (-x^2) = 0$.

Combine the expressions and write the result below the horizontal line: $(x^2 + 8x - 3) + (-x^2 + 2x) = 10x - 3$.

$$\begin{array}{r} x \\ x - 2 \overline{) x^2 + 8x - 3} \\ -x^2 + 2x \\ \hline 10x - 3 \end{array}$$

Now divide the leftmost term of the bottom expression by the leftmost term of the divisor: $\dfrac{10x}{x} = 10$. Write the result above the -3 term of the dividend because $+10$ and -3 are like terms.

$$\begin{array}{r} x + 10 \\ x - 2 \overline{) x^2 + 8x - 3} \\ -x^2 + 2x \\ \hline 10x - 3 \end{array}$$

Multiply the newly placed 10 by each term of the divisor, write the *opposites* of the results at the bottom of the division problem, and combine the expressions: $(10x - 3) + (-10x + 20) = 17$.

$$
\begin{array}{r}
x + 10 \\
x-2{\overline{\smash{\big)}\,x^2 + 8x - 3}} \\
\underline{-x^2 + 2x} \\
10x - 3 \\
\underline{-10x + 20} \\
+ 17
\end{array}
$$

The degree of the divisor is now greater than the degree of the bottommost expression, so the division problem is complete. Write the solution as the sum of the quotient ($x + 10$) and the fraction whose numerator is the remainder (17) and whose denominator is the divisor ($x - 2$).

$$
\left(x^2 + 8x - 3\right) \div \left(x - 2\right) = x + 10 + \frac{17}{x-2}
$$

> The degree of $x - 2$ is 1 because it's the highest power of x in the expression. The degree of 17 is 0 because it has no x's.

Note: Problems 1.41–1.42 refer to the expression ($x^2 + 8x - 3$) ÷ ($x - 2$).

1.42 Calculate the quotient and remainder using synthetic division to verify the solution generated by Problem 1.41.

The constant of the divisor $x - 2$ is –2. Write the *opposite* of that number in a box and list the coefficients of the dividend in a row to the right of that box. Draw a horizontal line below the entire row.

$$
\begin{array}{r|rrr}
2 & 1 & 8 & -3 \\
\hline
\end{array}
$$

Drop the first coefficient below the horizontal line.

$$
\begin{array}{r|rrr}
2 & 1 & 8 & -3 \\
\hline
 & 1 & &
\end{array}
$$

Multiply the number below the line by the number within the box ($1 \cdot 2 = 2$) and write the result above the line in the next column.

$$
\begin{array}{r|rrr}
2 & 1 & 8 & -3 \\
 & & 2 & \\
\hline
 & 1 & &
\end{array}
$$

Add the newly placed 2 to the other number in that column ($2 + 8 = 10$) and write the result below the horizontal line.

$$
\begin{array}{r|rrr}
2 & 1 & 8 & -3 \\
 & & 2 & \\
\hline
 & 1 & 10 &
\end{array}
$$

Multiply the number below the line by the number within the box ($10 \cdot 2 = 20$) and write the result above the line in the next column.

$$
\begin{array}{r|rrr}
\underline{2|} & 1 & 8 & -3 \\
 & & 2 & 20 \\
\hline
 & 1 & 10 &
\end{array}
$$

Combine the numbers in the final column and record the result below the horizontal line.

$$
\begin{array}{r|rrr}
\underline{2|} & 1 & 8 & -3 \\
 & & 2 & 20 \\
\hline
 & 1 & 10 & 17
\end{array}
$$

The numbers below the horizontal line represent the coefficients of the quotient and the remainder. ◂

$$
\left(x^2 + 8x - 3\right) \div (x - 2) = x + 10 + \frac{17}{x - 2}
$$

> The dividend had degree 2, and you're dividing by something with degree 1. Therefore, the quotient will have degree 2 – 1 = 1. That means the numbers 1, 10, and 17 are (from left to right) the coefficient of x^1, the constant, and the remainder.

Evaluating Expressions

Plug numbers into variables

1.43 Evaluate the expression $3x + 2y$, given $x = 5$ and $y = -4$.

Substitute $x = 5$ and $y = -4$ into the expression and simplify.

$$
\begin{aligned}
3x + 2y &= 3(5) + 2(-4) \\
&= 15 - 8 \\
&= 7
\end{aligned}
$$

1.44 Evaluate the expression $a - bc$, given $a = -3$, $b = 6$, and $c = \dfrac{1}{2}$.

Substitute $a = -3$, $b = 6$, and $c = \dfrac{1}{2}$ into the expression.

$$
a - bc = (-3) - (6)\left(\frac{1}{2}\right)
$$

According to the order of operations, multiplication must occur before addition and subtraction.

$$
\begin{aligned}
&= (-3) - \left(\frac{6}{2}\right) \\
&= (-3) - 3 \\
&= -6
\end{aligned}
$$

1.45 Evaluate the expression $x^2 - 6xy + 9y^2$, given $x = \dfrac{1}{3}$ and $y = -1$.

Substitute $x = \dfrac{1}{3}$ and $y = -1$ into the expression.

$$x^2 - 6xy + 9y^2 = \left(\frac{1}{3}\right)^2 - 6\left(\frac{1}{3}\right)(-1) + 9(-1)^2$$

According to the order of operations, exponents should be simplified first.

$$= \frac{1^2}{3^2} - 6\left(\frac{1}{3}\right)(-1) + 9(1)$$

$$= \frac{1}{9} - 6\left(\frac{1}{3}\right)(-1) + 9$$

Multiplication should be completed left to right before addition or subtraction may occur.

$$= \frac{1}{9} - \left(\frac{6}{1}\right)\left(\frac{1}{3}\right)(-1) + 9$$

$$= \frac{1}{9} - \left(\frac{6}{3}\right)(-1) + 9$$

$$= \frac{1}{9} - 2(-1) + 9$$

$$= \frac{1}{9} + 2 + 9$$

Use a common denominator to simplify the expression.

$$= \frac{1}{9} + \frac{18}{9} + \frac{81}{9}$$

$$= \frac{1 + 18 + 81}{9}$$

$$= \frac{100}{9}$$

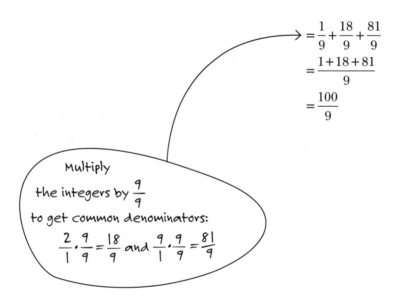

Multiply the integers by $\dfrac{9}{9}$ to get common denominators:

$\dfrac{2}{1} \cdot \dfrac{9}{9} = \dfrac{18}{9}$ and $\dfrac{9}{1} \cdot \dfrac{9}{9} = \dfrac{81}{9}$

1.46 Determine the value of y in the expression $y = mx + b$ if $m = -\dfrac{3}{5}$, $x = 2$, and $b = 8$.

Substitute $m = -\dfrac{3}{5}$, $x = 2$, and $b = 8$ into the expression and simplify.

$$y = mx + b$$
$$= \left(-\frac{3}{5}\right)(2) + 8$$
$$= \left(-\frac{3}{5}\right)\left(\frac{2}{1}\right) + 8$$
$$= -\frac{6}{5} + 8$$

Use the least common denominator to combine the rational numbers. ←

> Even though 8 isn't written as a fraction, it's a rational number. All integers are automatically rational numbers, too.

$$= -\frac{6}{5} + \frac{8}{1} \cdot \frac{5}{5}$$
$$= -\frac{6}{5} + \frac{40}{5}$$
$$= \frac{34}{5}$$

Therefore, $y = \dfrac{34}{5}$.

1.47 Evaluate the expression $\dfrac{a + b \cdot c}{d}$, given $a = 6$, $b = -3$, $c = 10$, and $d = -12$.

Substitute $a = 6$, $b = -3$, $c = 10$, and $d = -12$ into the expression.

$$\frac{a + b \cdot c}{d} = \frac{6 + (-3)(10)}{-12}$$

According to the order of operations, multiplication must be completed before addition.

$$= \frac{6 + (-30)}{-12}$$
$$= \frac{-24}{-12}$$
$$= 2$$

Chapter 2
LINEAR EQUATIONS

Solve for x in an equation or solve for (x,y) given two equations

Nearly all geometry problems can be classified in one of two categories: investigating logical statements (including proof) and calculating unknown values (such as an angle measurement in a diagram). When completing problems in the latter category, you are often required to solve linear equations and occasionally systems of linear equations. This chapter presents an opportunity to review these foundational skills.

This chapter reviews solving equations. The first half of the chapter focuses on simple equations that contain only one variable. To solve these, you manipulate both sides of the equation by adding, subtracting, multiplying, or dividing the same value in an attempt to isolate (solve for) the variable that's present.

The second half of the chapter focuses on systems of equations, two or more equations that you solve at the same time. You use either the substitution or the elimination method to find a single coordinate pair (x,y) that makes both equations true. Well, most of the time that's what happens. Sometimes systems have no solutions, and sometimes they have infinitely many. Usually, though, they have just one.

Basic Equation-Solving Strategies
Do the same thing to both sides of the equation

2.1 Solve the equation $x - 9 = 23$ and verify the solution.

If two things are equal (like the sides of this equation) and you add 9 to the left side only, the sides aren't equal anymore.

To solve an equation, you must isolate the variable within the equation on one side of the equals sign. In other words, one side of the equation will contain only the variable (in this case x). Though the left side of the equation currently contains x, it also contains the term -9.

To eliminate -9 from the left side of the equation, add its opposite, $+9$. However, adding 9 to the left side of the equation would invalidate the statement, so you must add 9 to the right side of the equation as well.

$$x - 9 + 9 = 23 + 9$$

Simplify both sides of the equation ($-9 + 9 = 0$ and $23 + 9 = 32$).

$$x + 0 = 32$$
$$x = 32$$

To verify that $x = 32$ is the correct solution, substitute $x = 32$ into the original equation.

$$x - 9 = 23$$
$$32 - 9 = 23$$
$$23 = 23 \quad \text{True}$$

Because the result is a true statement ($23 = 23$), the solution $x = 32$ is correct.

2.2 Solve the equation and verify the solution: $10 = 14 + x$.

The solution was $-4 = x$, but $x = -4$ means the same thing. You're allowed to swap the sides of an equation according to something called the symmetric property.

The variable x appears on the right side of this equation. Eliminate 14, the other term on that side of the equation, by subtracting 14 from both sides.

$$10 - 14 = 14 - 14 + x$$
$$-4 = 0 + x$$
$$-4 = x$$

To verify that $x = -4$ is the correct solution, substitute it into the original equation.

$$10 = 14 + x$$
$$10 = 14 + (-4)$$
$$10 = 14 - 4$$
$$10 = 10 \quad \text{True}$$

Because the resulting statement ($10 = 10$) is true, $x = -4$ is the solution to the equation.

2.3   Solve the equation: $9 + 6x = 7x$.

Both sides of this equation contain a variable; in fact, they contain the same variable, so $6x$ and $7x$ are like terms. To isolate the like terms on one side of the equals sign, subtract $6x$ from both sides of the equation. ←

$$9 + 6x - 6x = 7x - 6x$$
$$9 + 0 = 1x$$
$$9 = x$$

The solution to the equation is $x = 9$.

> $6x$ is smaller than $7x$, so if you subtract $6x$ from $7x$, you get $7x - 6x = 1x$, a positive result. This isn't the only way to solve the equation, but it's the fastest.

2.4   Solve the equation: $2(x - 1) = 3x$.

Apply the distributive property to the left side of the equation.

$$2(x) + 2(-1) = 3x$$
$$2x - 2 = 3x$$

Subtract $2x$ from both sides of the equation to isolate the x-terms right of the equals sign.

$$2x - 2x - 2 = 3x - 2x$$
$$0 - 2 = 1x$$
$$-2 = x$$

The solution to the equation is $x = -2$.

> Multiply everything inside the parentheses by the number outside the parentheses.

2.5   Solve the equation: $5x = 20$.

The x-term in this equation is already isolated left of the equals sign. To eliminate the coefficient of x, and thereby isolate x and solve the equation, divide both sides of the equation by 5.

$$\frac{5x}{5} = \frac{20}{5}$$

Reduce both fractions to lowest terms: $\frac{5}{5} = 1$ and $\frac{20}{5} = 4$.

$$x = 4$$

> When an equation is solved for x, one side of the equation is just an x, not $5x$ or x with any coefficient other than 1.

2.6 Solve the equation: $-4x = 26$.

To solve the equation for x, you must eliminate the x-coefficient. Divide both sides by -4 and reduce fractions to lowest terms.

$$\frac{-4x}{-4} = \frac{26}{-4}$$
$$x = -\frac{13}{2}$$

2.7 Solve the equation: $21 = \frac{1}{3}x$.

Like the equations in Problems 2.5–2.6, this equation has an x-term isolated on one side of the equation and its coefficient must be eliminated. To solve for x, divide both sides of the equation by the coefficient of x, $\frac{1}{3}$.

As stated in Problem 1.19, dividing by a fraction is equivalent to multiplying by its reciprocal. Therefore, dividing both sides of the equation by $\frac{1}{3}$ is equivalent to multiplying both sides of the equation by $\frac{3}{1}$.

RULE OF THUMB: To get rid of an x-coefficient that's a fraction, multiply both sides of the equation by the reciprocal.

$$\frac{3}{1}\left(\frac{21}{1}\right) = \frac{3}{1}\left(\frac{1}{3}x\right)$$
$$\frac{63}{1} = \frac{3}{3}x$$
$$63 = x$$

2.8 Solve the equation: $\frac{4}{9}y = 12$.

Multiply both sides of the equation by $\frac{9}{4}$, the reciprocal of the y-coefficient, to solve for y.

$$\frac{9}{4}\left(\frac{4}{9}y\right) = \frac{9}{4}\left(\frac{12}{1}\right)$$
$$\frac{36}{36}y = \frac{108}{4}$$

Reduce the fractions to lowest terms: $\frac{36}{36} = 1$ and $\frac{108 \div 4}{4 \div 4} = \frac{27}{1}$.

$$y = 27$$

2.9   Solve the equation: $3x - 2 = -20$.

Isolate the x-term left of the equals sign by adding 2 to both sides of the equation.

$$3x - 2 + 2 = -20 + 2$$
$$3x + 0 = -18$$
$$3x = -18$$

Divide both sides of the equation by 3 to eliminate the coefficient of x.

$$\frac{3x}{3} = \frac{-18}{3}$$
$$x = -6$$

Note: Problems 2.10–2.11 refer to the equation $5x + 9 = 8x - 1$.

2.10   Solve the equation by isolating x on the right side of the equals sign.

Subtract $5x$ from both sides of the equation to ensure that all the x-terms are right of the equals sign.

$$5x - 5x + 9 = 8x - 5x - 1$$
$$9 = 3x - 1$$

Isolate the x-term by adding 1 to both sides of the equation.

$$9 + 1 = 3x - 1 + 1$$
$$10 = 3x$$

Divide both sides of the equation by the coefficient of x.

$$\frac{10}{3} = \frac{3x}{3}$$
$$\frac{10}{3} = x$$

> It doesn't matter whether you isolate x on the left or right side of the equals sign. You'll get the same solution.

2.11   Solve the equation by isolating x on the left side of the equals sign. Verify that the solutions to Problems 2.10 and 2.11 are equal.

Move the x-terms left of the equals sign by subtracting $8x$ from both sides of the equation.

$$5x - 8x + 9 = 8x - 8x - 1$$
$$-3x + 9 = -1$$

Isolate the x-term by subtracting 9 from both sides of the equation.

$$-3x + 9 - 9 = -1 - 9$$
$$-3x = -10$$

Divide both sides of the equation by the coefficient of x to identify the solution.

$$\frac{-3x}{-3} = \frac{-10}{-3}$$
$$x = \frac{10}{3}$$

Note that the solution matches, and thereby verifies, the solution to Problem 2.10.

2.12   Solve the equation: $9(3 - x) + 4 = 6x + 11$.

Simplify the left side of the equation by applying the distributive property and combining like terms.

$$9(3) + 9(-x) + 4 = 6x + 11$$
$$27 - 9x + 4 = 6x + 11$$
$$(27 + 4) - 9x = 6x + 11$$
$$31 - 9x = 6x + 11$$

> Add 9x to both sides of the equation to combine the x-terms. Then subtract 11 from both sides to isolate the x-terms. Divide by the coefficient of x to finish. You can also isolate x on the left side of the equation, as in Problem 2.11.

Solve the equation for x.

$$31 - 9x + 9x = 6x + 9x + 11$$
$$31 = 15x + 11$$
$$31 - 11 = 15x + 11 - 11$$
$$20 = 15x$$
$$\frac{20}{15} = x$$
$$\frac{4}{3} = x$$

2.13 Solve the proportion: $\dfrac{x}{4} = \dfrac{1}{5}$.

Cross-multiply to eliminate fractions.

$$x \cdot 5 = 4 \cdot 1$$
$$5x = 4$$

Divide both sides of the equation by the coefficient of x to solve.

$$\frac{5x}{5} = \frac{4}{5}$$
$$x = \frac{4}{5}$$

Note: Problems 2.14–2.15 refer to the equation $\dfrac{4x}{3} = \dfrac{x+2}{9}$.

2.14 Eliminate fractions by cross-multiplying and then solve the equation.

Cross-multiply to eliminate the fractions from the equation.

$$(4x)(9) = 3(x+2)$$
$$36x = 3(x) + 3(2)$$
$$36x = 3x + 6$$

Isolate x on the left side of the equation.

$$36x - 3x = 3x - 3x + 6$$
$$33x = 6$$
$$\frac{33x}{33} = \frac{6}{33}$$
$$x = \frac{2}{11}$$

"Proportion" is a fancy word for an equation that contains two fractions set equal to each other.

To cross-multiply a proportion, multiply the top-left number by the bottom-right number (x · 5) and set that product equal to the bottom-left number times the top-right number (4 · 1).

If you can't remember how to find the least common denominator, flip back to Problems 1.15 and 1.16.

Note: Problems 2.14–2.15 refer to the equation $\dfrac{4x}{3} = \dfrac{x+2}{9}$.

2.15 Eliminate fractions using the least common denominator and then solve the equation.

Multiply each expression by 9, the least common denominator, to eliminate fractions.

$$\frac{4x}{3}\cdot\frac{9}{1} = \frac{x+2}{9}\cdot\frac{9}{1}$$
$$\frac{36x}{3} = \frac{9(x+2)}{9}$$
$$\frac{36}{3}x = \frac{9}{9}(x+2)$$
$$12x = 1(x+2)$$
$$12x = x+2$$

Solve the equation for x.

$$12x - x = x - x + 2$$
$$11x = 2$$
$$\frac{11x}{11} = \frac{2}{11}$$
$$x = \frac{2}{11}$$

2.16 Solve the equation and verify the solution: $\dfrac{2}{x} + \dfrac{1}{5} = \dfrac{11}{5x}$.

The denominator 5x is the LCD because the other denominators (x and 5) divide into it evenly.

Multiply each expression by $5x$, the least common denominator, to eliminate the fractions. Note that multiplying an entire equation by a variable expression might introduce false solutions, so complete the problem by testing all solutions that are generated.

$$\left(\frac{5x}{1}\right)\left(\frac{2}{x}\right) + \left(\frac{5x}{1}\right)\left(\frac{1}{5}\right) = \left(\frac{5x}{1}\right)\left(\frac{11}{5x}\right)$$
$$\frac{10x}{x} + \frac{5x}{5} = \frac{55x}{5x}$$
$$\frac{10}{1}\left(\frac{x}{x}\right) + \frac{5}{5}\left(\frac{x}{1}\right) = \frac{55}{5}\left(\frac{x}{x}\right)$$
$$10 + x = 11$$
$$10 - 10 + x = 11 - 10$$
$$x = 1$$

Substitute $x = 1$ into the original equation to verify that the result is a true statement, thereby affirming that the solution is valid.

$$\frac{2}{x} + \frac{1}{5} = \frac{11}{5x}$$

$$\frac{2}{1} + \frac{1}{5} = \frac{11}{5(1)}$$

$$\frac{2}{1} + \frac{1}{5} = \frac{11}{5}$$

Use common denominators to simplify the left side of the equation.

$$\frac{2}{1} \cdot \frac{5}{5} + \frac{1}{5} = \frac{11}{5}$$

$$\frac{10}{5} + \frac{1}{5} = \frac{11}{5}$$

$$\frac{10+1}{5} = \frac{11}{5}$$

$$\frac{11}{5} = \frac{11}{5} \quad \text{True}$$

Equations with Multiple Variables
Solve for one variable when others are present

2.17 Solve the equation $x - 4y + 3z = 1$ for x.

Isolate x left of the equals sign by adding $4y$ to, and subtracting $3z$ from, both sides of the equation.

$$x + (-4y + 4y) + (3z - 3z) = 1 + 4y - 3z$$
$$x = 1 + 4y - 3z$$

2.18 Solve the equation $F = \frac{9}{5}C + 32$ for C.

Isolate the term containing C on the right side of the equation.

$$F - 32 = \frac{9}{5}C + 32 - 32$$

$$F - 32 = \frac{9}{5}C$$

This is the formula that converts Celsius temperatures into Fahrenheit. By solving for C, you create the formula that does the opposite.

Multiply both sides of the equation by the reciprocal of the coefficient of C.

$$\frac{5}{9}(F-32) = \frac{5}{9}\left(\frac{9}{5}C\right)$$

$$\frac{5}{9}(F-32) = \frac{45}{45}C$$

$$\frac{5}{9}(F-32) = C$$

2.19 Solve the equation $S = 2\pi rh$ for h. Assume that $r \neq 0$.

The value of π is constant, so it should be treated like any other real number when solving equations. In this problem, the variable h is multiplied by the constants 2 and π, as well as the variable r. To solve for h, divide both sides of the equation by those three factors.

$$\frac{S}{2\pi r} = \frac{2\pi rh}{2\pi r}$$

$$\frac{S}{2\pi r} = \frac{2}{2} \cdot \frac{\pi}{\pi} \cdot \frac{r}{r} \cdot h$$

$$\frac{S}{2\pi r} = h$$

2.20 Solve the equation $x - xy = 1$ for y. Assume that $x \neq 0$.

> You want to isolate $-xy$ because it contains y, the variable you're solving for.

Isolate $-xy$ left of the equals sign by subtracting x from both sides of the equation.

$$x - x - xy = 1 - x$$

$$-xy = 1 - x$$

To solve for y, divide both sides of the equation by $-x$.

$$\frac{-xy}{-x} = \frac{1-x}{-x}$$

$$y = \frac{x-1}{x}$$

> Instead of writing the answer as $-\frac{1-x}{x}$, the book multiplies the fraction $\frac{1-x}{x}$ by the negative sign out front, $\frac{-1}{1}$:
>
> $$= \frac{-1(1-x)}{1 \cdot x}$$
>
> $$= \frac{-1+x}{x}$$
>
> $$= \frac{x-1}{x}$$

2.21 Solve the equation $V = 4lw + 2w^2$ for l. Assume that $w \neq 0$.

Isolate the term containing l on the right side of the equation.

$$V - 2w^2 = 4lw$$

Divide both sides of the equation by $4w$, the other factors of the product $4lw$.

$$\frac{V - 2w^2}{4w} = \frac{4lw}{4w}$$

$$\frac{V - 2w^2}{4w} = l$$

2.22 Solve the equation $Ax + By = C$ for y.

Isolate the y-term on the left side of the equals sign and eliminate its coefficient, B.

$$By = C - Ax$$

$$y = \frac{C - Ax}{B}$$

2.23 Solve the equation $y - y_1 = m(x - x_1)$ for x.

Apply the distributive property to the right side of the equation.

$$y - y_1 = mx - mx_1$$

Add mx_1 to both sides of the equation to isolate mx. ←

$$y - y_1 + mx_1 = mx$$

Divide every term in the equation by m to solve for x.

$$\frac{y - y_1 + mx_1}{m} = x$$

> You're not solving for x_1, so move it across the equals sign with everything else, leaving only mx behind.

Solving Systems of Equations
Using the substitution and elimination techniques

2.24 Demonstrate that the coordinate pair $(x,y) = (-1,5)$ is a solution to the following system of equations.

$$\begin{cases} 4x + y = 1 \\ 2x - 3y = -17 \end{cases}$$

> If $(x,y) = (-1,5)$, then $x = -1$ and $y = 5$.

If $(x,y) = (-1,5)$ is a solution to the system, then substituting the coordinate pair into each equation of the system should result in true statements. Plug the coordinate pair into the first equation, $4x + y = 1$.

$$4(-1) + 5 = 1$$

$$-4 + 5 = 1$$

$$1 = 1 \quad \text{True}$$

Now substitute $(x,y) = (-1,5)$ into the remaining equation of the system $(2x - 3y = -17)$ to verify that the result is true as well.

$$2(-1) - 3(5) = -17$$
$$-2 - 15 = -17$$
$$-17 = -17 \quad \text{True}$$

2.25 For what value of c is the coordinate pair $(c,-7)$ a solution to the following system of equations?

$$\begin{cases} 2x - y = 8 \\ 6x + 2y = -11 \end{cases}$$

If $(x,y) = (c,-7)$ is a solution to the system, then substituting $x = c$ and $y = -7$ into either equation of the system must produce a true statement. Here, $(c,-7)$ is substituted into the first equation, $2x - y = 8$.

$$2c - (-7) = 8$$
$$2c + 7 = 8$$
$$2c = 8 - 7$$
$$2c = 1$$
$$c = \frac{1}{2}$$

You get the same answer if you plug into the second equation:

$$6c + 2(-7) = -11$$
$$6c - 14 = -11$$
$$6c = -11 + 14$$
$$6c = 3$$
$$c = \frac{3}{6}$$
$$c = \frac{1}{2}$$

Note: Problems 2.26–2.27 refer to the following system of equations.

$$\begin{cases} 4x + y = 19 \\ 9x + 2y = 40 \end{cases}$$

2.26 Solve the system using substitution.

The first step in the substitution method is to solve one of the equations for either x or y. Of the four variables in this system, it is easiest to solve for y in the equation $4x + y = 19$ because its coefficient is 1.

$$y = -4x + 19$$

Substitute the expression $-4x + 19$ for y in the other equation of the system.

$$9x + 2y = 40$$
$$9x + 2(-4x + 19) = 40$$

Solve the equation for x.

$$9x + 2(-4x) + 2(19) = 40$$
$$9x - 8x + 38 = 40$$
$$x + 38 = 40$$
$$x = 40 - 38$$
$$x = 2$$

The solution to this system of equations is a coordinate pair (x,y). Calculate the corresponding value of y by substituting $x = 2$ into the equation you first solved for y.

$$y = -4x + 19$$
$$= -4(2) + 19$$
$$= -8 + 19$$
$$= 11$$

The solution to the system of equations is $(x,y) = (2,11)$.

Most systems of linear equations have a single coordinate pair solution like this. See Problems 2.38–2.41 for the exceptions.

Note: Problems 2.26–2.27 refer to the following system of equations.
$$\begin{cases} 4x + y = 19 \\ 9x + 2y = 40 \end{cases}$$

2.27 Verify the solution generated by Problem 2.26.

To verify a solution to a system of linear equations, substitute the coordinate pair into each equation of the system. In this case, substitute $(x,y) = (2,11)$ into the equations $4x + y = 19$ and $9x + 2y = 40$.

$$4(2) + 11 = 19 \qquad\qquad 9(2) + 2(11) = 40$$
$$8 + 11 = 19 \qquad\qquad 18 + 22 = 40$$
$$19 = 19 \ \ \text{True} \qquad\qquad 40 = 40 \ \ \text{True}$$

Because each of the resulting statements is true, $(x,y) = (2,11)$ is the solution to the system of equations.

Note: Problems 2.28–2.29 refer to the following system of equations.
$$\begin{cases} x - 3y = -11 \\ 2x + 21y = -4 \end{cases}$$

2.28 Solve the system of equations using substitution.

Solve $x - 3y = -11$ for x by adding $3y$ to both sides of the equation.

$$x = 3y - 11$$

Substitute $x = 3y - 11$ into the second equation of the system and solve for y.

$$2x + 21y = -4$$
$$2(3y - 11) + 21y = -4$$
$$6y - 22 + 21y = -4$$
$$6y + 21y = -4 + 22$$
$$27y = 18$$
$$y = \frac{18}{27}$$
$$y = \frac{2}{3}$$

Divide the top and bottom by 9 to reduce the fraction.

Substitute $y = \dfrac{2}{3}$ into the equation previously solved for x to calculate the corresponding x-value.

$$x = 3y - 11$$
$$= \frac{3}{1}\left(\frac{2}{3}\right) - 11$$
$$= \frac{6}{3} - 11$$
$$= 2 - 11$$
$$= -9$$

Be sure to write the numbers in order. The x-value -9 must come before the y-value $\frac{2}{3}$.

The solution to the system of equations is $(x, y) = \left(-9, \dfrac{2}{3}\right)$.

> **Note: Problems 2.28–2.29 refer to the following system of equations.**
> $$\begin{cases} x - 3y = -11 \\ 2x + 21y = -4 \end{cases}$$

2.29 Verify the solution generated by Problem 2.28.

Substitute the coordinate pair $(x, y) = \left(-9, \dfrac{2}{3}\right)$ into both equations of the system and verify that the resulting statements are true.

$$x - 3y = -11 \qquad\qquad 2x + 21y = -4$$
$$-9 - \frac{3}{1}\left(\frac{2}{3}\right) = -11 \qquad 2(-9) + \frac{21}{1}\left(\frac{2}{3}\right) = -4$$
$$-9 - \frac{6}{3} = -11 \qquad\qquad -18 + \frac{42}{3} = -4$$
$$-9 - 2 = -11 \qquad\qquad -18 + 14 = -4$$
$$-11 = -11 \ \text{ True} \qquad\qquad -4 = -4 \ \text{ True}$$

> **Note: Problems 2.30–2.31 refer to the following system of equations:**
> $$\begin{cases} 2x + 3y = 1 \\ 3x - y = 29 \end{cases}$$

2.30 Solve one of the equations for y and solve the system using substitution.

If you solve 2x + 3y = 1 for y, you have to divide everything by 3.

Solve the second equation for y, because the coefficient of y is -1.

$$3x - y = 29$$
$$3x = 29 + y$$
$$3x - 29 = y$$

Substitute $y = 3x - 29$ into the remaining equation of the system, $2x + 3y = 1$, and solve for x.

$$2x + 3(3x - 29) = 1$$
$$2x + 9x - 87 = 1$$
$$11x = 1 + 87$$
$$11x = 88$$
$$x = \frac{88}{11}$$
$$x = 8$$

Substitute $x = 8$ into the equation already solved for y to complete the ordered pair.

$$y = 3x - 29$$
$$= 3(8) - 29$$
$$= 24 - 29$$
$$= -5$$

The solution to the system is $(x,y) = (8,-5)$.

Note: Problems 2.30–2.31 refer to the following system of equations:

$$\begin{cases} 2x + 3y = 1 \\ 3x - y = 29 \end{cases}$$

2.31 Solve one of the equations for x and solve the system using substitution. Verify that the solutions to Problems 2.30 and 2.31 are equal.

Begin by solving either of the equations for x. In the following solution, $2x + 3y = 1$ is chosen, but solving $3x - y = 29$ for x and proceeding accordingly produces an identical coordinate pair solution.

$$2x + 3y = 1$$
$$2x = -3y + 1$$
$$\frac{2}{2}x = \frac{-3}{2}y + \frac{1}{2}$$
$$x = -\frac{3}{2}y + \frac{1}{2}$$

> Don't expect to get the same equation when you solve $3x - y = 29$ for x. The final answer will match, but almost none of the other steps will.

Substitute $x = -\dfrac{3}{2}y + \dfrac{1}{2}$ into the remaining equation of the system, $3x - y = 29$.

$$3\left(-\frac{3}{2}y + \frac{1}{2}\right) - y = 29$$
$$\frac{3}{1}\left(-\frac{3}{2}y\right) + \frac{3}{1}\left(\frac{1}{2}\right) - y = 29$$
$$-\frac{9}{2}y + \frac{3}{2} - y = 29$$

Multiply the equation by 2, the least common denominator, to eliminate the fractions.

$$\frac{2}{1}\left(-\frac{9}{2}y\right)+\frac{2}{1}\left(\frac{3}{2}\right)+2\left(-y\right)=2\left(29\right)$$

$$-\frac{18}{2}y+\frac{6}{2}-2y=58$$

$$-9y+3-2y=58$$

Solve for y.

$$-9y-2y=58-3$$

$$-11y=55$$

$$y=\frac{55}{-11}$$

$$y=-5$$

Substitute $y=-5$ into the equation already solved for x to complete the solution.

$$x=-\frac{3}{2}y+\frac{1}{2}$$

$$=-\frac{3}{2}\left(-\frac{5}{1}\right)+\frac{1}{2}$$

$$=\frac{15}{2}+\frac{1}{2}$$

$$=\frac{16}{2}$$

$$=8$$

> Substitution works best when one of the equations of the system has an x- or y-term with a coefficient of 1 or –1 (as in Problem 2.30). Otherwise, you end up dealing with fractions, which complicates things.

The solution to the system of equations is $(x,y)=(8,-5)$. Although the solution matches the coordinate pair generated by Problem 2.30, the volume of work required to reach that solution is significantly greater in Problem 2.31.

Note: *Problems 2.32–2.32 refer to the following system of equations:*

$$\begin{cases} 2x+y=9 \\ 5x-y=19 \end{cases}$$

2.32 Solve the system using the elimination technique.

To apply the elimination technique, ensure that the sum of one set of corresponding coefficients is 0. In other words, if you added the x- or y-coefficients together, one of the sums should be 0. In this case, the coefficients of the y-terms have a zero sum. Add the equations by combining like terms.

> The coefficient of y in the first equation is 1, and in the second equation the coefficient is –1. They're opposites, so their sum is 0.

$$
\begin{array}{rcrcr}
2x & + & y & = & 9 \\
5x & - & y & = & 19 \\
\hline
7x & + & 0 & = & 28
\end{array}
$$

Solve the resulting equation.

$$\frac{7x}{7} = \frac{28}{7}$$
$$x = 4$$

Calculate the corresponding y-value (and, therefore, complete the solution) by substituting $x = 4$ into either of the original equations of the system. Here, $x = 4$ is substituted into the first equation, $2x + y = 9$.

$$2(4) + y = 9$$
$$8 + y = 9$$
$$y = 9 - 8$$
$$y = 1$$

The solution to the system of equations is $(x,y) = (4,1)$.

Note: Problems 2.32–2.32 refer to the following system of equations:
$$\begin{cases} 2x + y = 9 \\ 5x - y = 19 \end{cases}$$

2.33 Check the solution generated by Problem 2.32.

Substitute $(x,y) = (4,1)$ into each equation of the system. If both of the resulting statements are true, the solution is valid.

$$2x + y = 9 \qquad\qquad 5x - y = 19$$
$$2(4) + 1 = 9 \qquad\qquad 5(4) - 1 = 19$$
$$8 + 1 = 9 \qquad\qquad 20 - 1 = 19$$
$$9 = 9 \ \text{ True} \qquad\qquad 19 = 19 \ \text{ True}$$

2.34 Solve the following system using the elimination technique.
$$\begin{cases} 4x + 3y = -48 \\ 2x - 5y = 28 \end{cases}$$

The most efficient way to eliminate a variable from the system is to multiply the second equation by –2. The result, $-4x + 10y = -56$, contains an x-term that is the opposite of the x-term in the first equation of the system. Combine the first equation with the newly modified second equation.

$$\begin{array}{rrrcr} 4x & + & 3y & = & -48 \\ -4x & + & 10y & = & -56 \\ \hline 0 & + & 13y & = & -104 \end{array}$$

Solve the resulting equation for y.

$$\frac{13y}{13} = \frac{-104}{13}$$

$$y = -8$$

> You could also plug $y = -8$ into the equation $-4x + 10y = -56$, which you got when you multiplied by –2 in the very beginning of the problem.

Substitute $y = -8$ into either of the original equations to calculate the corresponding value of x.

$$2x - 5y = 28$$
$$2x - 5(-8) = 28$$
$$2x + 40 = 28$$
$$2x = 28 - 40$$
$$2x = -12$$
$$x = \frac{-12}{2}$$
$$x = -6$$

The solution to the system of equations is $(x,y) = (-6,-8)$.

Note: Problems 2.35–2.36 refer to the following system of equations:

$$\begin{cases} 2x + 6y = 11 \\ 7x - 2y = 27 \end{cases}$$

2.35 Solve the system by eliminating y.

> This is the second equation after you multiply each of its terms by 3.

To eliminate y, the y-coefficients of the equations must be opposites. Notice that multiplying the second equation by 3 generates a y-coefficient of $3(-2) = -6$, the opposite of the y-coefficient in the first equation of the system. Combine $2x + 6y = 11$ with the modified second equation.

$$\begin{array}{rrrrr} 2x & + & 6y & = & 11 \\ 21x & - & 6y & = & 81 \\ \hline 23x & + & 0 & = & 92 \end{array}$$

> By plugging x into either $2x + 6y = 11$ or $7x - 2y = 27$.

Solve the equation for x.

$$\frac{23x}{23} = \frac{92}{23}$$
$$x = 4$$

Calculate the corresponding value of y.

$$2x + 6y = 11$$
$$2(4) + 6y = 11$$
$$8 + 6y = 11$$
$$6y = 11 - 8$$
$$6y = 3$$
$$\frac{6y}{6} = \frac{3}{6}$$
$$y = \frac{1}{2}$$

The solution to the system is $(x, y) = \left(4, \frac{1}{2}\right)$.

Note: Problems 2.35–2.36 refer to the following system of equations:

$$\begin{cases} 2x + 6y = 11 \\ 7x - 2y = 27 \end{cases}$$

2.36 Verify the solution generated in Problem 2.35 by solving the system once more, this time by eliminating x.

Unlike Problems 2.34 and 2.35, neither of the coefficients to be eliminated is a multiple of the other. In such cases, the most straightforward method to eliminate the terms is to multiply each equation by the coefficient of the other.

$$\begin{cases} 7[2x + 6y = 11] \\ 2[7x - 2y = 27] \end{cases} = \begin{cases} 14x + 42y = 77 \\ 14x - 4y = 54 \end{cases}$$

> If one of the coefficients divides evenly into the other, you simply multiply one of the equations by a number when eliminating. In this case, 2 and 7 don't divide into each other evenly, so both equations must be multiplied by something.

Now multiply one of the equations by –1 so that the x-terms are opposite values instead of equal values.

$$\begin{cases} 14x + 42y = 77 \\ [-1][14x - 4y = 54] \end{cases} = \begin{cases} 14x + 42y = 77 \\ -14x + 4y = -54 \end{cases}$$

Combine the equations and solve for y.

$$46y = 23$$
$$y = \frac{23}{46}$$
$$y = \frac{1}{2}$$

> Add the x-terms, the y-terms, and the constants: 14x – 14x = 0, 42y + 4y = 46y, and 77 – 54 = 23. You end up with 0 + 46y = 23.

Calculate the corresponding value of x.

$$2x + 6y = 11$$

$$2x + \frac{6}{1}\left(\frac{1}{2}\right) = 11$$

$$2x + \frac{6}{2} = 11$$

$$2x + 3 = 11$$

$$2x = 11 - 3$$

$$2x = 8$$

$$x = \frac{8}{2}$$

$$x = 4$$

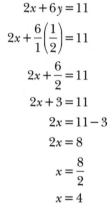

It doesn't matter whether you eliminate x or y to solve a system. Either way, you get the same answer.

The solution to the system of equations is $(x, y) = \left(4, \frac{1}{2}\right)$, the same solution generated by Problem 2.35.

2.37 Solve the following system of equations using either the substitution or elimination technique.

$$\begin{cases} 5x - 2y = -18 \\ 3x + y = 9 \end{cases}$$

The second equation is easily solved for y: $y = -3x + 9$. Substitute this expression into the first equation and solve for x.

$$5x - 2(-3x + 9) = -18$$

$$5x + 6x - 18 = -18$$

$$5x + 6x = -18 + 18$$

$$11x = 0$$

$$\frac{11x}{11} = \frac{0}{11}$$

$$x = 0$$

Determine the corresponding value of y.

$$y = -3x + 9$$

$$= -3(0) + 9$$

$$= 0 + 9$$

$$= 9$$

The solution to the system of equations is $(x, y) = (0, 9)$.

2.38 Solve the following system of equations using either the substitution or elimination technique.

$$\begin{cases} 6x - 2y = -1 \\ 3x - y = 4 \end{cases}$$

Multiply the second equation by –2 to eliminate the *x*-terms.

$$\begin{array}{rrrrr} 6x & - & 2y & = & -1 \\ -6x & + & 2y & = & -8 \\ \hline 0 & + & 0 & = & -9 \end{array}$$

> This is the equation 3x – y = 4 multiplied by –2.

Eliminating the *x*-terms from the system eliminates the *y*-terms as well, producing the false statement 0 = –9. If solving a system produces a false statement, that system has no solution and is described as "inconsistent."

Consider this visual rationale for the behavior of inconsistent systems. The solution to a system of linear equations is the point at which the graphs of the lines intersect. However, inconsistent systems are comprised of parallel lines, which do not intersect; therefore, such systems have no solution.

> Chapter 6 covers parallel lines in great detail.

2.39 Solve the following system of equations using either the substitution or elimination technique.

$$\begin{cases} x - 2y = 4 \\ y = \dfrac{1}{2}x - 2 \end{cases}$$

Substitute the second equation, which is solved for *y*, into the first to calculate the *x*-value of the solution.

$$x - 2\left(\frac{1}{2}x - 2\right) = 4$$

$$x - \frac{2}{1}\left(\frac{1}{2}x\right) - 2(-2) = 4$$

$$x - \frac{2}{2}x + 4 = 4$$

$$x - x + 4 = 4$$

$$4 = 4$$

> That's because the equations are different-looking versions of each other. If you solve the first equation for y, you get the second one.

The resulting equation, 4 = 4, is true regardless of the values of *x* and *y*. Therefore, any point (*x*,*y*) that satisfies one equation of the system satisfies the other equation as well.

When attempting to solve a system results in a true statement (like 4 = 4), the system has an infinite number of solutions and is described as "dependent." The infinite collection of solutions consists of the ordered pairs that satisfy either (and, therefore, both) of the equations.

> In set notation, you can write the solution like this: {(x,y), x – 2y = 4}. This means that the solution includes any point (x,y), as long as x – 2y = 4.

2.40 For what value of c is the following system of equations inconsistent? Assume that $x \neq 0$ and $y \neq 0$.
$$\begin{cases} 4x + y = 3 \\ 6x + cy = -4 \end{cases}$$

If the system is inconsistent, eliminating one of the variables eliminates the other as well. Multiply the first equation by -3 and the second equation by 2 to eliminate the x-terms.

$$\begin{cases} -12x - 3y = -9 \\ 12x + 2cy = -8 \end{cases}$$

Combine the equations of the system.

> Remember, the y-terms are supposed to cancel out just as the x-terms did. This means that they must add up to 0.

$$
\begin{array}{rcrcr}
-12x & - & 3y & = & -9 \\
12x & + & 2cy & = & -8 \\
\hline
0 & + & \left(-3y + 2cy\right) & = & -17
\end{array}
$$

Calculate c.

$$-3y + 2cy = 0$$
$$2cy = 0 + 3y$$
$$\frac{2cy}{2y} = \frac{3y}{2y}$$
$$c = \frac{3}{2}$$

2.41 For what values of a and b is the following system of equations dependent? Assume that $x \neq 0$ and $y \neq 0$.
$$\begin{cases} 3x - 2y = -7 \\ ax + by = 2 \end{cases}$$

A dependent system of equations consists of equivalent equations. Both of the equations in this system are written in the same form (a sum of an x- and a y-term are set equal to a constant). Calculate the real number k by which the first equation must be multiplied to generate the second equation.

> The only way these two equations could be equal is if you multiply one of them by a number to get the other. Each of the terms is multiplied by the same mystery number (called k here), so multiply the constant −7 in the first equation by k and set it equal to 2, the constant in the second equation.

$$-7k = 2$$
$$\frac{-7k}{-7} = \frac{2}{-7}$$
$$k = -\frac{2}{7}$$

To generate the x- and y-terms of the second equation, multiply the x- and y-terms of the first equation by $-\frac{2}{7}$.

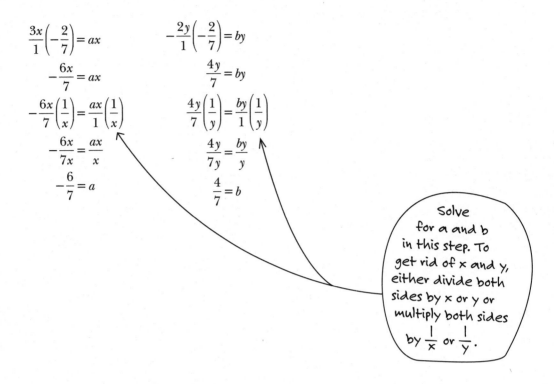

$$\frac{3x}{1}\left(-\frac{2}{7}\right) = ax \qquad\qquad -\frac{2y}{1}\left(-\frac{2}{7}\right) = by$$

$$-\frac{6x}{7} = ax \qquad\qquad \frac{4y}{7} = by$$

$$-\frac{6x}{7}\left(\frac{1}{x}\right) = \frac{ax}{1}\left(\frac{1}{x}\right) \qquad\qquad \frac{4y}{7}\left(\frac{1}{y}\right) = \frac{by}{1}\left(\frac{1}{y}\right)$$

$$-\frac{6x}{7x} = \frac{ax}{x} \qquad\qquad \frac{4y}{7y} = \frac{by}{y}$$

$$-\frac{6}{7} = a \qquad\qquad \frac{4}{7} = b$$

Solve for a and b in this step. To get rid of x and y, either divide both sides by x or y or multiply both sides by $\frac{1}{x}$ or $\frac{1}{y}$.

Chapter 3
QUADRATIC EQUATIONS

And some other related skills

The techniques applied in Chapter 2 to solve linear equations are not sufficient for solving equations of higher degree. Quadratic equations, for instance, are typically solved using the quadratic formula or by factoring the expressions within. However, before you can investigate those methods, you must master a number of prerequisite arithmetic and algebraic skills. This chapter begins by highlighting those foundational skills, including factoring and the manipulation of radical expressions.

The title of the chapter is "Quadratic Equations," but you don't actually solve them until the very end. Before then, you review factoring real numbers, factoring things with variables in them, and simplifying radicals (such as square roots). The techniques include familiar concepts, such as greatest common factor, the fundamental theorem of arithmetic, and the difference of perfect squares.

Factoring Real Numbers

The fundamental theorem of arithmetic and greatest common factors

> The natural numbers are 1, 2, 3, 4, 5, and so on. They are positive and don't contain fractions or decimals.

3.1 Complete the following statement.

According to the fundamental theorem of arithmetic, all natural numbers greater than 1 are equal to a _____ product of _____.

According to the fundamental theorem of arithmetic, all natural numbers greater than 1 are equal to a <u>unique</u> product of <u>prime numbers</u>. This unique set of factors is called the "prime factorization" for that natural number.

3.2 Identify the prime factorization of 14.

The natural number 14 is equal to the product of two prime numbers: $14 = 2 \cdot 7$. Although $1 \cdot 14$ also equals 14, 1 is not a prime number.

Note: Problems 3.3–3.4 demonstrate that the prime factorization of a number is unchanged regardless of the initial product you choose.

3.3 Identify the prime factorization of 250, beginning with the initial product $250 = 25 \cdot 10$.

Neither of the factors identified by the problem is prime, because 10 and 25 are divisible by natural numbers other than 1 and the numbers themselves. Rewrite the factorization of 250 by explicitly identifying the factors of 10 and 25.

$$250 = 25 \cdot 10$$
$$250 = (5 \cdot 5) \cdot (2 \cdot 5)$$

> The prime factorization consists of three 5's. Instead of writing $5 \cdot 5 \cdot 5$, write 5^3.
> $250 = 2 \cdot 5^3$

The resulting factors (2 and the repeated factor 5) are prime. Complete the prime factorization, using an exponent to denote the repeated factor: $250 = 2 \cdot 5^3$.

Note: Problems 3.3–3.4 demonstrate that the prime factorization of a number is unchanged regardless of the initial product you choose.

3.4 Identify the prime factorization of 250, starting with the initial product $250 = 125 \cdot 2$.

Although the factor 2 is prime, 125 is not. Notice that 125 is evenly divisible by 25: $125 = 5 \cdot 25$.

$$250 = 125 \cdot 2$$
$$250 = (5 \cdot 25) \cdot 2$$

All the resulting factors are prime except 25, which should be written as a factor of prime numbers: $25 = 5 \cdot 5$.

$$250 = 5 \cdot (5 \cdot 5) \cdot 2 \;\longleftarrow$$

Use exponents to indicate repeated factors in the prime factorization. It is conventional to write the factors in order, from least to greatest.

$$250 = 2 \cdot 5^3$$

Notice that the prime factorizations in Problems 3.3 and 3.4 are equal, despite the different initial products used to generate them.

> You might have noticed right away that $125 = 5^3$ and skipped the $25 = 5 \cdot 5$ step, but it's good to know that you don't have to. Just keep breaking down the composite (factorable) numbers until all the numbers are prime.

3.5 Identify the prime factorization of 324.

Notice that 324 is evenly divisible by 4.

$$324 = 4 \cdot 81$$

Neither 4 nor 81 is a prime number, but both are perfect squares of prime numbers.

$$324 = (2 \cdot 2) \cdot (9 \cdot 9)$$

One of the repeated factors is not prime. Express it as a product of prime numbers: $9 = 3 \cdot 3$.

$$324 = 2 \cdot 2 \cdot (3 \cdot 3) \cdot (3 \cdot 3)$$
$$324 = 2^2 \cdot 3^4$$

> If you multiply a number times itself, the result is a perfect square. In this case, 2 times itself equals 4, and 9 times itself equals 81.

3.6 Identify the greatest common factor of 12 and 36.

One of the numbers is evenly divisible by the other ($36 \div 12$ has no remainder). Therefore, 12 is the greatest common factor of 12 and 36. $\longleftarrow$

3.7 Identify the greatest common factor of 10 and 25.

Generate the prime factorizations of 10 and 25, using the technique modeled in Problems 3.2–3.5.

$$10 = 2 \cdot 5 \qquad 25 = 5^2$$

The prime factorizations have one factor in common: 5. More specifically, 5^1 is a factor of 10, and 5^2 is a factor of 25. The greatest common factor, therefore, is the common factor 5 raised to the lesser of the two powers: 5^1. It is unnecessary to include an exponent of 1, so it is also correct (and more common) to report that the greatest common factor is 5.

> The greatest common factor can't be bigger than the smallest number. In rare cases, like this one, it actually is the smaller number.

3.8 Identify the greatest common factor of 48 and 160.

Generate the prime factorizations of 48 and 160.

$$48 = 4 \cdot 12 \qquad\qquad 160 = 2 \cdot 80$$
$$48 = (2 \cdot 2) \cdot (4 \cdot 3) \qquad 160 = 2 \cdot (4 \cdot 20)$$
$$48 = 2 \cdot 2 \cdot (2 \cdot 2) \cdot 3 \qquad 160 = 2 \cdot (2 \cdot 2) \cdot (10 \cdot 2)$$
$$48 = 2^4 \cdot 3 \qquad\qquad 160 = 2 \cdot 2 \cdot 2 \cdot (2 \cdot 5) \cdot 2$$
$$160 = 2^5 \cdot 5$$

The prime factorizations share one common factor: 2. More specifically, 2^4 is a factor of 48 and 2^5 is a factor of 160. The greatest common factor, therefore, is the common factor raised to the lesser exponent: $2^4 = 16$.

3.9 Identify the greatest common factor of 375 and 1,575.

Generate the prime factorizations of 375 and 1,575.

$$375 = 3 \cdot 125 \qquad\qquad 1{,}575 = 9 \cdot 175$$
$$375 = 3 \cdot (5 \cdot 25) \qquad 1{,}575 = (3 \cdot 3) \cdot (7 \cdot 25)$$
$$375 = 3 \cdot 5 \cdot (5 \cdot 5) \qquad 1{,}575 = 3 \cdot 3 \cdot 7 \cdot (5 \cdot 5)$$
$$375 = 3 \cdot 5^3 \qquad\qquad 1{,}575 = 3^2 \cdot 5^2 \cdot 7$$

> Here's a trick: add the digits of 375: $3 + 7 + 5 = 15$. The sum (15) is divisible by 3, so the actual number 375 is divisible by 3. Caution: This doesn't work for all numbers. For example, 32 is NOT divisible by 5 even though $3 + 2 = 5$.

The numbers 375 and 1,575 have two common factors: 3 and 5. More specifically, 3^1 and 5^3 are factors of 375, and 3^2 and 5^2 are factors of 1,575. The greatest common factor, therefore, is the product of 3 and 5 when each is raised to its lowest power: $3^1 \cdot 5^2 = 3 \cdot 25 = 75$.

Simplifying Radicals
Cleaning up square roots and cube roots

3.10 Simplify the radical expression: $\sqrt[3]{8}$.

> This expression asks the question, "What is x if $x \cdot x \cdot x = 8$?" In other words, if three x's are multiplied to get 8, what must x be?

To simplify a radical expression, remove factors of the radicand that are raised to the same exponent as the index of the expression. In this expression, the radicand is 8 and the index is 3; therefore, you should identify factors of 8 that are raised to the third power.

The prime factorization of 8 is 2^3. Write the radical expression, explicitly identifying the cubed factor.

$$\sqrt[3]{8} = \sqrt[3]{2^3}$$

A factor raised to a power that is equal to the index of the radical expression may be removed from the radicand.

$$\sqrt[3]{2^3} = 2$$

Therefore, $\sqrt[3]{8} = 2$.

3.11 Simplify the radical expression: $\sqrt{49}$.

This radical expression has no explicitly stated index. Any such expression has an implicit index of 2. Therefore, to simplify the expression, you must identify factors of 49 that are raised to the second power. Notice that the prime factorization of 49 is 7^2. Rewrite the radical, explicitly identifying the squared factor.

$$\sqrt{49} = \sqrt{7^2} = 7$$

The exponent of the factor 7^2 matches the index of the radical (2), so pull 7 out of the radical and drop the matching exponent.

3.12 Simplify the radical expression: $\sqrt{18}$.

This expression, like the expression $\sqrt{49}$ in Problem 3.11, has an implied index of 2 and is described as a "square root." Identify the prime factorization of 18 and substitute it into the expression.

$$\sqrt{18} = \sqrt{2 \cdot 9} = \sqrt{2 \cdot 3^2}$$

Simplify the expression by removing the factor 3^2 from the radicand, because its exponential power is equal to the index of the expression.

$$\sqrt{2 \cdot 3^2} = 3 \cdot \sqrt{2}$$

Therefore, $\sqrt{18} = 3\sqrt{2}$.

The expression $\sqrt[3]{8}$ in Problem 3.10 has an index of 3. Radicals with that index are referred to as "cube roots."

3.13 Verify the solution generated by Problem 3.12.

If $3\sqrt{2} = \sqrt{18}$, then squaring both sides of the equation should produce a true statement that contains no radical expressions.

$$\left(3\sqrt{2}\right)^2 = \left(\sqrt{18}\right)^2$$
$$3^2\left(\sqrt{2}\right)^2 = \left(\sqrt{18}\right)^2$$
$$9 \cdot \sqrt{2} \cdot \sqrt{2} = \sqrt{18} \cdot \sqrt{18}$$

You raise both sides to the second power because the index of the radicals is 2.

According to a property of radicals, the product of two radical expressions with the same index is equal to the radical of the product: $\sqrt[n]{a} \cdot \sqrt[n]{b} = \sqrt[n]{ab}$. In this problem, $\sqrt{2} \cdot \sqrt{2} = \sqrt{2 \cdot 2}$ and $\sqrt{18} \cdot \sqrt{18} = \sqrt{18 \cdot 18}$.

$$9\sqrt{2 \cdot 2} = \sqrt{18 \cdot 18}$$
$$9\sqrt{2^2} = \sqrt{18^2}$$

Simplify the radical expressions.

$$9 \cdot 2 = 18 \quad \text{True}$$

Because the final result is a true statement, the solution to Problem 3.12 is verified.

3.14 Simplify the radical expression: $\sqrt[3]{135}$.

Identify the prime factorization of 135.

$$135 = 5 \cdot 27$$
$$135 = 5 \cdot (3 \cdot 9)$$
$$135 = 5 \cdot 3 \cdot (3 \cdot 3)$$
$$135 = 3^3 \cdot 5$$

Rewrite the radical expression, explicitly identifying the factor raised to the third power, and simplify the expression.

$$\sqrt[3]{135} = \sqrt[3]{3^3 \cdot 5} = 3 \cdot \sqrt[3]{5}$$

3.15 Simplify the radical expression: $\sqrt[3]{x^7 y^8}$.

The index of the radical expression is 3, so divide each of the exponents by 3: $7 \div 3 = 2$ remainder 1 and $8 \div 3 = 2$ remainder 2. According to these results, x^7 has the perfect cube factor $\left(x^2\right)^3$, and y^8 has the perfect cube factor $\left(y^2\right)^3$.

$$x^7 = \left(x^2\right)^3 \cdot x^1$$
$$y^8 = \left(y^2\right)^3 \cdot y^2$$

You're dividing seven x's and eight y's into groups of three, because you need something to the third power to simplify the expression. You can form two groups of three x's and two groups of three y's, leaving one x and two y's ungrouped.

$7 \div 3 = 2$ remainder 1, so x to the quotient (2) is raised to the third power and multiplied by x to the remainder (1); $8 \div 3 = 2$ remainder 2, so y to the quotient (2) is raised to the third power and multiplied by y to the remainder (2).

Rewrite the radical expression using the factored forms of x^7 and y^8 that contain the factors raised to the third power.

$$\sqrt[3]{x^7 y^8} = \sqrt[3]{\left(x^2\right)^3 \cdot x^1 \cdot \left(y^2\right)^3 \cdot y^2}$$

Simplify the expression by removing the factors raised to the third power from the radicand.

$$\sqrt[3]{x^7 y^8} = \left(x^2 y^2\right)\sqrt[3]{xy^2}$$

3.16 Simplify the radical expression: $\sqrt{x^4 y^{11}}$.

The implied index of the radical is 2, so divide each of the exponents by 2 to rewrite the expression using squared factors: $4 \div 2 = 2$ and $11 \div 2 = 5$ remainder 1.

$$\sqrt{x^4 y^{11}} = \sqrt{\left(x^2\right)^2 \cdot \left(y^5\right)^2 \cdot y^1}$$

Simplify the expression by removing the squared factors from the radicand. If a variable is raised to an even power that is equal to the index of the radical, the factor is placed in absolute values: $\sqrt[n]{x^n} = |x|$ if n is even.

$$= \left|x^2 y^5\right|\sqrt{y}$$

Even roots always have to be positive, and absolute values make sure they are, no matter what x and y are equal to.

Note that $x^2 \geq 0$ for all x, so placing x^2 within absolute values is unnecessary.

$$= x^2 \left| y^5 \right| \sqrt{y}$$

> x^2 (or x to any even power) isn't negative even if x is negative. There's no need to force x^2 to be non-negative using absolute values.

3.17 Simplify the expression: $\left(\sqrt{6xy} \right)\left(\sqrt{2x} \right)$.

The product of two radicals with the same index is equal to the radical of the product.

$$\left(\sqrt{6xy} \right)\left(\sqrt{2x} \right) = \sqrt{6 \cdot 2 \cdot x \cdot x \cdot y}$$
$$= \sqrt{12x^2 y}$$

Rewrite the radicand, explicitly identifying the factors that are raised to the second power; remove those factors from the radicand.

$$= \sqrt{2^2 \cdot 3 \cdot x^2 \cdot y}$$
$$= 2\left| x \right| \sqrt{3y}$$

> When you take a variable with an even exponent out of a radical with a matching index, you must use absolute values. See Problem 3.16.

3.18 Simplify the expression: $\sqrt{27x} + \sqrt{3x}$.

You cannot add or subtract radical expressions unless they have like radicands (the same quantity within the radical). Initially, these two terms do not share a common radicand, but they will after $\sqrt{27x}$ is simplified.

$$\sqrt{27x} + \sqrt{3x} = \sqrt{9 \cdot 3 \cdot x} + \sqrt{3x}$$
$$= \sqrt{3^2 \cdot 3 \cdot x} + \sqrt{3x}$$
$$= 3\sqrt{3x} + \sqrt{3x}$$

> $\sqrt{3x}$ doesn't have an explicitly stated coefficient, so its implied coefficient is 1: $3\sqrt{3x} + 1\sqrt{3x} = 4\sqrt{3x}$.

The terms now contain the common radical $\sqrt{3x}$, so their coefficients may be combined.

$$= 4\sqrt{3x}$$

3.19 Simplify the expression: $\sqrt[5]{64x^9} - \sqrt[5]{486x^4}$.

To simplify radicals of index 5, it is helpful to consider several natural numbers raised to the fifth power: $2^5 = 32$, $3^5 = 243$, $4^5 = 1,024$, and $5^5 = 3,125$. Notice that 32 and 243 are factors of the left and right radical expressions, respectively. Rewrite the radicands, explicitly identifying the factors raised to the fifth power.

$$\sqrt[5]{64x^9} - \sqrt[5]{486x^4} = \sqrt[5]{32 \cdot 2 \cdot x^5 \cdot x^4} - \sqrt[5]{243 \cdot 2 \cdot x^4}$$
$$= \sqrt[5]{2^5 \cdot 2 \cdot x^5 \cdot x^4} - \sqrt[5]{3^5 \cdot 2 \cdot x^4}$$
$$= 2x\sqrt[5]{2x^4} - 3\sqrt[5]{2x^4}$$
$$= (2x - 3)\left(\sqrt[5]{2x^4}\right)$$

> The coefficients of the radicals are 2x and –3. They're not like terms (one has an x and one doesn't), so "combining" them just means writing them together in front of the common radical.

3.20 Rationalize the denominator: $\dfrac{5x}{\sqrt{2}}$.

A rationalized denominator does not contain a radical expression. To eliminate $\sqrt{2}$, multiply the numerator and denominator of the fraction by $\sqrt{2}$.

$$\frac{5x}{\sqrt{2}} \cdot \frac{\sqrt{2}}{\sqrt{2}} = \frac{5x\sqrt{2}}{\sqrt{2 \cdot 2}}$$
$$= \frac{5x\sqrt{2}}{\sqrt{2^2}}$$
$$= \frac{5x\sqrt{2}}{2}$$

> Multiply the bottom AND top of the fraction by the same value.

> To eliminate a cube root, you need things to the third power. You've got x^2 already, so you need only one more x. However, there's only one 7, so you need two more: 7^2.

3.21 Rationalize the denominator: $\sqrt[3]{\dfrac{4x}{7x^2}}$.

The radical of a quotient is equal to the quotient of the radicals.

$$\sqrt[3]{\frac{4x}{7x^2}} = \frac{\sqrt[3]{4x}}{\sqrt[3]{7x^2}}$$

To create a perfect cube in the radicand of the denominator, multiply by $\sqrt[3]{7^2 x}$. Multiply the numerator by the same value to ensure that the value of the fraction remains unchanged.

$$= \frac{\sqrt[3]{4x}}{\sqrt[3]{7x^2}} \cdot \frac{\sqrt[3]{7^2 \cdot x}}{\sqrt[3]{7^2 \cdot x}}$$
$$= \frac{\sqrt[3]{4 \cdot 7^2 \cdot x \cdot x}}{\sqrt[3]{7^3 \cdot x^3}}$$
$$= \frac{\sqrt[3]{196x^2}}{7x}$$

Note that $\sqrt[3]{196x^2}$ cannot be further simplified because its prime factorization $(2^2 \cdot 7^2)$ contains no perfect cubes.

Factoring Algebraic Expressions

Using the GCF and common factoring patterns

3.22 Identify the greatest common factor of $20x^2y^5$ and $30x^3y^4$.

Express the coefficients 20 and 30 as prime factorizations.

$$20x^2y^5 = 2^2 \cdot 5 \cdot x^2 \cdot y^5$$
$$30x^3y^4 = 2 \cdot 3 \cdot 5 \cdot x^3 \cdot y^4$$

The terms share four common factors: 2, 5, x, and y. The greatest common factor of $20x^2y^5$ and $30x^3y^4$ is the product of the common factors raised to their lowest powers: $2^1 \cdot 5^1 \cdot x^2 \cdot y^4 = 10x^2y^4$.

> In other words, use 2^1 instead of 2^2, x^2 instead of x^3, and y^4 instead of y^5.

Note: Problems 3.23–3.24 refer to the expression $24xy^7 + 60x^9y^5z$.

3.23 Identify the greatest common factor of the terms in the expression.

Express the coefficients 24 and 60 as prime factorizations.

$$24xy^7 = 2^3 \cdot 3 \cdot x \cdot y^7$$
$$60x^9y^5z = 2^2 \cdot 3 \cdot 5 \cdot x^9 \cdot y^5 \cdot z$$

The terms share common factors: 2, 3, x, and y. The greatest common factor of $24xy^7$ and $60x^9y^5z$ is the product of the common factors raised to their lowest powers: $2^2 \cdot 3 \cdot x \cdot y^5 = 12xy^5$.

Note: Problems 3.23–3.24 refer to the expression $24xy^7 + 60x^9y^5z$.

3.24 Factor the expression.

According to Problem 3.23, the greatest common factor of the expression is $12xy^5$. Divide each of the terms by the greatest common factor.

> When you divide exponential expressions with the same base, subtract the powers. For example, $\dfrac{y^7}{y^5} = y^{7-5}$.

$$\frac{24xy^7}{12xy^5} = \frac{24}{12}x^{1-1}y^{7-5} = 2y^2$$
$$\frac{60x^9y^5z}{12xy^5} = \frac{60}{12}x^{9-1}y^{5-5}z = 5x^8z$$

Write each term as the product of the greatest common factor $12xy^5$ and the quotients identified above.

$$24xy^7 + 60x^9y^5z = (12xy^5)(2y^2) + (12xy^5)(5x^8z)$$

Factor the expression by removing $12xy^5$ from each term and multiplying it by the sum of the remaining factors.

$$= 12xy^5(2y^2 + 5x^8z)$$

> Write the GCF. Then write what's left over from each term (after the GCF is removed) inside a set of parentheses. You can check your answer by distributing $12xy^5$ to see if you get the original expression.

3.25 Factor the expression: $6x^3 - 3x^2 + 21x$.

Divide each of the terms by $3x$, the greatest common factor.

$$\frac{6x^3}{3x} = \frac{6}{3}x^{3-1} = 2x^2$$

$$\frac{-3x^2}{3x} = -\frac{3}{3}x^{2-1} = -x$$

$$\frac{21x}{3x} = \frac{21}{3}x^{1-1} = 7$$

Write the expression as a product of the greatest common factor and the sum of the quotients.

$$6x^3 - 3x^2 + 21x = 3x(2x^2 - x + 7)$$

3.26 Factor by grouping: $x^3 + 5x^2 - 2x - 10$.

The four terms of the expression share no common factor. Instead of considering the expression as a whole, group it into two smaller expressions; place the two leftmost terms into one group and the two rightmost terms into another.

$$(x^3 + 5x^2) + (-2x - 10)$$

> If you factor -2 out of both terms instead of just 2, the numbers inside both sets of parentheses will match.

The terms in the left quantity have a greatest common factor of x^2; the terms in the right quantity have a greatest common factor of -2. Factor both quantities individually.

$$= x^2(x + 5) + (-2)(x + 5)$$

The expression now consists of two terms, $x^2(x + 5)$ and $(-2)(x + 5)$. Factor $(x + 5)$ out of both terms.

$$= (x + 5)(x^2 + (-2))$$

$$= (x + 5)(x^2 - 2)$$

> Just as you factored x^2 out of $x^3 + 5x^2$ to get $x^2(x + 5)$, factor $(x + 5)$ out of both terms here, which leaves behind x^2 in the first term and -2 in the second.

Therefore, $x^3 + 5x^2 - 2x - 10 = (x + 5)(x^2 - 2)$.

3.27 Factor by grouping: $14x^3 + 21x^2 + 8x + 12$.

Use the method described in Problem 3.26: group the expression into two quantities and then factor the greatest common factor out of each.

$$\left(14x^3 + 21x^2\right) + \left(8x + 12\right)$$
$$= 7x^2\left(2x + 3\right) + 4\left(2x + 3\right)$$

The two resulting terms share the common factor $(2x + 3)$, which should be factored out of each.

$$= (2x + 3)\left(7x^2 + 4\right)$$

3.28 Factor the expression: $x^2 - 64$.

Notice that the expression is a difference of perfect squares—one perfect square is subtracted from another. The perfect square $a^2 - b^2$ can be factored using the formula $(a + b)(a - b)$. In this case, $a^2 = x^2$ and $b^2 = 64$; therefore, $a = x$ and $b = 8$.

$$a^2 - b^2 = (a + b)(a - b)$$
$$(x)^2 - (8)^2 = (x + 8)(x - 8)$$
$$x^2 - 64 = (x + 8)(x - 8)$$

You get a perfect square by multiplying something times itself. In this case, $x \cdot x = x^2$ and $8 \cdot 8 = 64$.

3.29 Factor the expression: $8x^2 - 50$.

The greatest common factor of $8x^2$ and -50 is 2; factor it out of the expression.

$$8x^2 - 50 = 2(4x^2 - 25)$$

The parenthetical quantity $(4x^2 - 25)$ is a difference of perfect squares. Apply the formula $a^2 - b^2 = (a + b)(a - b)$, as described in Problem 3.28.

$$= 2(2x + 5)(2x - 5) \longleftarrow$$

$4x^2$ is a perfect square because $(2x)(2x) = 4x^2$, and 25 is a perfect square because $5 \cdot 5 = 25$. Set $a = 2x$ and $b = 5$.

3.30 Factor the expression: $64x^3 - 27$.

This expression is a difference of perfect cubes, because $(4x)^3 = (4x)(4x)(4x) = 64x^3$ and $3^3 = (3)(3)(3) = 27$. Differences of perfect *cubes* are factored according to the formula $a^3 - b^3 = (a - b)(a^2 + ab + b^2)$; in this problem, $a^3 = 64x^3$ and $b^3 = 27$, so $a = 4x$ and $b = 3$.

$$a^3 - b^3 = (a - b)\left(a^2 + ab + b^2\right)$$
$$(4x)^3 - (3)^3 = (4x - 3)\left[(4x)^2 + (4x)(3) + (3)^2\right]$$
$$64x^3 - 27 = (4x - 3)\left(16x^2 + 12x + 9\right)$$

3.31 Factor the expression: $x^3 + 1,000$.

This expression is the sum of perfect cubes. Unlike the sum of perfect squares, sums of perfect cubes may be factored by means of the formula $a^3 + b^3 = (a + b)(a^2 - ab + b^2)$. In this problem, $a^3 = x^3$ and $b^3 = 1,000$, so $a = x$ and $b = 10$.

$$a^3 + b^3 = (a + b)(a^2 - ab + b^2)$$
$$(x)^3 + (10)^3 = (x + 10)\left[(x)^2 - (x)(10) + (10)^2\right]$$
$$x^3 + 1,000 = (x + 10)(x^2 - 10x + 100)$$

> So you can't factor $x^2 + 2^2$ (a sum of perfect squares), but you CAN factor $x^3 + 2^3$ (a sum of perfect cubes).

3.32 Factor the expression $x^2 + 9x + 14$ and verify your answer.

To factor the quadratic trinomial $x^2 + ax + b$, you must identify two numbers m and n such that $m + n = a$ and $mn = b$.

$$x^2 + ax + b = (x + m)(x + n)$$

In this problem, m and n must have a sum of 9 and a product of 14, so $m = 2$ and $n = 7$. (Note that it is equally correct to set $m = 7$ and $n = 2$. The values themselves are important, but the variable names of the values are not.)

$$x^2 + 9x + 14 = (x + 2)(x + 7)$$

To verify the factored form of the expression, distribute each of the terms in the left quantity to the terms in the right quantity, and combine the results. In other words, simplify the expression $x(x + 7) + 2(x + 7)$.

$$x(x + 7) + 2(x + 7) = x(x) + x(7) + 2(x) + 2(7)$$
$$= x^2 + 7x + 2x + 14$$
$$= x^2 + 9x + 14$$

Because the result is the original trinomial expression, the factored form of the expression is correct.

> Trinomial means "containing three terms," and quadratic means "the highest exponent is 2."

3.33 Factor the expression: $x^2 - 8x + 7$.

As explained in Problem 3.32, to factor this quadratic trinomial with a leading coefficient of 1, you must identify two numbers m and n such that $m + n = -8$ and $mn = 7$. Notice that the product mn is positive, so m and n are either both positive numbers or both negative numbers. Furthermore, the sum of m and n is negative, so m and n must be negative numbers.

If $m = -1$ and $n = -7$, then $m + n = -1 + (-7) = -8$ and $mn = (-1)(-7) = 7$. Factor the quadratic.

$$x^2 - 8x + 7 = (x + m)(x + n)$$
$$= (x + (-1))(x + (-7))$$
$$= (x - 1)(x - 7)$$

> The leading coefficient is the number in front of the variable that's raised to the highest exponent. In this case, x^2 has the highest exponent and its coefficient is 1.

3.34 Factor the expression: $x^2 + 11x - 26$.

Using the technique described in Problems 3.32–3.33, the quadratic trinomial $x^2 + 11x - 26$ is factored $(x + m)(x + n)$ when $m + n = 11$ and $mn = -26$. The product mn is negative, so m and n have opposite signs; one is positive and the other is negative. Furthermore, the sum is positive, so the positive number has a greater absolute value than the negative number.

> This shortcut works only when x^2 has a coefficient of 1. If the coefficient of x^2 is anything else, use the factoring by decomposition technique described in Problems 3.35–3.36.

If $m = -2$ and $n = 13$, then $m + n = -2 + 13 = 11$ and $mn = (-2)(13) = -26$. Therefore, the factored form of $x^2 + 11x - 26$ is $(x - 2)(x + 13)$.

3.35 Factor the expression: $2x^2 + 3x - 5$.

This problem differs from Problems 3.32–3.34, because the coefficient of x^2 is not 1. To factor this expression, use a modified version of the technique described in the preceding problems, known as factoring by decomposition. The goal remains the same: identify two numbers, m and n, based upon the coefficients of the expression. However, the factored form of the expression will *not* be $(x + m)(x + n)$, and n is identified in a slightly different manner.

> In other words, if you ignore the signs, the positive number is "bigger" than the negative.

To factor the expression $ax^2 + bx + c$, identify two numbers m and n such that $m + n = b$ and $mn = ac$. In this problem, m and n should have a sum of 3 and a product of $2(-5) = -10$. The product mn is negative, so m and n must have different signs. The sum $m + n$ is positive, so the absolute value of the positive number must be greater than the absolute value of the negative number.

If $m = -2$ and $n = 5$, then $m + n = -2 + 5 = 3$ and $mn = (-2)(5) = -10$. Rewrite the coefficient $b = 3$ in the expression as $m + n = -2 + 5$.

$$2x^2 + 3x - 5 = 2x^2 + (-2 + 5)x - 5$$

Distribute x through the adjacent quantity $(-2 + 5)$.

$$= 2x^2 - 2x + 5x - 5$$

Factor the expression by grouping, as described in Problems 3.26–3.27.

$$= \left(2x^2 - 2x\right) + \left(5x - 5\right)$$
$$= 2x(x - 1) + 5(x - 1)$$
$$= (x - 1)(2x + 5)$$

3.36 Factor the expression: $6x^2 - 23x + 21$.

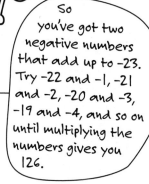

The coefficient of x^2 does not equal 1, so the quadratic trinomial should be factored by decomposition, the technique outlined in Problem 3.35. Begin by identifying two numbers, m and n, such that $m + n = -23$ and $mn = 6(21) = 126$. The product mn is positive, so m and n must have the same sign; the sum $m + n$ is negative, so m and n must be negative numbers.

> So you've got two negative numbers that add up to –23. Try –22 and –1, –21 and –2, –20 and –3, –19 and –4, and so on until multiplying the numbers gives you 126.

If $m = -14$ and $n = -9$, then $m + n = -14 - 9 = -23$ and $mn = (-14)(-9) = 126$. Replace the x-coefficient in the expression $6x^2 - 23x + 21$ with $m + n = -23$ and factor the resulting expression by grouping.

Factor
−3 out of
(−9x + 21) so
that the result
is (3x − 7), which
matches the other
set of parentheses.
If you factor out 3
instead, you get
(−3x + 7), which
is the opposite
of (3x − 7).

$$6x^2 - 23x + 21 = 6x^2 + (-14 - 9)x + 21$$
$$= 6x^2 - 14x - 9x + 21$$
$$= \left(6x^2 - 14x\right) + \left(-9x + 21\right)$$
$$= 2x(3x - 7) + (-3)(3x - 7)$$
$$= (3x - 7)(2x - 3)$$

Solving Quadratic Equations

By factoring and applying the quadratic formula

3.37 Complete the following statement and justify your answer.

According to the zero product property, if xy = 0, then _____.

No two values have a product of 0 unless at least one of the values is equal to 0. Therefore, according to the zero product property, if $xy = 0$, then $\underline{x = 0 \text{ or } y = 0}$.

Note: Problems 3.38–3.39 refer to the equation x(x − 5) = 0.

3.38 Solve the equation.

Solve the
equation x − 5 = 0
(by adding 5 to both
sides) to get the
solution x = 5.

According to the zero product property, if the product of x and $(x - 5)$ equals 0, then either $x = 0$ or $x - 5 = 0$. Therefore, the solution is $x = 0$ or $x = 5$.

Note: Problems 3.38–3.39 refer to the equation x(x − 5) = 0.

3.39 Explain why the equation has two possible solutions, whereas the linear equations investigated in Chapter 2 have only one possible solution.

Complex
numbers
include both
real and imaginary
numbers. Because
geometry deals
with measurement,
you'll deal almost
exclusively with real
numbers—measuring
things using
imaginary numbers
can be as hard
as it sounds.

According to the fundamental theorem of algebra, an equation of degree n has exactly n complex roots. Linear equations have degree 1 and therefore have only one solution. Apply the distributive property to the equation $x(x - 5) = 0$, and the result is $x^2 - 5x = 0$, a quadratic equation. Quadratic equations have degree 2 and therefore possess exactly two complex roots.

3.40 Solve the equation: $(x-4)(2x+9)(4x-1) = 0$.

According to the zero product property, when a product equals 0, at least one of the factors must equal 0. Set each of the factors equal to 0 and solve the three resulting equations.

$$
\begin{array}{ccc}
x - 4 = 0 & 2x + 9 = 0 & 4x - 1 = 0 \\
& \text{or} \quad 2x = -9 & \text{or} \quad 4x = 1 \\
x = 4 & x = -\dfrac{9}{2} & x = \dfrac{1}{4}
\end{array}
$$

The solution to the equation is $x = -\dfrac{9}{2}$, $x = \dfrac{1}{4}$, or $x = 4$.

3.41 Solve the equation by factoring: $x^2 - 15x + 56 = 0$.

Factor the quadratic expression using the technique described in Problems 3.32–3.34.

$$(x - 7)(x - 8) = 0$$

Set each of the factors equal to 0 and solve the resulting equations.

$$
\begin{array}{cc}
x - 7 = 0 & x - 8 = 0 \\
\text{or} & \\
x = 7 & x = 8
\end{array}
$$

The solution to the equation is $x = 7$ or $x = 8$.

3.42 Solve the equation by factoring: $20x^3 + 13x^2 = 15x$.

To solve a quadratic equation by factoring, apply the zero product property.

$$20x^3 + 13x^2 - 15x = 0$$

Factor x, the greatest common factor, out of the terms left of the equals sign.

$$x(20x^2 + 13x - 15) = 0$$

Factor the quadratic trinomial by decomposition.

$$
\begin{aligned}
x\left[20x^2 + (25 - 12)x - 15\right] &= 0 \\
x\left[(20x^2 + 25x) + (-12x - 15)\right] &= 0 \\
x\left[5x(4x + 5) - 3(4x + 5)\right] &= 0 \\
x(4x + 5)(5x - 3) &= 0
\end{aligned}
$$

One side of the equation has to equal 0, so subtract 15x from both sides. The last step has to set all the factors (on one side of the equation) equal to the 0 on the other side.

You need two numbers that add up to 13 and multiply to give you (20)(-15) = -300. Those numbers are +25 and -12.

Set each factor equal to 0 and solve the resulting equations.

$$4x + 5 = 0 \qquad\qquad 5x - 3 = 0$$

$$x = 0 \quad \text{or} \qquad 4x = -5 \quad \text{or} \qquad 5x = 3$$

$$x = -\frac{5}{4} \qquad\qquad x = \frac{3}{5}$$

The solution to the equation is $x = -\dfrac{5}{4}$, $x = 0$, or $x = \dfrac{3}{5}$.

3.43 Solve the equation $x^2 - 3x - 28 = 0$ using the quadratic formula.

According to the quadratic formula, the solution to the equation $ax^2 + bx + c = 0$ is $x = \dfrac{-b \pm \sqrt{b^2 - 4ac}}{2a}$. The values a, b, and c are the coefficients of the expression; in this problem, $a = 1$, $b = -3$, and $c = -28$.

> You can use the quadratic formula only when one side of a quadratic equation equals 0.

$$x = \frac{-(-3) \pm \sqrt{(-3)^2 - 4(1)(-28)}}{2(1)}$$

$$= \frac{3 \pm \sqrt{9 + 112}}{2}$$

$$= \frac{3 \pm \sqrt{121}}{2}$$

$$= \frac{3 \pm 11}{2}$$

$$= \frac{3 + 11}{2} \text{ or } \frac{3 - 11}{2}$$

$$= \frac{14}{2} \text{ or } -\frac{8}{2}$$

$$= 7 \text{ or } -4$$

> This equation is MUCH easier to solve if you factor it instead. However, some equations can't be factored and you have to use the quadratic formula.

The solution to the equation is $x = -4$ or $x = 7$.

Note: Problems 3.44–3.45 refer to the equation $6x^2 - 13x - 8 = 0$.

3.44 Solve the equation by factoring.

Factor the quadratic trinomial by decomposition.

> The two numbers with a sum of –13 and a product of –48 are –16 and +3.

$$6x^2 + (3 - 16)x - 8 = 0$$

$$\left(6x^2 + 3x\right) + (-16x - 8) = 0$$

$$3x(2x + 1) - 8(2x + 1) = 0$$

$$(2x + 1)(3x - 8) = 0$$

Set both factors equal to 0 and solve the resulting equations.

$$2x + 1 = 0 \qquad 3x - 8 = 0$$
$$2x = -1 \quad \text{or} \quad 3x = 8$$
$$x = -\frac{1}{2} \qquad\qquad x = \frac{8}{3}$$

The solution to the equation is $x = -\dfrac{1}{2}$ or $x = \dfrac{8}{3}$.

Note: Problems 3.44–3.45 refer to the equation $6x^2 - 13x - 8 = 0$.

3.45 Verify the solution to Problem 3.44 by solving the equation again, this time using the quadratic formula.

Substitute $a = 6$, $b = -13$, and $c = -8$ into the quadratic formula.

$$
\begin{aligned}
x &= \frac{-b \pm \sqrt{b^2 - 4ac}}{2a} \\[2mm]
&= \frac{-(-13) \pm \sqrt{(-13)^2 - 4(6)(-8)}}{2(6)} \\[2mm]
&= \frac{13 \pm \sqrt{169 + 192}}{12} \\[2mm]
&= \frac{13 \pm \sqrt{361}}{12} \\[2mm]
&= \frac{13 \pm 19}{12} \\[2mm]
&= \frac{13 - 19}{12} \ \text{or}\ \frac{13 + 19}{12} \\[2mm]
&= -\frac{6}{12} \ \text{or}\ \frac{32}{12} \\[2mm]
&= -\frac{1}{2} \ \text{or}\ \frac{8}{3}
\end{aligned}
$$

3.46 Solve the equation: $x^2 + 5x = 8$.

Before applying the quadratic formula (or solving a quadratic equation by factoring), one side of the equation must equal 0. Subtract 8 from both sides of the equation.

$$x^2 + 5x - 8 = 0$$

Substitute $a = 1$, $b = 5$, and $c = -8$ into the quadratic formula.

This equation can't be solved by factoring. You can tell because the final answer has square roots in it.

$$x = \frac{-b \pm \sqrt{b^2 - 4ac}}{2a}$$

$$= \frac{-5 \pm \sqrt{5^2 - 4(1)(-8)}}{2(1)}$$

$$= \frac{-5 \pm \sqrt{25 + 32}}{2}$$

$$= \frac{-5 \pm \sqrt{57}}{2}$$

The solution to the equation is $x = -\frac{5}{2} - \frac{\sqrt{57}}{2}$ or $x = -\frac{5}{2} + \frac{\sqrt{57}}{2}$.

3.47 Solve the equation: $4x = 2x^2 + 1$.

Before the quadratic formula can be applied, one side of the equation must equal 0, so subtract $4x$ from both sides.

$$0 = 2x^2 - 4x + 1$$

Substitute $a = 2$, $b = -4$, and $c = 1$ into the quadratic formula.

$$x = \frac{-b \pm \sqrt{b^2 - 4ac}}{2a}$$

$$= \frac{-(-4) \pm \sqrt{(-4)^2 - 4(2)(1)}}{2(2)}$$

$$= \frac{4 \pm \sqrt{16 - 8}}{4}$$

$$= \frac{4 \pm \sqrt{8}}{4}$$

$$= \frac{4 \pm 2\sqrt{2}}{4}$$

Reduce the rational expression to lowest terms.

$$= \frac{4}{4} - \frac{2\sqrt{2}}{4} \quad \text{or} \quad \frac{4}{4} + \frac{2\sqrt{2}}{4}$$

$$= \frac{4}{4} - \frac{\cancel{2}\sqrt{2}}{\cancel{2} \cdot 2} \quad \text{or} \quad \frac{4}{4} + \frac{\cancel{2}\sqrt{2}}{\cancel{2} \cdot 2}$$

$$= 1 - \frac{\sqrt{2}}{2} \quad \text{or} \quad 1 + \frac{\sqrt{2}}{2}$$

The solution to the equation is $x = 1 - \frac{\sqrt{2}}{2}$ or $1 + \frac{\sqrt{2}}{2}$.

3.48 Identify all complex solutions of the equation $5x^3 - 40 = 0$.

Factor 5, the greatest common factor, out of the terms left of the equals sign.

$$5(x^3 - 8) = 0$$

According to the zero product property, if $5(x^3 - 8) = 0$, then either $5 = 0$ or $x^3 - 8 = 0$. The statement $5 = 0$ is false, so discard it.

$$x^3 - 8 = 0$$

The expression $x^3 - 8$ is a difference of perfect cubes. Apply the formula from Problem 3.30 to factor it.

$$(x - 2)(x^2 + 2x + 4) = 0$$

Set each factor equal to 0 and solve the corresponding equations. The left factor produces the equation $x - 2 = 0$, which has the solution $x = 2$. Solve the remaining equation, $x^2 + 2x + 4 = 0$, using the quadratic formula.

$$x = \frac{-b \pm \sqrt{b^2 - 4ac}}{2a}$$

$$= \frac{-(2) \pm \sqrt{2^2 - 4(1)(4)}}{2(1)}$$

$$= \frac{-2 \pm \sqrt{4 - 16}}{2}$$

$$= \frac{-2 \pm \sqrt{-12}}{2}$$

> There's no way that $5 = 0$ can be a true statement, so you can drop it. You can also start the problem by dividing the entire equation by 5 to get $x^3 - 8 = 0$.

> The formula is $a^3 - b^3 = (a - b)(a^2 + ab + b^2)$. In this problem, $a = x$ and $b = 2$.

Simplify the radical expression using the complex value $i = \sqrt{-1}$.

$$= \frac{-2 \pm \sqrt{-1 \cdot 4 \cdot 3}}{2}$$

$$= \frac{-2 \pm \sqrt{-1} \cdot \sqrt{4} \cdot \sqrt{3}}{2}$$

$$= \frac{-2 \pm i \cdot 2\sqrt{3}}{2}$$

$$= \frac{-2 \pm 2i\sqrt{3}}{2}$$

> Don't forget the solution to the equation $x - 2 = 0$.

Reduce the fraction to lowest terms.

$$= -\frac{2}{2} \pm \frac{2i\sqrt{3}}{2}$$

$$= -1 \pm i\sqrt{3}$$

The solution to the equation is $x = 2$, $x = -1 - i\sqrt{3}$, or $x = -1 + i\sqrt{3}$.

> The solutions containing i are imaginary numbers, because they exist only when you're allowed to take the square root of a negative number. In geometry, you focus primarily on real numbers, not imaginary numbers.

Chapter 4
FOUNDATIONAL GEOMETRIC CONCEPTS

Postulates, lines, and angles

This chapter explores the basic concepts of geometry, beginning with the core assumptions about points, lines, and planes. These axioms, or postulates, must be assumed true in order to lay the logical foundation of geometry. By combining a few core concepts, new concepts arise. Defining a fixed length of line based upon two points creates line segments, introducing only one endpoint to a line creates a ray, and joining two rays at a common endpoint creates an angle.

Geometry involves two major skills: working with diagrams and proving things. In this chapter, you focus on the first one (proofs start in Chapter 5). All geometric diagrams are made up of the same pieces: points, lines, planes, segments, and rays. This chapter helps you review what they are, how you're supposed to identify them (notation is extremely important in geometry), and what you can do with them.

Geometric Axioms

Obvious things that don't need proof

4.1 Complete the following statement and justify your answer.

Two distinct points define a _____.

> Points are usually indicated by capital letters, such as C, R, and S in this problem.

Two distinct points define a <u>line</u>. An infinite number of lines can be drawn through the single point C below, but only one line can be drawn connecting points R and S.

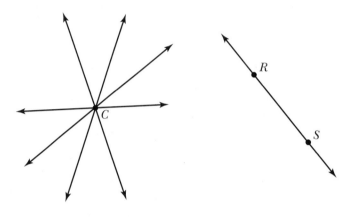

4.2 Complete the following statement and justify your answer.

_____ *distinct,* _____ *points define a plane.*

> Noncollinear means "not all lying on the same line."

<u>Three</u> distinct, <u>noncollinear</u> points define a plane. According to Problem 4.1, two distinct points are required to define a one-dimensional line. A third point not belonging to that line (and, therefore, noncollinear) is required to define a plane. An infinite number of planes may be drawn through line *l* below (including the two planes illustrated in the diagram), but only plane *M* contains points *Q* and *R* (on line *j*) and noncollinear point *P*.

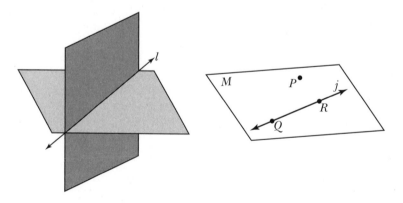

Note that the surface of plane *M* extends infinitely, as does line *j*, which lies on the plane.

4.3   Draw intersecting lines *m* and *n* and describe their intersection. Assume that *m* and *n* do not describe the same line.

If two lines intersect, they intersect at exactly one point. In the diagram below, line *m* contains point *A* and line *n* contains point *B*. The lines intersect at point *P*.

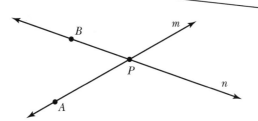

Lines are named either with a lowercase letter (such as m) or by two points on the line (such as $\overleftrightarrow{AP}$).

4.4   Draw intersecting planes *C* and *D* and describe their intersection.

A line defines the intersection of two planes. Consider the walls of a rectangular room. Each vertical wall meets its adjacent walls at a straight line, the corner of the room. In the following diagram, planes *C* and *D* intersect at line *l*.

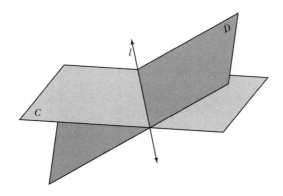

4.5   Construct a diagram containing a line $\overleftrightarrow{XY}$ and a noncollinear point *Z*. How many planes may be drawn that contain both? Justify your answer.

Draw a line passing through points *X* and *Y*. As long as both points appear on the line, their placement is irrelevant. Construct a point *Z* that does not belong to that line. According to the axiom described in Problem 4.2, three distinct and noncollinear points (like *X*, *Y*, and *Z*) define exactly one plane, labeled *M* in the diagram below.

X and Y ARE collinear, but the line doesn't pass through Z. That means that all three points (X, Y, and Z) are NOT collinear.

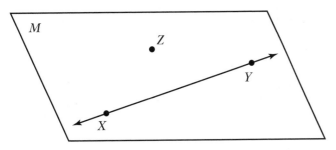

4.6 Construct a diagram illustrating coplanar, nonintersecting lines p and q. What term best describes lines p and q?

Two lines in the same plane that do not intersect are described as "parallel." Parallel lines have the same algebraic slope; therefore, they remain the same distance apart for the entirety of their infinite lengths. Consider the parallel lines illustrated by the rails of a railroad track—they stay the same distance apart (as defined by the railroad ties between them) for the entire length of the track. The notation "∥" is used to indicate parallel lines, so in plane N below, $p \parallel q$.

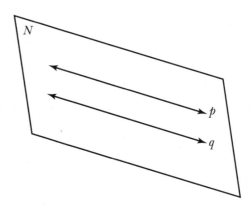

4.7 Construct a diagram illustrating noncoplanar, nonintersecting lines r and s. What term best describes those lines?

Nonintersecting lines that do not belong to the same plane are described as "skew." In the diagram below, line r lies on plane M, line s lies on plane N, and the lines do not intersect. Notice that planes M and N need not be parallel to contain nonintersecting lines.

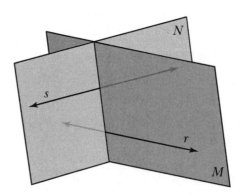

Segments, Rays, and Lines

Two endpoints, one endpoint, and no endpoints, respectively

4.8 According to the ruler postulate, points along a line may be assigned real number values called coordinates, constructing what is most commonly described as a "number line." If A has coordinate 3 and B has coordinate 8 on $\overleftrightarrow{AB}$, what is the length of line segment $\overline{AB}$?

To determine the length of the line segment with endpoints A and B, take the absolute value of the difference of the coordinates. In this problem, A has coordinate 3 and B has coordinate 8, so $\overline{AB}$ has length $|3-8| = |-5| = 5$. ←

> Absolute value bars erase negative values, so simplify the expression inside the bars and then make the result positive. If it's already positive, absolute values won't affect it.

4.9 Describe the difference between the notations $\overleftrightarrow{AB}$, $\overline{AB}$, and AB.

The notation $\overleftrightarrow{AB}$ indicates a line that passes through points A and B, whereas $\overline{AB}$ indicates a segment of that line. Specifically, $\overline{AB}$ is the portion of the line that lies between (and includes) points A and B. The *length* of line segment $\overline{AB}$ is described by the value AB. Therefore, AB is unique among the three notations because it represents a real number value.

Note: Problems 4.10–4.15 refer to the number line below.

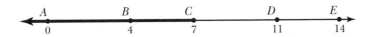

4.10 Which of the labeled points (A, B, C, D, and E) are contained within $\overline{BD}$?

The line segment $\overline{BD}$ is the portion of the number line extending from point B to point D. Both of those endpoints belong to the segment, as does point C, which lies between the endpoints.

Note: Problems 4.10–4.15 refer to the number line in Problem 4.10.

4.11 Which of the labeled points (A, B, C, D, and E) are contained within $\overrightarrow{CB}$?

The notation $\overrightarrow{CB}$ indicates the ray with endpoint C that extends in the direction of point B, the darkened portion of the number line in the diagram below.

A   B   C   D   E
0   4   7   11   14

The ray contains its endpoint C and all the points left of C on the number line, including B and A.

> A line segment stops at both endpoints, but a ray stops only at the first point in its name. That means $\overrightarrow{CB}$ doesn't stop at B, but extends through it, passing through B and all points left of B.

Note: Problems 4.10–4.15 refer to the number line in Problem 4.10.

4.12 Calculate *CD*.

The notation *CD* represents the length of the line segment $\overline{CD}$. Point *C* has coordinate 7 and point *D* has coordinate 11. Apply the ruler postulate (described in Problem 4.8) to calculate *CD* by taking the absolute value of the difference of the coordinates.

$$CD = |7 - 11| = |-4| = 4$$

Note: Problems 4.10–4.15 refer to the number line in Problem 4.10.

4.13 Given *M* is the midpoint of $\overline{AB}$, identify the coordinate of *M*.

The midpoint of a segment is the point equidistant from the segment's endpoints—the point that bisects the segment. If *M* is the midpoint of $\overline{AB}$, then *AM* = *MB*.

> Bisects means "cut in half."

Apply the ruler postulate to calculate *AB*.

$$AB = |0 - 4| = |-4| = 4$$

The segments $\overline{AM}$ and $\overline{MB}$ are half as long as $\overline{AB}$.

$$AM = MB = \frac{AB}{2} = \frac{4}{2} = 2$$

> M is also 2 units left of B, so you could subtract 2 from B's coordinate to get the same answer: 4 – 2 = 2.

If *AM* = 2, then *M* is 2 units right of point *A* along the number line. Add 2 to the coordinate of *A* to identify the coordinate of *M*.

$$M = A + 2 = 0 + 2 = 2$$

Note: Problems 4.10–4.15 refer to the number line in Problem 4.10.

4.14 Given *M* is the midpoint of $\overline{AB}$, calculate *ME*.

According to Problem 4.13, *M* has coordinate 2. Apply the ruler postulate to determine the length of $\overline{ME}$.

$$ME = |2 - 14| = |-12| = 12$$

Note: Problems 4.10–4.15 refer to the number line in Problem 4.10.

4.15 Identify all the congruent segments in the diagram.

The diagram contains two segments that are 3 units in length: $\overline{BC}$ and $\overline{DE}$.

$$BC = |4 - 7| = |-3| = 3$$
$$DE = |11 - 14| = |-3| = 3$$

Two segments have a length of 4 units: $\overline{AB}$ and $\overline{CD}$.

$$AB = |0 - 4| = |-4| = 4$$
$$CD = |7 - 11| = |-4| = 4$$

Finally, three segments have a length of 7 units: $\overline{AC}$, $\overline{BD}$, and $\overline{CE}$.

$$AC = |0 - 7| = |-7| = 7$$
$$BD = |4 - 11| = |-7| = 7$$
$$CE = |7 - 14| = |-7| = 7$$

> Congruent means "has the same measure." Congruent segments have the same length.

4.16 Given distinct, collinear points A and B, which of the following statements must be false? Justify your answer.

$$\overline{AB} \cong \overline{BA} \qquad AB = BA \qquad \overrightarrow{AB} \cong \overrightarrow{BA}$$

A segment is named using the endpoints that define it. If a segment has endpoints A and B, then $\overline{AB}$ and $\overline{BA}$ both identify the same segment. Because $\overline{AB}$ and $\overline{BA}$ are merely different ways of naming the same segment, $\overline{AB} \cong \overline{BA}$ and therefore $AB = BA$. However, $\overrightarrow{AB}$ and $\overrightarrow{BA}$ are not congruent, because they begin at different endpoints and extend in opposite directions, as illustrated below.

> The postulate that states "Everything is congruent to itself" is called the reflexive property of equality.

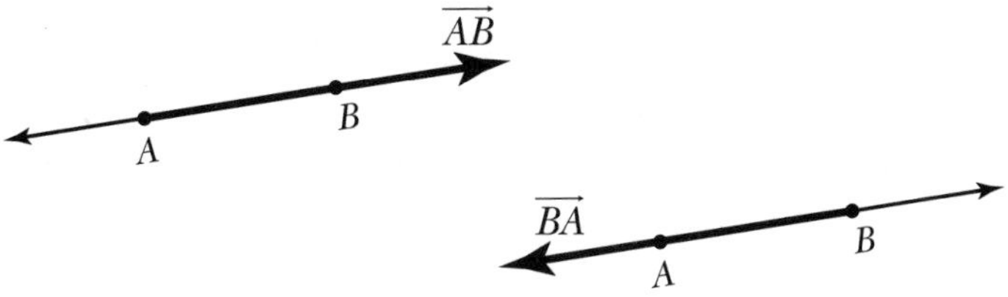

Note: Problems 4.17–4.20 refer to the number line below. Assume that K is the midpoint of $\overline{JL}$.

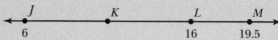

4.17 Complete the following statement.

If K is the midpoint of $\overline{JL}$, then _____ $\cong$ _____ and _____ = _____.

A segment is bisected by its midpoint, resulting in two congruent segments that are half as long as the original segment. In this problem, if K is the midpoint of $\overline{JL}$, then $\overline{JK} \cong \overline{KL}$ and $JK = KL$.

Note: Problems 4.17–4.20 refer to the number line illustrated in Problem 4.17. Assume that K is the midpoint of $\overline{JL}$.

4.18 Complete the following statement.

According to the _____ postulate, KL + LM = _____.

The segment addition postulate allows you to glue segments together to make bigger segments.

Segments $\overline{KL}$ and $\overline{LM}$ belong to the same line and share the common endpoint L. Whereas L serves as the right endpoint of $\overline{KL}$, it serves as the left endpoint of $\overline{LM}$. Combining those segments creates the longer segment $\overline{KM}$. According to the segment addition postulate, $KL + LM = \underline{KM}$.

Note: Problems 4.17–4.20 refer to the number line illustrated in Problem 4.17. Assume that K is the midpoint of $\overline{JL}$.

4.19 What is the coordinate of K?

Apply the ruler postulate to calculate JL.

$$JL = |6 - 16| = |-10| = 10$$

Because K bisects $\overline{JL}$, JK and KL are congruent segments, each half the length of $\overline{JL}$.

$$JK = KL = \frac{1}{2} JL$$
$$= \frac{1}{2}(10)$$
$$= 5$$

Point K is 5 units to the right of endpoint J (and 5 units left of endpoint L); add 5 to the coordinate of J (or subtract 5 from the coordinate of L) to calculate the coordinate of K.

$$K = J + 5 = 6 + 5 = 11$$
$$K = L - 5 = 16 - 5 = 11$$

Note: Problems 4.17–4.20 refer to the number line illustrated in Problem 4.17.

4.20 If *L* is the midpoint of $\overline{PM}$, what is the coordinate of *P*?

Every segment has a unique midpoint that is equidistant from the endpoints of the segment. If the midpoint of $\overline{PM}$ is *L*, then the distances between *L* and the endpoints of the segment (*PL* and *LM*) must be equal. Apply the ruler postulate to calculate *LM*.

$$LM = |16 - 19.5| = |-3.5| = 3.5$$

If *L* is 3.5 units *left* of point *M*, then *L* must be 3.5 units *right* of *P*. Subtract 3.5 from the coordinate of *L* to identify the coordinate of *P*.

$$P = L - 3.5 = 16 - 3.5 = 12.5$$

The coordinate of *P* is 12.5. ←

> The fractional version of the answer is also correct:
>
> $12.5 = 12\frac{1}{2} = \frac{25}{2}$

Note: Problems 4.21–4.23 refer to the diagram below, in which AB = BC, AD = 8, and BE = 2.5.

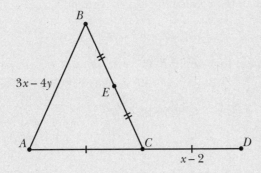

4.21 Identify three pairs of congruent segments based on the diagram and the given information.

The diagram contains four tick marks (sometimes called hash marks or hatch marks) that designate congruent segments. A single set of tick marks passes through $\overline{AC}$ and $\overline{CD}$, so $\overline{AC} \cong \overline{CD}$. A double set of tick marks indicates a second set of congruent segments: $\overline{BE} \cong \overline{CE}$. Finally, the problem itself (in the information listed preceding the diagram) indicates that $\overline{AB} \cong \overline{BC}$.

Note: Problems 4.21–4.23 refer to the diagram in Problem 4.21. Assume that AB = BC, AD = 8, and BE = 2.5.

4.22 Calculate x.

According to Problem 4.21, $\overline{AC} \cong \overline{CD}$, so C is the midpoint of $\overline{AD}$, bisecting it into two congruent segments. You are given $AD = 8$, so both AC and CD must equal 4. Set CD equal to 4 and solve for x.

$$x - 2 = 4$$
$$x = 4 + 2$$
$$x = 6$$

According to the diagram, the length of $\overline{CD}$ is $x - 2$. Now that you know $CD = 4$, you can figure out what x is.

Note: Problems 4.21–4.23 refer to the diagram in Problem 4.21. Assume that AB = BC, AD = 8, and BE = 2.5.

4.23 Calculate y.

Point E is the midpoint of $\overline{BC}$ because E bisects the segment into two congruent segments, $\overline{BE}$ and $\overline{EC}$. According to the segment addition postulate, the smaller segments combine to produce the larger segment.

$$BE + CE = BC$$

The problem states that $BE = 2.5$. Because $\overline{BE}$ and $\overline{EC}$ are congruent segments, $EC = 2.5$ as well. Substitute these values into the previous segment addition equation.

$$2.5 + 2.5 = BC$$
$$5 = BC$$

The problem also stipulates that $AB = BC$, so $AB = 5$. According to the diagram, $AB = 3x - 4y$. Set the lengths equal to calculate y.

$$3x - 4y = 5$$

According to Problem 4.22, $x = 6$.

$$3(6) - 4y = 5$$
$$18 - 4y = 5$$

Solve for y.

$$-4y = 5 - 18$$
$$-4y = -13$$
$$\frac{-4y}{-4} = \frac{-13}{-4}$$
$$y = \frac{13}{4}$$

Angles

Made by two rays with a common endpoint

4.24   List three alternate names for ∠1 in the diagram that follows.

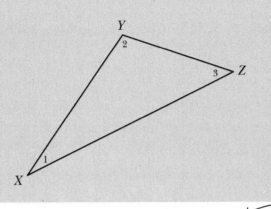

In this diagram, ∠1 has vertex *X*, so it can be named ∠*X*. The angle may also be described using three points: ∠*YXZ* and ∠*ZXY*. In each of those angle notations, the vertex of the angle (*X*) appears in the middle position. The order of the other two points (*Y* and *Z*) is irrelevant; the only condition they must meet is that each point must belong to a different side of the angle.

> If multiple angles have the same vertex, you must use more than a single letter to indicate which angle you're talking about. In this diagram, each angle has a different vertex.

Note: Problems 4.25–4.31 refer to the diagram that follows, in which $\overline{OC} \perp \overline{EA}$.

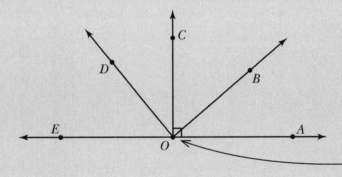

> Look at ∠COA in the diagram. It has a little square drawn inside it that's used to indicate right angles.

4.25   Calculate *m*∠*COA*.

The notation *m*∠*COA* is read "the measure of ∠*COA*." Its value is a real number, in the same way that *AB* is a real number value that represents the length of $\overline{AB}$.

The problem indicates that the sides of ∠*COA* are perpendicular ($\overline{OC} \perp \overline{EA}$). Perpendicular lines form right angles, which measure 90°.

> The protractor postulate enables you to measure angles, as the ruler postulate enables you to measure segments.

Note: Problems 4.25–4.31 refer to the diagram in Problem 4.25, in which $\overline{OC} \perp \overline{EA}$.

4.26   Classify $\angle BOA$.

According to the protractor postulate, the rays extending from O may each be assigned a unique real number between 0° (at $\overrightarrow{OA}$) and 180° (at $\overrightarrow{OE}$). Because $\overrightarrow{OC}$ is perpendicular to $\overleftrightarrow{EA}$, $\overrightarrow{OC}$ is assigned the measurement 90°.

Angles are classified by their measures:

♦ Acute angles have measures greater than 0° and less than 90°.

♦ Right angles measure 90°.

♦ Obtuse angles have measures greater than 90° and less than 180°.

♦ Straight angles measure 180°.

♦ Reflex angles have measures greater than 180° and less than 360°.

Because $\overrightarrow{OB}$ lies between $\overrightarrow{OA}$ and $\overrightarrow{OC}$, the value that the protractor postulate assigns to it is between 0° and 90°. Therefore, $\angle BOA$ is an acute angle.

Note: Problems 4.25–4.31 refer to the diagram in Problem 4.25, in which $\overline{OC} \perp \overline{EA}$.

4.27   Classify $\angle AOD$.

According to the protractor postulate, each ray extending from O may be assigned a real number, beginning with 0° at $\overrightarrow{OA}$ and 180° at $\overrightarrow{OE}$. Furthermore, Problem 4.25 states that $\overrightarrow{OC}$ is assigned the value 90°.

> In Problem 4.29, you subtract these values to calculate angle measures (just like you subtract coordinates to calculate the lengths of segments).

Because $\overrightarrow{OD}$ falls between $\overrightarrow{OC}$ and $\overrightarrow{OE}$, its value is greater than 90° and less than 180°, so $\angle AOD$ is obtuse.

Note: Problems 4.25–4.31 refer to the diagram in Problem 4.25, in which $\overline{OC} \perp \overline{EA}$.

4.28   Complete the following statement and justify your answer.

According to the _____ postulate, $m\angle BOD + m\angle DOE =$ _____.

Just as the segment addition postulate enables you to combine collinear, connected line segments to form larger line segments, the angle addition postulate enables you to combine adjacent angles to form larger angles.

> Adjacent angles share the same vertex, have one overlapping side, and no common interior points. They're basically two pieces of pie that make an even bigger pie piece when you push them together.

According to the <u>angle addition</u> postulate, $m\angle BOD + m\angle DOE = \underline{m\angle BOE}$.

Note: Problems 4.25–4.31 refer to the diagram in Problem 4.25, in which $\overline{OC} \perp \overline{EA}$.

4.29   If $\overrightarrow{OB}$ represents 40°, apply the protractor postulate to calculate $m\angle COB$.

Consider the ruler postulate, as explained in Problem 4.8. It defines the length of a segment according to the coordinates of its endpoints—length is equal to the absolute value of their difference. Similarly, the protractor postulate defines the measure of an angle according to the degree values of its sides—the measure is once again equal to the absolute value of their difference.

The sides of $\angle COB$ are $\overrightarrow{OC}$ (which corresponds to 90° according to Problem 4.26) and $\overrightarrow{OB}$ (which corresponds to 40°).

$$m\angle COB = |90° - 40°| = |50°| = 50°$$

Note: Problems 4.25–4.31 refer to the diagram in Problem 4.25, in which $\overline{OC} \perp \overline{EA}$. Assume that $m\angle BOA = 40°$.

4.30 Calculate $m\angle DOC$, assuming $\overline{OD} \perp \overline{OB}$.

If $\overline{OD} \perp \overline{OB}$, then $m\angle DOB = 90°$. According to the angle addition postulate, the sum of the measures of two adjacent angles is equal to the measure of the angle formed by their nonadjacent sides. ◄

$$m\angle DOC + m\angle COB = m\angle DOB$$

According to Problem 4.29, $m\angle COB = 50°$. Substitute the known measurements into the equation and solve for $\angle DOC$.

$$m\angle DOC + 50° = 90°$$
$$m\angle DOC = 90° - 50°$$
$$m\angle DOC = 40°$$

> In this case, $\angle DOC$ and $\angle COB$ share side $\overrightarrow{OC}$, so if you add up the smaller angles, it describes the angle formed by the nontouching sides: $\overrightarrow{OD}$ and $\overrightarrow{OB}$, which form angle $\angle DOB$.

Note: Problems 4.25–4.31 refer to the diagram in Problem 4.25, in which $\overline{OC} \perp \overline{EA}$. Assume that $m\angle BOA = 40°$ and $\overline{OD} \perp \overline{OB}$.

4.31 Draw $\overrightarrow{OF}$, the angle bisector of $\angle EOD$ and calculate $m\angle EOF$.

Perpendicular lines form 90° angles. In this diagram, $\overline{OC} \perp \overline{EA}$, so $m\angle EOC = m\angle COA = 90°$. According to Problem 4.30, $m\angle DOC = 40°$. Apply the angle addition postulate to calculate $m\angle EOD$.

$$m\angle EOD + m\angle DOC = m\angle EOC$$
$$m\angle EOD + 40° = 90°$$
$$m\angle EOD = 90° - 40°$$
$$m\angle EOD = 50°$$

In the diagram below, $\overrightarrow{OF}$ bisects $\angle EOD$, forming $\angle EOF$ and $\angle DOF$.

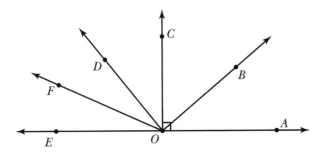

An angle bisector divides an angle into two congruent angles, each half the measure of the original angle. Here, $m\angle EOF = \frac{1}{2} m\angle EOD$.

$$m\angle EOF = \frac{1}{2} m\angle EOD = \frac{1}{2}(50°) = 25°$$

Therefore, $m\angle EOF = 25°$.

Note: Problems 4.32–4.36 refer to the diagram below.

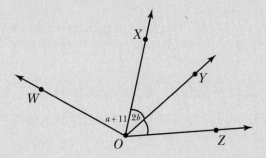

4.32 Name two pairs of adjacent angles and one pair of nonadjacent angles.

Because they share side $\overrightarrow{OX}$, $\angle WOX$ and $\angle XOY$ are adjacent. Similarly, $\angle XOY$ and $\angle YOZ$ are adjacent because they share side $\overrightarrow{OY}$. However, $\angle WOX$ and $\angle YOZ$ are nonadjacent angles because they share no common side.

Note: Problems 4.32–4.36 refer to the diagram in Problem 4.32.

4.33 Are $\angle WOY$ and $\angle XOZ$ adjacent angles? Explain your answer.

Adjacent angles overlap only at a shared side, but $\angle WOY$ and $\angle XOZ$ overlap between $\overrightarrow{OX}$ and $\overrightarrow{OY}$. Therefore, the angles are not adjacent.

Note: Problems 4.32–4.36 refer to the diagram in Problem 4.32.

4.34 Given $m\angle XOZ = 70°$, calculate b.

According to the diagram, $\angle XOY \cong \angle YOZ$. Just as tick marks are used to indicate congruent segments (see Problem 4.21), small arcs are used to indicate congruent angles. Notice that $\overrightarrow{OY}$ divides $\angle XOZ$ into two congruent angles. Therefore, $\overrightarrow{OY}$ is the angle bisector of $\angle XOZ$ and $m\angle XOY = m\angle YOZ = \frac{1}{2} m\angle XOZ$.

$$m\angle XOY = \frac{1}{2} m\angle XOZ$$
$$2b = \frac{1}{2}(70)$$
$$2b = 35$$

Solve the equation for b.

$$b = \frac{35}{2} = 17.5$$

Note: Problems 4.32–4.36 refer to the diagram in Problem 4.32. Assume that $m\angle XOZ = 70°$.

4.35 Draw $\overrightarrow{OA}$ such that $\overrightarrow{OX}$ and $\overrightarrow{OY}$ trisect $\angle AOZ$ and express $m\angle WOA$ in terms of a.

In the diagram below, $\angle AOZ$ is trisected by $\overrightarrow{OX}$ and $\overrightarrow{OY}$.

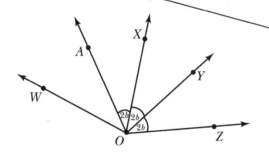

Trisecting an angle means splitting it into three equal angles. The value $b = 17.5$ in the equation that follows comes from Problem 4.34.

$$m\angle AOX = m\angle XOY = m\angle YOZ = 2b = 2(17.5°) = 35°$$

Apply the angle addition postulate.

$$m\angle WOA + m\angle AOX = m\angle WOX$$

Substitute $m\angle WOX = a + 11$ and $m\angle AOX = 35°$ into the equation and solve for $m\angle WOA$.

$$m\angle WOA + 35° = (a+11)°$$
$$m\angle WOA = (a+11-35)°$$
$$m\angle WOA = (a-24)°$$

Note: Problems 4.32–4.36 refer to the diagram in Problem 4.32. Assume that $m\angle XOZ = 70°$.

4.36 For what value of a does $\overrightarrow{OX}$ bisect $\angle WOZ$?

If $\overrightarrow{OX}$ bisects $\angle WOZ$, then $m\angle WOX = m\angle XOZ$. Recall that $m\angle WOX = (a+11)°$ and $m\angle XOZ = 70°$.

$$m\angle WOX = m\angle XOZ$$
$$(a+11)° = 70°$$
$$a = 70-11$$
$$a = 59$$

Note: Problems 4.37–4.41 refer to the diagram below, in which $\overleftrightarrow{AE} \perp \overleftrightarrow{CG}$.

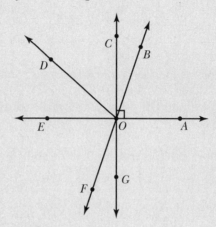

4.37 Identify the adjacent complement of ∠AOB.

A pair of angles is complementary when the sum of their measures equals 90°. Because $\overleftrightarrow{AE} \perp \overleftrightarrow{CG}$, the lines intersect at right angles.

$$m\angle AOC = m\angle COE = m\angle EOG = m\angle GOA = 90°$$

Apply the angle addition postulate.

$$m\angle AOB + m\angle BOC = m\angle AOC$$
$$m\angle AOB + m\angle BOC = 90°$$

Because the sum of their measures is 90°, ∠AOB and ∠BOC are complementary. Note that ∠FOG and ∠AOB are also complementary but do not share a side, so they are not adjacent angles.

∠FOG is also a complement of ∠AOB because ∠COB and ∠FOG are vertical angles (and, therefore, congruent). See Problem 4.41 for details.

Note: Problems 4.37–4.41 refer to the diagram in Problem 4.36, in which $\overleftrightarrow{AE} \perp \overleftrightarrow{CG}$.

4.38 Identify the supplement of ∠AOB.

A pair of angles is supplementary when the sum of their measures equals 180°. In other words, two adjacent angles are supplementary when their unshared sides lie on the same straight line.

Apply the angle addition postulate.

$$m\angle AOB + m\angle BOE = m\angle AOE$$

Points *A*, *O*, and *E* are collinear, so ∠AOE is a straight angle and $m\angle AOE = 180°$.

$$m\angle AOB + m\angle BOE = 180°$$

Because the sum of their measures equals 180°, ∠AOB and ∠BOE are supplementary.

Note: Problems 4.37–4.41 refer to the diagram in Problem 4.36, in which $\overline{AE} \perp \overline{CG}$.

4.39 Identify the complement of $\angle DOE$.

Apply the angle addition postulate.

$$m\angle DOE + m\angle DOC = m\angle EOC$$

The sides of $\angle COE$ are perpendicular, so $m\angle COE = 90°$.

$$m\angle DOE + m\angle DOC = 90°$$

Because the sum of their measures is $90°$, $\angle DOE$ and $\angle DOC$ are complementary.

Note: Problems 4.37–4.41 refer to the diagram in Problem 4.36, in which $\overline{AE} \perp \overline{CG}$.

4.40 Identify the two adjacent supplements of $\angle AOG$. ←

Because its sides are perpendicular, $\angle AOG$ is a right angle. Supplementary right angles are adjacent to both sides of $\angle AOG$.

$$m\angle AOG + m\angle AOC = 180°$$
$$m\angle AOG + m\angle GOE = 180°$$

Therefore, $\angle AOC$ and $\angle GOE$ are both adjacent supplements of $\angle AOG$.

> Unless it's specifically stated, never assume that problems refer to reflex (over 180°) angles. That means that $\angle AOG$ is NOT the 270° angle that starts at $\overrightarrow{OA}$ and goes counterclockwise to $\overrightarrow{OG}$.

Note: Problems 4.37–4.41 refer to the diagram in Problem 4.36, in which $\overline{AE} \perp \overline{CG}$.

4.41 Identify two pairs of vertical angles that don't measure $90°$.

Vertical angles are formed by intersecting lines. In this diagram, $\overrightarrow{CG}$ and $\overrightarrow{BF}$ form vertical angles $\angle COB$ and $\angle GOF$; $\overrightarrow{AE}$ and $\overrightarrow{BF}$ form vertical angles $\angle AOB$ and $\angle EOF$.

> Vertical angles are congruent.

4.42 Identify the word that best completes the statement that follows and justify your answer.

Complementary angles measuring more than $0°$ must be _____.

If $\angle A$ and $\angle B$ are complementary, then $m\angle A + m\angle B = 90°$. If the sum of two angles is $90°$ and neither of the angles measures $0°$, then each of the angles must measure less than $90°$. Angle measurements are positive values, so neither $\angle A$ nor $\angle B$ may measure more than $90°$, or the sum of the measures of the angles will be greater than $90°$. Thus, complementary angles measuring more than $90°$ must be <u>acute</u>.

> So you can't have one angle that measures 120° and another one that measures −30° for a total of 90°.

4.43 Given $\angle A$ is its own complement and $\angle B$ is its own supplement, calculate $m\angle A + m\angle B$.

If $\angle A$ is its own complement, then $m\angle A + m\angle A = 90°$. Solve the equation.

$$2(m\angle A) = 90°$$
$$\frac{2(m\angle A)}{2} = \frac{90°}{2}$$
$$m\angle A = 45°$$

If $\angle B$ is its own supplement, then $m\angle B + m\angle B = 180°$. Solve this equation as well.

$$2(m\angle B) = 180°$$
$$\frac{2(m\angle B)}{2} = \frac{180°}{2}$$
$$m\angle B = 90°$$

Calculate the sum of the angle measures.

$$m\angle A + m\angle B = 45° + 90° = 135°$$

Note: Problems 4.44–4.45 refer to the diagram below.

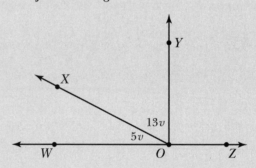

4.44 For what value of v are $\angle WOX$ and $\angle XOY$ complementary?

According to the diagram, $m\angle WOX = 5v$ and $m\angle XOY = 13v$. For the angles to be complementary, their measures must have a sum of 90°.

$$m\angle WOX + m\angle XOY = 90°$$
$$5v + 13v = 90°$$
$$18v = 90°$$

Solve the equation for v.

$$\frac{18v}{18} = \frac{90°}{18}$$
$$v = 5$$

If $v = 5$, then $m\angle WOX = 5(5) = 25°$ and $m\angle XOY = 13(5) = 65°$. Note that $25° + 65° = 90°$, so the angles are complementary.

Note: Problems 4.44–4.45 refer to the diagram in Problem 4.44.

4.45 Given the value of v generated by Problem 4.44, calculate $m\angle XOZ$.

Problem 4.44 states that $\angle WOX$ and $\angle XOY$ are complementary when $v = 5$, so $m\angle WOY = 90°$. Note that $\angle WOY$ and $\angle YOZ$ are supplementary because they are adjacent angles whose nonadjacent sides lie on the straight line $\overleftrightarrow{WZ}$. ←

$$m\angle WOY + m\angle YOZ = 180°$$

Calculate $m\angle YOZ$.

$$90° + m\angle YOZ = 180°$$
$$m\angle YOZ = 180° - 90°$$
$$m\angle YOZ = 90°$$

Apply the angle addition postulate to calculate $m\angle XOZ$.

$$m\angle XOZ = m\angle XOY + m\angle YOZ$$
$$= 65° + 90°$$
$$= 155°$$

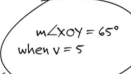

The angles are supplementary because combining them gives you a straight line (sort of like combining complementary angles gives you a right angle).

$m\angle XOY = 65°$ when $v = 5$

Note: Problems 4.46–4.47 refer to distinct, coplanar lines $\overleftrightarrow{AB}$ and $\overleftrightarrow{CD}$ that intersect at point M, such that M is the midpoint of $\overline{AB}$ and $\overline{CD}$.

4.46 Identify two pairs of vertical angles formed by the intersection of the lines.

Begin by constructing a diagram that reflects the given information.

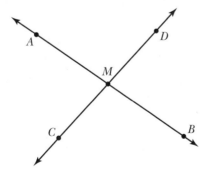

Although your diagram may differ from this illustration, points A and B must be collinear and equidistant from point M; points C and D must be collinear with each other (but not with points A and B) and must be equidistant from M as well. The intersection point M belongs to both lines.

The diagram contains two pairs of vertical angles: $\angle AMC$ and $\angle BMD$ comprise the first pair, and $\angle AMD$ and $\angle BMC$ comprise the second.

Note: Problems 4.46–4.47 refer to distinct, coplanar lines $\overleftrightarrow{AB}$ and $\overleftrightarrow{CD}$ that intersect at point M, such that M is the midpoint of $\overline{AB}$ and $\overline{CD}$.

4.47 Given $m\angle AMD$ is 5 times as large as $m\angle AMC$, calculate $m\angle BMC$. Express your answer as a real number in degrees.

Although the given diagram does not necessarily reflect this newest piece of information (the comparative size of the angles was not known in Problem 4.46), the diagram can still be used to identify the relationships between the angles.

Let x represent the measure of the smaller angle: $x = m\angle AMC$. If $\angle AMD$ is 5 times as large as $\angle AMC$, then $m\angle AMD = 5(m\angle AMC) = 5x$. Note that $\angle AMC$ and $\angle AMD$ are supplementary angles, so their measures must have a sum of 180°.

$$m\angle AMC + m\angle AMD = 180$$
$$x + 5x = 180$$
$$6x = 180$$
$$x = \frac{180}{6}$$
$$x = 30$$

> $\angle AMC$ and $\angle AMD$ are adjacent (they share vertex M and side $\overrightarrow{MA}$). Their nonadjacent sides ($\overrightarrow{MC}$ and $\overrightarrow{MD}$) lie on the same line ($\overleftrightarrow{CD}$), so the angles are supplementary.

Hence, $m\angle AMC = x = 30°$ and $m\angle AMD = 5x = 5(30) = 150°$. According to Problem 4.46, $\angle AMD$ and $\angle BMC$ are vertical angles. Vertical angles are congruent, so $m\angle BMC = m\angle AMD = 150°$.

Chapter 5
LOGIC AND PROOF

If-then statements and two-column proofs

The study of geometry consists of two major tasks: measurement and proof. This chapter begins the investigation of proof, the logical method by which all geometric theorems are verified. Though all of mathematics is structured in a similar logical fashion, geometry is unique in that the majority of the theorems do not require advanced mathematics to complete.

Geometry problems fall into two major categories: calculating something (such as length, angle measure, area, perimeter, volume, or surface area) and proving something. This chapter focuses on the second category.

You prove things called theorems, which are statements like "The diagonals of a rectangle are congruent." Before you can jump into proving that statement true, you need a few foundational skills. One of those skills is understanding how conditional statements like that rectangle theorem work.

You also need to justify things. Think of the last time you argued with someone. If you can prove that you're right using some sort of evidence, then you're more convincing. You use two-column proofs: one column containing your arguments and one column presenting your evidence.

Conditional Statements

If-then statements

Note: Problems 5.1–5.6 refer to the following true statement: "All tiger sharks are carnivores."

5.1 Rewrite the sentence as a conditional statement, identifying its hypothesis and conclusion.

> You can construct the conditional statement in different ways, but they all pretty much look like this. Yours doesn't have to match word for word, but it should be pretty close.

A conditional statement is commonly known as an "if-then" statement. The classic example of a conditional statement is "if *p*, then *q*," in which you may conclude *q* about anything that meets condition *p*. Thus, the condition statement *p* is the hypothesis and the conclusion statement is *q*.

One correct conditional statement is "If the organism is a tiger shark, then it is a carnivore." The hypothesis in the statement is "the organism is a tiger shark," because it states the only condition the organism in question must meet. The conclusion summarizes what is known about any subject that meets that criterion: "it is a tiger shark."

Note: Problems 5.1–5.6 refer to the following true statement: "All tiger sharks are carnivores."

5.2 Construct a Venn diagram illustrating the conditional statement generated in Problem 5.1.

A Venn diagram uses circles to represent collections of items. In this case, two circles are necessary: one that represents all tiger sharks and one that represents all carnivores. Note that the entire set of tiger sharks must be contained within the set of carnivores because the given statement stipulates that every member of the tiger shark circle must also be a member of the carnivore circle.

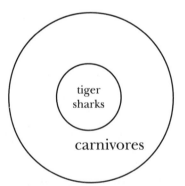

The diagram is meant to present the statement visually, but not to represent it with scientific accuracy. For instance, the area of the tiger shark circle does not represent the actual percentage of carnivores that are tiger sharks.

Note: Problems 5.1–5.6 refer to the following true statement: "All tiger sharks are carnivores."

5.3   Identify the converse of the statement and determine whether it is true or false using a Venn diagram.

The converse of the statement "if *p*, then *q*" is "if *q*, then *p*." In other words, replace the hypothesis with the conclusion and vice versa. The converse of this conditional statement is "If the organism is a carnivore, then it is a tiger shark." The statement is false, because many carnivores aren't tiger sharks.

The following Venn diagram demonstrates that the converse is false by identifying three counterexamples: a lion, a great white shark, and a dog. A counterexample proves a statement false by identifying something that satisfies the hypothesis but not the conclusion.

To disprove a statement, you need to give only a single counterexample. This book gives three to illustrate the point better, but it could have stopped at one.

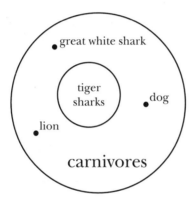

Note: Problems 5.1–5.6 refer to the following true statement: "All tiger sharks are carnivores."

5.4   Identify the contrapositive of the statement and determine whether it is true or false using a Venn diagram.

The contrapositive of the statement "if *p*, then *q*" is "if not *q*, then not *p*." Like the converse, the contrapositive is constructed by reversing the hypothesis and conclusion statements. Once reversed, both statements are then negated, hence the notation "not *p*" and "not *q*." The symbols "¬" and "~" are most commonly used to indicate negation, so "if ¬*q*, then ¬*p*" is an acceptable way to communicate the contrapositive.

The contrapositive of this conditional statement is "If the organism is not a carnivore, then it is not a tiger shark." This statement is true because any organism that does not belong to the carnivore circle in the following Venn diagram is automatically excluded from the tiger shark circle as well.

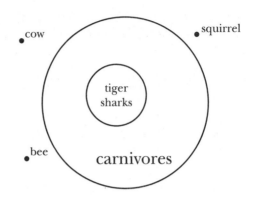

Even though one counterexample can prove a statement false, all the examples in the world won't prove something true. The contrapositive is definitely true, but not because you found three reasons why.

Three illustrative examples of herbivores (noncarnivores) are provided. They aren't counterexamples, because they serve to demonstrate that the contrapositive is true rather than prove it false.

Note: Problems 5.1–5.6 refer to the following true statement: "All tiger sharks are carnivores."

5.5 Identify the inverse of the statement and determine whether it is true or false.

The inverse of the statement "if p, then q" is "if $\neg p$, then $\neg q$." Both the hypothesis and conclusion statements are negated, but unlike the converse and contrapositive statements, their order does not change.

In this case, the inverse is "If the organism is not a tiger shark, then it is not a carnivore." This statement is false, because there are numerous organisms that aren't tiger sharks but are still carnivores. Consider the Venn diagram in Problem 5.3—great white sharks, dogs, and lions aren't tiger sharks, but they are most definitely carnivores.

Note: Problems 5.1–5.6 refer to the following true statement: "All tiger sharks are carnivores."

5.6 Is the condition stated by the hypothesis necessary and/or sufficient to reach the conclusion?

If you're confused about the question, do Problem 5.7 first and then come back to Problem 5.6.

To determine whether a condition is necessary or sufficient, you must determine the accuracy of the statement and its converse. The statement itself is true, so classifying an organism as a tiger shark provides sufficient evidence to then conclude that the organism is a carnivore.

According to Problem 5.3, the converse is false, so the condition is not necessary. In other words, it is not necessary that an organism be a tiger shark to be a carnivore.

5.7   Describe how to classify a condition as necessary and/or sufficient to reach a conclusion.

As Problem 5.6 states, to classify conditions as sufficient and/or necessary, you must know whether the statement and its converse are true. If both the statement and its converse are true, then the conditions are sufficient and necessary. If a statement is true but its converse is false (as in Problem 5.6), then the conditions are sufficient but not necessary.

If a statement and its converse are both false, then the conditions are neither sufficient nor necessary to reach the conclusion. However, if a false statement has a true converse, then the conditions are necessary but not sufficient.

> True statement = sufficient
> False statement = not sufficient
> True converse = necessary
> False converse = not necessary

5.8   Is the condition in the following statement sufficient and/or necessary to reach the conclusion? Explain your answer.

If you mix blue and yellow paint, then the result is green paint.

The given statement is true—mixing blue and yellow paint produces green paint. The converse of the statement, "If a paint mixture is green, then it was produced by combining blue and yellow paint," is also true because it describes the only way to produce a green mixture.

According to Problem 5.7, if a statement and its converse are both true, then the condition is both sufficient (mixing blue and yellow will always make green) and necessary (mixing blue and yellow is the *only* way to make green).

Note: Problems 5.9–5.11 refer to the following assertion: "All birds with wings can fly."

5.9   Write the sentence as a conditional statement, identifying the hypothesis and conclusion, and determine whether the statement is true or false.

The statement "All birds with wings can fly" is equivalent to the conditional statement "If a bird has wings, then it can fly." The hypothesis is that the bird in question has wings, and the conclusion is that it can fly. The statement is false because a number of flightless birds have wings, including ostriches, penguins, emus, and kiwis.

> This statement is restricted to birds, so a statement like "If something can fly, then it is a bird with wings" is no good.

Note: Problems 5.9–5.11 refer to the following assertion: "All birds with wings can fly."

5.10   Construct the converse, contrapositive, and inverse of the conditional statement generated in Problem 5.9. Determine whether each is true or false.

The conditional statement generated in Problem 5.9 is "If a bird has wings, then it can fly." The converse of "if p, then q" is "if q, then p": if a bird can fly, then it has wings. This is a true statement because all birds capable of flight are propelled by wings.

> Not sure how else they'd fly. Rocket packs?

The contrapositive of "if p, then q" is "if not q, then not p." If a bird cannot fly, then it does not have wings. This statement is false. Problem 5.9 lists a number of birds that cannot fly but do have wings.

The inverse of "if p, then q" is "if not p, then not q." If a bird doesn't have wings, then it cannot fly. This statement is true; without wings, a bird cannot fly.

Note: Problems 5.9–5.11 refer to the following assertion: "All birds with wings can fly."

5.11   Is the condition necessary and/or sufficient to reach the conclusion? Explain your answer.

The original statement is false (according to Problem 5.9) and its converse is true (according to Problem 5.10). Therefore, the condition of having wings is not sufficient to allow flight, but the presence of wings is necessary for flight to occur.

5.12   Complete the statement below and justify your answer using the following conditional statement: "If ∠A and ∠B are complementary, then $m\angle A + m\angle B = 90°$."

A conditional statement and its _____ are logically equivalent.

A conditional statement and its <u>contrapositive</u> are logically equivalent—if one statement is true, then the other must be true as well. Similarly, if one statement is false, so is the other.

The contrapositive of the given statement is "If $m\angle A + m\angle B \neq 90°$, then ∠A and ∠B are not complementary." This statement is true because two angles are complementary only if the sum of their measures is 90°.

5.13   Describe the logical equivalency of a conditional statement, its converse, and its inverse.

In other words, you can't tell whether the inverse or converse will be true based on the original statement, but you do know that the inverse and converse will both be true or both be false.

Just as a statement and its contrapositive are logically equivalent (as explained in Problem 5.12), the inverse and converse of a statement are logically equivalent. There is no standard logical equivalence relationship between a statement and either its inverse or converse.

5.14   Describe the difference between an "if-then" statement and an "if and only if" statement. Use the following definition to explain your answer: "Two lines are perpendicular iff they intersect at right angles."

Iff is an abbreviation for "if and only if."

The converse of an "if and only if" statement is also true. Therefore, "If lines are perpendicular, then they intersect at right angles" and "If lines intersect at right angles, then they are perpendicular" are both true statements. For lines to be perpendicular, it is both necessary and sufficient for the lines to intersect at 90° angles.

Algebraic Properties

The official reasons you're allowed to do things in algebra

5.15 Identify the algebraic property that justifies the statement $AB = AB$, and explain the implications of the property.

According to the reflexive property of equality, all values are equal to themselves. This does not represent a significant leap of faith and is perhaps the most obvious of all algebraic properties.

5.16 Identify the algebraic property that justifies the following statement and explain the implications of the property.

If $a = b$, then $b = a$.

According to the symmetric property of equality, two equal quantities may be placed on either side of an equals sign. Similarly, two congruent quantities may be placed on either side of a congruency symbol: if $\angle A \cong \angle B$, then $\angle B \cong \angle A$.

5.17 Identify the algebraic property that justifies the following statement and explain the implications of the property.

If $a = b$ and $b = c$, then $a = c$.

The transitive property of equality states that two values equal to a common value are equal to each other. In this case, a and c are both equal to b, so a and c must be equal. If you have difficulty understanding this property, assign a real number value to b, such as $b = 5$: if $a = 5$ and $5 = c$, then $a = c$ because $5 = 5$.

5.18 Identify the algebraic property that justifies the following statement and explain the implications of the property.

If $x + y + 9 = 12$, then $y + 9 + x = 12$.

The terms of a sum may be written in any order, according to the commutative property of addition. To demonstrate the validity of the property, assign x and y real number values such as $x = 1$ and $y = 2$.

$$x + y + 9 = 12 \qquad y + 9 + x = 12$$
$$1 + 2 + 9 = 12 \qquad 2 + 9 + 1 = 12$$
$$3 + 9 = 12 \qquad 11 + 1 = 12$$
$$12 = 12 \text{ True} \qquad 12 = 12 \text{ True}$$

There's also a commutative property of multiplication, which allows you to multiply in any order: $5(3) = 3(5) = 15$. Subtraction and division do NOT have a commutative property because order matters: $10 - 6 \neq 6 - 10$ and $4 \div 2 \neq 2 \div 4$.

5.19 Identify the algebraic property that justifies the following statement and explain the implications of the property.

If (rs)t = −1, then r(st) = −1.

There's also an associative property for addition: $(x + y) + z = x + (y + z)$

The only difference between the expressions $(rs)t$ and $r(st)$ is the placement of the parentheses, which changes the order in which the expressions are evaluated; the product in parentheses must be calculated first. The associative property of multiplication states that the order in which you multiply factors is irrelevant; the product remains unchanged.

To demonstrate the validity of the property, substitute $r = 2$, $s = 5$, and $t = -\dfrac{1}{10}$ into the equations.

$$(rs)t = -1 \qquad\qquad r(st) = -1$$

$$(2 \cdot 5)\left(-\frac{1}{10}\right) = -1 \qquad\qquad 2\left(5 \cdot -\frac{1}{10}\right) = -1$$

$$(10)\left(-\frac{1}{10}\right) = -1 \qquad\qquad 2\left(-\frac{5}{10}\right) = -1$$

$$\left(-\frac{10}{10}\right) = -1 \qquad\qquad 2\left(-\frac{1}{2}\right) = -1$$

$$-1 = -1 \quad \text{True} \qquad\qquad -1 = -1 \quad \text{True}$$

To solve for x, you usually have to add something to, subtract something from, multiply something by, or divide something into both sides of an equation. When you do, the equation stays true, even though it may look different.

5.20 Identify the algebraic property that justifies the following statement and explain the implications of the property.

If a = b and c ≠ 0, then $\dfrac{a}{c} = \dfrac{b}{c}$.

The division property of equality states that two equal values will remain equal if both are divided by the same nonzero quantity. Similarly, the addition, subtraction, and multiplication properties of equality allow you to add, subtract, and multiply equal values by the same quantity, and the values will remain equal.

This family of properties is most often used when solving equations. To solve for a variable, you often need to manipulate both sides of an equation.

5.21 Identify the algebraic property that justifies the following statement and explain the implications of the property.

If a + 1 = b and a = 2c, then 2c + 1 = b.

The substitution property of equality allows you to replace one expression with a different, equivalent expression without affecting the value of the expression or equation. In this case, $a = 2c$, so you can replace a in the equation $a + 1 = b$ with the expression $2c$.

$$\boxed{a} + 1 = b$$
$$\boxed{2c} + 1 = b$$

Substituting equivalent values (in this case, $2c$ replaces a) maintains the equality of the statement.

5.22 Identify the algebraic property that justifies the following statement and explain the implications of the property.

x · 1 = 1 · x = x

The product of any *x*-value and 1 is the original *x*-value. This guarantee comprises the multiplicative identity property; multiplying any number by 1 does not change the value or identity of the number. Similarly, the additive identity property states that the sum of 0 and a real number *y* is equal to *y*.

$$y + 0 = 0 + y = y$$

Any number times 1 equals itself.

5.23 Identify the algebraic property that justifies the following statement and explain the implications of the property.

For any real number x, $x \cdot \dfrac{1}{x} = \dfrac{1}{x} \cdot x = 1$.

This statement is more commonly known as the multiplicative inverse property. It states that the product of any real number x and its reciprocal $\left(\dfrac{1}{x}\right)$ is 1. You would apply the multiplicative inverse property to solve the equation $\dfrac{2}{3}x = 6$, multiplying both sides of the equation by $\dfrac{3}{2}$ so that the resulting coefficient of x is 1.

An inverse property also exists for addition. The additive inverse property states that the sum of any number and its opposite is 0.

$$a + (-a) = -a + a = 0$$

The reciprocal of a fraction is the fraction turned upside down, so the reciprocal of $\dfrac{2}{3}$ is $\dfrac{3}{2}$. The reciprocal of an integer is 1 over that integer, so the reciprocal of 8 is $\dfrac{1}{8}$.

5.24 Identify the algebraic property that justifies the following statement and explain the implications of the property.

If a > b, then a + c > b + c.

According to the addition property of inequality, adding the same value to both sides of an inequality maintains the inequality of the statement. In this example, the left side of the inequality is greater ($a > b$), and after adding c to both sides of the inequality, the left side remains the larger side.

Similarly, the subtraction property of inequality states that an inequality remains true after the same value is subtracted from both sides of the inequality statement: if $a > b$, then $a - c > b - c$.

5.25 Identify the algebraic property that justifies the following statement and explain the implications of the property.

If x ≥ y and z < 0, then xz ≤ yz.

The multiplicative property of inequality asserts that multiplying both sides of an inequality (in this case, $x \geq y$) by a negative number (z) reverses the inequality (in this case, $\geq$ becomes $\leq$). Substitute $x = 3$, $y = 2$, and $z = -4$ into the inequality to demonstrate the validity of the property.

$$x \geq y$$
$$3 \geq 2 \quad \text{True}$$

Multiply both sides of the inequality by $z = -4$.

$$3(-4) \geq 2(-4)$$
$$-12 \geq -8 \quad \text{False}$$

Multiplying the larger number (3) by a negative number produces a number that is more negative. Similarly, dividing both sides of an inequality reverses the inequality relationship. However, multiplying or dividing both sides of an inequality by a *positive* number maintains the inequality statement. Therefore, if $x \geq y$ and $z > 0$, then $xz \geq yz$.

z is negative because the problem states that z < 0. Only negative numbers are less than 0.

5.26 Identify the algebraic property that justifies the following statement and explain the implications of the property.

2(x + 9) = 2x + 18

According to the distributive property, if an entire quantity (such as $x + 9$) is multiplied by a value (such as 2), then each of the terms in the quantity may be multiplied by that value. For more information about the distributive property, see Problem 1.33.

Deductive Reasoning

Apply logic and properties to diagrams and equations

Note: Problems 5.27–5.31 refer to the figure below, in which M bisects $\overline{AB}$ and $\overline{CD}$.

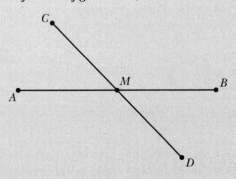

5.27 Justify the following statement: $AM = MB$.

A midpoint, by definition, divides a segment into two segments of equal length. Thus, M is the midpoint of $\overline{AB}$ and $AM = MB$.

> Justify means "explain why it's true."

Note: Problems 5.27–5.31 refer to the figure in Problem 5.27, in which M bisects $\overline{AB}$ and $\overline{CD}$.

5.28 Justify the statement: $AM + MB = AB$.

Points A, M, and B are collinear, and $\overline{AM}$ and $\overline{MB}$ share common endpoint M. According to the segment addition postulate, the sum of the lengths of $\overline{AM}$ and $\overline{MB}$ is equal to the length of $\overline{AB}$.

Note: Problems 5.27–5.31 refer to the figure in Problem 5.27, in which M bisects $\overline{AB}$ and $\overline{CD}$.

5.29 Given $AM + MB = AB$, justify the following statement: $MB = AB - AM$.

The subtraction property of equality states that subtracting the same number from two equal values maintains the equality of those values. In this case, $AM + MB = AB$. Subtracting MB from both sides of the equation maintains the equality of the statement.

$$AM + MB = AB$$
$$AM + MB - MB = AB - MB$$
$$AM = AB - MB$$

Note: Problems 5.27–5.31 refer to the figure in Problem 5.27, in which M bisects $\overline{AB}$ and $\overline{CD}$.

5.30   Given $AM + MB = AB$, justify the following statement: $2MB = AB$.

According to Problem 5.27, $AM = MB$. Apply the substitution property of equality, replacing AM in the equation $AM + MB = AB$ with the equivalent value MB.

$$AM + MB = AB$$
$$MB + MB = AB$$
$$2MB = AB$$

> You can add one MB to another MB and get 2MB, just as you can add one x to another x and get 2x.

Note: Problems 5.27–5.31 refer to the figure in Problem 5.27, in which M bisects $\overline{AB}$ and $\overline{CD}$.

5.31   Justify the following statement: $m\angle CMA = m\angle DMB$.

Vertical angles are pairs of angles whose sides are formed by intersecting lines. Thus, $\angle CMA$ and $\angle DMB$ are vertical angles because they are formed by intersecting lines $\overleftrightarrow{AB}$ and $\overleftrightarrow{CD}$. Vertical angles are congruent, so $\angle CMA \cong \angle DMB$ and $m\angle CMA = m\angle DMB$.

> See Problems 4.41 and 4.47.

Note: Problems 5.32–5.34 refer to the figure below, in which $\angle 1$ and $\angle 3$ are complementary.

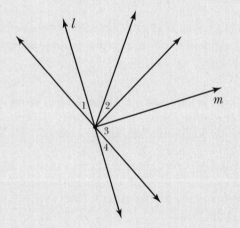

5.32   Justify the following statement: $\angle 3$ and $\angle 4$ are complementary.

The problem states that $\angle 1$ and $\angle 3$ are complementary, so the sum of their measures is $90°$.

$$m\angle 1 + m\angle 3 = 90°$$

Because $\angle 1$ and $\angle 4$ are formed by intersecting lines, they are vertical angles and therefore congruent. Apply the substitution property of equality to replace $m\angle 1$ with $m\angle 4$ in the previous equation.

> The sides of vertical angles are collinear rays that point in opposite directions.

$$m\angle 4 + m\angle 3 = 90°$$

The sum of the measures of $\angle 3$ and $\angle 4$ is $90°$, so $\angle 3$ and $\angle 4$ are complementary.

Note: Problems 5.32–5.34 refer to the figure in Problem 5.32, in which $\angle 1$ and $\angle 3$ are complementary.

5.33 Justify the following statement: $l \perp m$.

Lines l and m are perpendicular if and only if they intersect at right angles. One of the angles formed by the intersection of l and m has measure $m\angle 4 + m\angle 3$. Problem 5.32 states that $m\angle 4 + m\angle 3 = 90°$, so l and m intersect at right angles. Therefore, l and m must be perpendicular lines.

Note: Problems 5.32–5.34 refer to the figure in Problem 5.32, in which $\angle 1$ and $\angle 3$ are complementary.

5.34 Justify the following statement: If $\angle 4 \cong \angle 2$, then $\angle 1 \cong \angle 2$.

As Problem 5.32 states, $\angle 1$ and $\angle 4$ are vertical angles, so $\angle 1 \cong \angle 4$. According to the transitive property of equality, if $\angle 1 \cong \angle 4$ and $\angle 4 \cong \angle 2$, then $\angle 1 \cong \angle 2$.

See Problem 5.14.

Technically, the angle addition postulate states that you can combine $\angle 4$ and $\angle 3$ to get the right angle in the lower-right part of the diagram.

Two-Column Proofs
Prove that theorems are true using statements and reasons

Note: Problems 5.35–5.36 refer to the equation $4(x – 3) = 8$.

5.35 Solve the equation for x.

Distribute 4 through the quantity $(x – 3)$.

$$4(x) + 4(-3) = 8$$
$$4x - 12 = 8$$

Isolate $4x$ on the left side of the equation by adding 12 to both sides.

$$4x - 12 + 12 = 8 + 12$$
$$4x = 20$$

Divide both sides of the equation by the coefficient of x and express the fractions in lowest terms.

$$\frac{4}{4}x = \frac{20}{4}$$
$$x = 5$$

The left column lists the steps you took to solve the equation, and the right column identifies the properties you used in each step.

Note: Problems 5.35–5.36 refer to the equation $4(x - 3) = 8$.

5.36 Rewrite the solution generated in Problem 5.35 as a two-column proof.

A two-column proof begins with the given information and ends with the desired result. In this case, the equation is given and the goal is to prove $x = 5$.

Statement	Reason
1. $4(x - 3) = 8$	1. Given
2. $4x - 12 = 8$	2. Distributive property
3. $4x = 20$	3. Addition property of equality
4. $x = 5$	4. Division property of equality

Multiply every term by 6, the least common denominator, to get rid of the fractions:

$$\frac{6}{1}\left(\frac{2}{3}\right)x - \frac{6}{1}\left(\frac{1}{2}\right) = \frac{6}{1}\left(\frac{5}{6}\right)x$$

$$\frac{12}{3}x - \frac{6}{2} = \frac{30}{6}x$$

$$4x - 3 = 5x$$

5.37 Solve the equation $\frac{2}{3}x - \frac{1}{2} = \frac{5}{6}x$ using a two-column proof.

Statement	Reason
1. $\frac{2}{3}x - \frac{1}{2} = \frac{5}{6}x$	1. Given
2. $4x - 3 = 5x$	2. Multiplication property of equality
3. $-3 = x$	3. Subtraction property of equality

Subtract 4x from both sides of the equation.

5.38 Complete the following table to prove that supplements of the same angle are congruent.

Statement	Reason
1. $\angle B$ and $\angle C$ are supplements of $\angle A$	1. Given
2. $m\angle A + m\angle B = 180°$; $m\angle A + m\angle C = 180°$	2.
3. $m\angle A + m\angle B = m\angle A + m\angle C$	3.
4. $m\angle A = m\angle A$	4.
5. $m\angle B = m\angle C$	5.

$m\angle A + m\angle C = 180°$ and $m\angle A + m\angle C$ both equal 180°, so the sums equal each other.

Subtract $m\angle A$ from both sides of the equation in Step 3.

Statement	Reason
1. $\angle B$ and $\angle C$ are supplements of $\angle A$	1. Given
2. $m\angle A + m\angle B = 180°$; $m\angle A + m\angle C = 180°$	2. The sum of the measures of supplementary angles is 180°
3. $m\angle A + m\angle B = m\angle A + m\angle C$	3. Transitive property of equality
4. $m\angle A = m\angle A$	4. Reflexive property
5. $m\angle B = m\angle C$	5. Subtraction property of equality

Two angles with the same measures are congruent, so the statement $m\angle B = m\angle C$ is logically equivalent to the statement $\angle B \cong \angle C$.

5.39 Given $\angle A \cong \angle B$, $\angle C$ is a supplement of $\angle A$, and $\angle D$ is a supplement of $\angle B$, prove $\angle C \cong \angle D$.

Statement	Reason
1. $\angle C$ is a supplement of $\angle A$; $\angle D$ is a supplement of $\angle B$	1. Given
2. $m\angle C + m\angle A = 180°$; $m\angle D + m\angle B = 180°$	2. The sum of the measures of supplementary angles is $180°$
3. $m\angle C + m\angle A = m\angle D + m\angle B$	3. Transitive property of equality
4. $\angle A \cong \angle B$; $m\angle A = m\angle B$	4. Given
5. $m\angle C = m\angle D$	5. Subtraction property of equality

Because $\angle A$ and $\angle B$ are congruent, you're allowed to subtract $m\angle A$ from the left side of the equation in Step 3 and $m\angle B$ from the right side.

5.40 Given M is the midpoint of $\overline{XY}$ and $XM = 6$, in the following diagram, complete the proof to prove $XY = 12$.

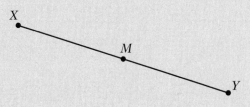

Statement	Reason
1. M is the midpoint of $\overline{XY}$	1. Given
2.	2. The midpoint of a segment divides it into two congruent segments
3. $XM + MY = XY$	3.
4.	4. Substitution property of equality (Steps 2, 3)
5. $2(MY) = XY$	5. Addition of like terms
6.	6. Given
7.	7. Transitive property of equality
8. $MY = 6$	8.

When you cite the substitution property of equality, indicate the steps that include the equations you're using.

Statement	Reason
1. M is the midpoint of $\overline{XY}$	1. Given
2. $XM = MY$	2. The midpoint of a segment divides it into two congruent segments
3. $XM + MY = XY$	3. Segment addition postulate
4. $MY + MY = XY$	4. Substitution property of equality (Steps 2, 3)
5. $2(MY) = XY$	5. Addition of like terms
6. $XY = 12$	6. Given
7. $2(MY) = 12$	7. Transitive property of equality
8. $MY = 6$	8. Division property of equality

Because XM = MY (according to Step 2), you can replace XM in Step 3 with MY.

Note: Problems 5.41–5.42 demonstrate two different ways to prove that the vertical angles in the following diagram are congruent.

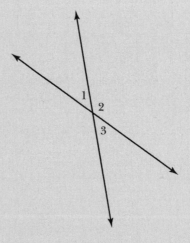

5.41 Prove $\angle 1 \cong \angle 3$ using the substitution property of equality.

Statement	Reason
1. $m\angle 1 + m\angle 2 = 180°$; $m\angle 2 + m\angle 3 = 180°$	1. Angle addition postulate
2. $m\angle 1 + m\angle 2 = m\angle 2 + m\angle 3$	2. Transitive property of equality
3. $m\angle 2 = m\angle 2$	3. Reflexive property of equality
4. $m\angle 1 = m\angle 3$	4. Subtraction property of equality

Note: Problems 5.41–5.42 demonstrate two different ways to prove that the vertical angles in Problem 5.41 are congruent.

5.42 Prove that $\angle 1 \cong \angle 3$ using the properties of supplementary angles.

Statement	Reason
1. $m\angle 1 + m\angle 2 = 180°$; $m\angle 2 + m\angle 3 = 180°$	1. Angle addition postulate
2. $\angle 1$ and $\angle 2$ are supplementary; $\angle 2$ and $\angle 3$ are supplementary	2. The sum of the measures of supplementary angles is 180°
3. $m\angle 2 = m\angle 2$	3. Reflexive property of equality
4. $m\angle 1 = m\angle 3$	4. Supplements of the same ← angle are congruent

Problem 5.38 proves this statement. After you prove something, you can use it in the "Reason" column of subsequent proofs.

5.43 Given $\angle A$, which is its own complement, prove that $m\angle A = 45°$.

Statement	Reason
1. $\angle A$ is its own complement	1. Given
2. $m\angle A + m\angle A = 90°$	2. Definition of complementary angles
3. $2(m\angle A) = 90°$	3. Addition of like terms
4. $m\angle A = 45°$	4. Division property of equality

5.44 Given $\overrightarrow{OB}$ and $\overrightarrow{OC}$ trisect $\angle AOD$ in the following diagram, prove $m\angle AOC = m\angle BOD$.

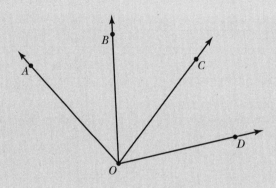

∠AOB and ∠COD are also congruent to ∠BOC, so it's okay to say m∠AOB = m∠BOC = m∠COD.

Statement	Reason
1. $\overrightarrow{OB}$ and $\overrightarrow{OC}$ trisect $\angle AOD$	1. Given
2. $m\angle AOB = m\angle COD$	2. Trisecting an angle divides it into three congruent angles
3. $m\angle BOC = m\angle BOC$	3. Reflexive property of equality
4. $m\angle AOB + m\angle BOC = m\angle BOC + m\angle COD$	4. Addition property of equality
5. $m\angle AOB + m\angle BOC = m\angle AOC$; $m\angle BOC + m\angle COD = m\angle BOD$	5. Angle addition postulate
6. $m\angle AOC = m\angle BOD$	6. Substitution property of equality (Steps 4, 5)

Chapter 6
PARALLEL AND PERPENDICULAR LINES

Including angles formed by a transversal

Parallel and perpendicular lines, though introduced in preceding chapters, deserve additional focused attention. This chapter is primarily dedicated to investigating parallel lines intersected by a transversal and the pairs of angles that result. In the problems that follow, you identify the angles, define the manner in which their measures are related, and use them to prove lines parallel. The chapter concludes with a series of theorems about perpendicular lines, including the behavior of points along a perpendicular bisector.

When two parallel lines are intersected by a third line (not parallel to the other two), that line is called a transversal. A transversal carves parallel lines into sets of angles, including corresponding angles, same-side interior angles, and alternate interior angles (just to name a few). When you figure out how the pairs of angles are related, you use them to calculate angle measurements and eventually move on to proving that lines are parallel.

At the end of the chapter, you return to the distant cousin of parallel lines—perpendicular lines, which intersect each other at 90° angles. You also learn about the perpendicular bisector, which intersects a segment at its midpoint AND is perpendicular to the segment, and angle bisectors.

Angles Formed by a Transversal

Corresponding, alternate interior, same-side interior angles

Note: Problems 6.1–6.10 refer to the following diagram in which parallel lines m and n are intersected by a transversal.

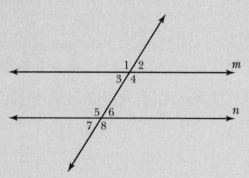

6.1 Identify all pairs of corresponding angles.

Corresponding angles occupy the same relative position on each parallel line. For instance, ∠1 is the upper-left angle formed by the intersection of line *m* and the transversal. It corresponds with ∠5, the upper-left angle formed by the intersection of line *n* and the transversal. You can identify three other pairs of corresponding angles: ∠2 and ∠6, ∠3 and ∠7, and ∠4 and ∠8.

Note: Problems 6.1–6.10 refer to the diagram in Problem 6.1, in which parallel lines m and n are intersected by a transversal.

6.2 Complete the following statement and explain your answer.

When parallel lines are intersected by a transversal, the corresponding angles formed are _____.

When parallel lines are intersected by a transversal, the corresponding angles formed are <u>congruent</u>. To visually verify this postulate, imagine that line *n* is shifted upward so that it coincides with line *m*, intersecting line *m* at every point along its length. The transversal intersects *n* at precisely the same angle it intersects *n*.

> You can't prove that corresponding angles are congruent; it's a postulate. Think of line n as a copy of line m that's just moved down a little, so the transversal cuts through both lines at the same angle.

Note: Problems 6.1–6.10 refer to the diagram in Problem 6.1, in which parallel lines m and n are intersected by a transversal.

6.3 Identify all pairs of alternate interior angles.

Alternate interior angles are nonadjacent angles located between the parallel lines on alternate sides of the transversal. In this diagram, ∠3 and ∠6 are alternate interior angles, as are ∠4 and ∠5.

Note: Problems 6.1–6.10 refer to the diagram in Problem 6.1, in which parallel lines m and n are intersected by a transversal.

6.4   Complete the following statement and prove it using the diagram and one pair of alternate interior angles identified in Problem 6.3.

When two parallel lines are intersected by a transversal, the alternate interior angles formed are _____.

When two parallel lines are intersected by a transversal, the alternate interior angles formed are <u>congruent</u>. Use a two-column proof to justify this statement; given $m \parallel n$, prove that $\angle 4 \cong \angle 5$. ←

> You can prove that $\angle 3 \cong \angle 6$ using the same technique.

Statement	Reason
1. $m \parallel n$	1. Given
2. $\angle 1 \cong \angle 5$	2. Corresponding angles formed by parallel lines and a transversal are congruent
3. $\angle 1 \cong \angle 4$	3. Vertical angles are congruent
4. $\angle 4 \cong \angle 5$	4. Transitive property of equality ←

> Instead of the transitive property, you can write "Substitution property of equality (Steps 2, 3)." ($\angle 1$ and $\angle 4$ are congruent, so you can replace $\angle 1$ in Step 2's equation with $\angle 4$.)

Note: Problems 6.1–6.10 refer to the diagram in Problem 6.1, in which parallel lines m and n are intersected by a transversal.

6.5   Identify all pairs of same-side interior angles.

Same-side interior angles are located between the parallel lines on the same side of the transversal. In this diagram, $\angle 3$ and $\angle 5$ are same-side interior angles, as are $\angle 4$ and $\angle 6$.

Note: Problems 6.1–6.10 refer to the diagram in Problem 6.1, in which parallel lines m and n are intersected by a transversal.

6.6   Complete the following statement and prove it using the diagram and one pair of same-side interior angles identified in Problem 6.5.

When two parallel lines are intersected by a transversal, the same-side interior angles formed are _____.

When two parallel lines are intersected by a transversal, the same-side interior angles formed are <u>supplementary</u>. Use a two-column proof to justify this statement; given $m \parallel n$, prove that $\angle 3$ and $\angle 5$ are supplementary.

Statement	Reason
1. $m \parallel n$	1. Given
2. $\angle 1 \cong \angle 5$	2. Corresponding angles formed by parallel lines and a transversal are congruent
3. $m\angle 1 + m\angle 3 = 180°$	3. Angle addition postulate
4. $m\angle 5 + m\angle 3 = 180°$	4. Substitution property of equality (Steps 2, 3)
5. $\angle 3$ and $\angle 5$ are supplementary	5. The sum of the measures of supplementary angles is 180°

Step 2 says ∠1 ≅ ∠5, so m∠1 = m∠5. Replace m∠1 in Step 3 with m∠5.

Note: Problems 6.1–6.10 refer to the diagram in Problem 6.1, in which parallel lines m and n are intersected by a transversal.

6.7 Identify all pairs of alternate exterior angles.

Alternate exterior angles are *not* located between the parallel lines. In the diagram, $\angle 1$, $\angle 2$, $\angle 7$, and $\angle 8$ are exterior angles. *Alternate* exterior angles are nonadjacent angles located on alternate sides of the transversal. In this diagram, $\angle 1$ and $\angle 8$ are alternate exterior angles, as are $\angle 2$ and $\angle 7$.

∠3, ∠4, ∠5, and ∠6 are interior angles because they are nestled between the parallel lines.

Note: Problems 6.1–6.10 refer to the diagram in Problem 6.1, in which parallel lines m and n are intersected by a transversal.

6.8 Complete the following statement and prove it using the diagram and one pair of alternate exterior angles identified in Problem 6.7.

When two parallel lines are intersected by a transversal, the alternate exterior angles formed are _____.

When two parallel lines are intersected by a transversal, the alternate exterior angles formed are <u>congruent</u>. Use a two-column proof to justify this statement; given $m \parallel n$, prove $\angle 2 \cong \angle 7$.

You can subtract m∠1 from the left side of the equation in Step 3 and subtract m∠5 from the right side of the equation, because the values are equal.

Statement	Reason
1. $m \parallel n$	1. Given
2. $m\angle 1 + m\angle 2 = 180°$; $m\angle 5 + m\angle 7 = 180°$	2. Angle addition postulate
3. $m\angle 1 + m\angle 2 = m\angle 5 + m\angle 7$	3. Transitive property of equality
4. $m\angle 1 = m\angle 5$	4. Corresponding angles formed by parallel lines and a transversal are congruent
5. $m\angle 2 = m\angle 7$	5. Subtraction property of equality

Note: Problems 6.1–6.10 refer to the diagram in Problem 6.1, in which parallel lines m and n are intersected by a transversal.

6.9 Identify all pairs of same-side exterior angles.

As explained in Problem 6.7, exterior angles are not located between the parallel lines; same-side angles are located on the same side of the transversal. Therefore, $\angle 1$ and $\angle 7$ are same-side exterior angles, as are $\angle 2$ and $\angle 8$.

Note: Problems 6.1–6.10 refer to the diagram in Problem 6.1, in which parallel lines m and n are intersected by a transversal.

6.10 Complete the following statement and prove it using the diagram and one pair of same-side exterior angles identified in Problem 6.9.

When two parallel lines are intersected by a transversal, the same-side exterior angles formed are _____.

When two parallel lines are intersected by a transversal, the same-side exterior angles formed are <u>supplementary</u>. Use a two-column proof to justify this statement.

Statement	Reason
1. $m \parallel n$	1. Given
2. $m\angle 6 + m\angle 8 = 180°$	2. Angle addition postulate
3. $m\angle 2 = m\angle 6$	3. Corresponding angles formed by parallel lines and a transversal are congruent
4. $m\angle 2 + m\angle 8 = 180°$	4. Substitution property of equality (Steps 2, 3)
5. $\angle 2$ and $\angle 8$ are supplementary	5. The sum of the measures of supplementary angles is 180°

6.11   Use the diagram below to prove the following statement.

If two parallel lines are intersected by a transversal and the transversal is perpendicular to one of the lines, then it is perpendicular to the other line as well.

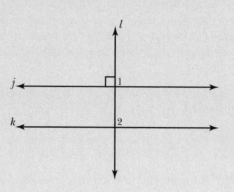

The little square above and to the left of the point where j and l intersect indicates a right angle, making the lines perpendicular.

Given $j \parallel k$ and $j \perp l$, prove $k \perp l$.

Statement	Reason
1. $j \parallel k$; $j \perp l$	1. Given
2. $m\angle 1 = 90°$	2. Perpendicular lines intersect at 90° angles
3. $m\angle 1 = m\angle 2$	3. Corresponding angles formed by parallel lines and a transversal are congruent
4. $m\angle 2 = 90°$	4. Transitive property of equality
5. $k \perp l$	5. Lines intersecting at 90° angles are perpendicular

Note: Problems 6.12–6.13 refer to the diagram below.

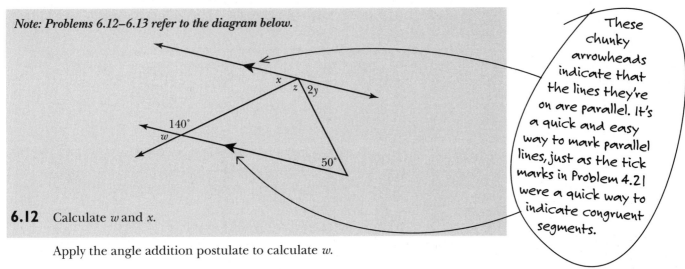

These chunky arrowheads indicate that the lines they're on are parallel. It's a quick and easy way to mark parallel lines, just as the tick marks in Problem 4.21 were a quick way to indicate congruent segments.

6.12 Calculate w and x.

Apply the angle addition postulate to calculate w.

$$w + 140 = 180$$
$$w = 180 - 140$$
$$w = 40$$

When parallel lines are intersected by a transversal, the corresponding angles formed are congruent. Therefore, $w = x = 40°$. It may be necessary for you to redraw the diagram, excluding the extraneous information and extending the transversal, to visualize the parallel lines and the transversal that form the corresponding angles.

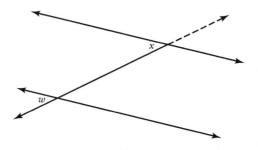

Note: Problems 6.12–6.13 refer to the diagram in Problem 6.12.

6.13 Calculate y and z.

When two parallel lines are intersected by a transversal, the alternate interior angles formed are congruent. Therefore, the angle measurements $2y$ and $50°$ are equal.

$$2y = 50$$
$$y = \frac{50}{2}$$
$$y = 25$$

The angle measures x, z, and $2y$ combine to form a straight line.

The x-value from Problem 6.12

$$x + z + 2y = 180$$

Substitute $y = 25$ and $x = 40$ into the equation and solve for z.

$$40 + z + 2(25) = 180$$
$$40 + z + 50 = 180$$
$$z + 90 = 180$$
$$z = 180 - 90$$
$$z = 90$$

Because they intersect at a 90° angle, the transversals in the diagram are perpendicular.

Note: Problems 6.14–6.15 refer to the diagram below.

This diagram has two sets of arrowheads to indicate parallel lines. Lines with single arrowheads are parallel to each other, and lines with double arrowheads are parallel to each other.

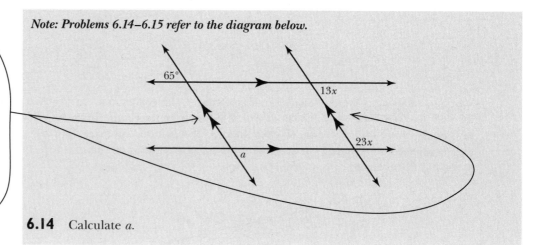

6.14 Calculate a.

The measures of two alternate exterior angles are given: 65° and a. According to Problems 6.7 and 6.8, alternate exterior angles are congruent. Therefore, $a = 65$.

Note: Problems 6.14–6.15 refer to the diagram in Problem 6.14.

6.15 Calculate x.

The angles in the diagram with measures that contain x are same-side interior angles. Therefore, the sum of their measures is 180°.

$$13x + 23x = 180$$

Solve the equation for x.

$$36x = 180$$
$$x = \frac{180}{36}$$
$$x = 5$$

Note: Problems 6.16–6.17 refer to the diagram below.

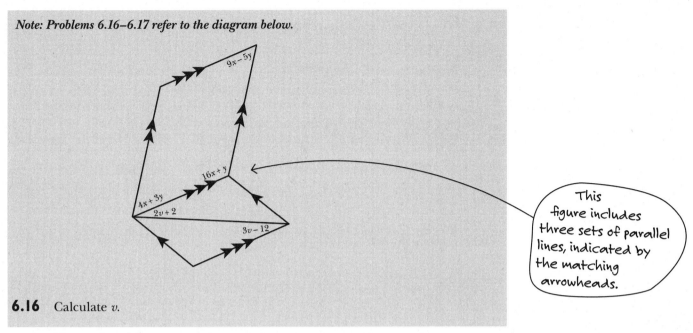

This figure includes three sets of parallel lines, indicated by the matching arrowheads.

6.16 Calculate v.

The angles with measures that contain v are congruent because they are alternate interior angles formed by parallel lines intersected by a transversal.

$$2v + 2 = 3v - 12$$

Solve the equation for v.

$$2v + 2 + 12 = 3v$$
$$2v + 14 = 3v$$
$$14 = 3v - 2v$$
$$14 = v$$

Note: Problems 6.16–6.17 refer to the diagram in Problem 6.16.

6.17 Calculate *x* and *y*.

The diagram contains two pairs of same-side interior angles. It may be necessary to redraw the diagram, excluding extraneous information and extending parallel lines, to identify the pairs of angles.

Think of the same-side interior angles as saying "hi." That's mighty friendly.

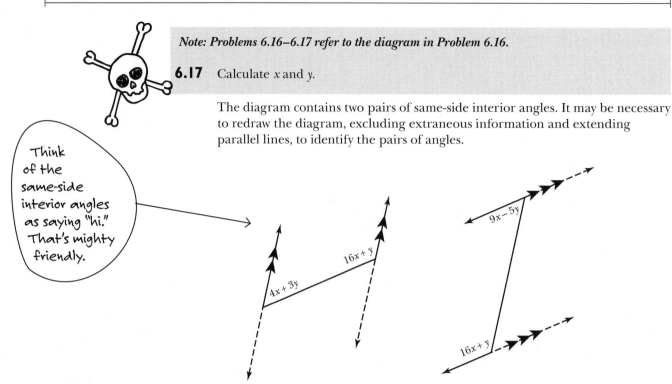

Each pair of angle measurements has a sum of 180° because same-side interior angles are supplementary.

$$(4x + 3y) + (16x + y) = 180 \qquad (9x - 5y) + (16x + y) = 180$$
$$20x + 4y = 180 \qquad\qquad 25x - 4y = 180$$

Look at Problems 2.32–2.36 if you need help.

Solve the system of liner equations by eliminating *y*.

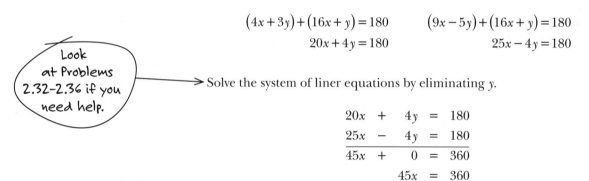

Solve the equation for *x*.

$$x = \frac{360}{45}$$
$$x = 8$$

Calculate the corresponding value of *y* by substituting *x* = 8 into an equation of the system.

$$20x + 4y = 180$$
$$20(8) + 4y = 180$$
$$160 + 4y = 180$$
$$4y = 180 - 160$$
$$4y = 20$$
$$y = \frac{20}{4}$$
$$y = 5$$

Therefore, $x = 8$ and $y = 5$.

Proving Lines Parallel

Converses of the angle theorems from the last section

6.18 Construct the converse of the foundational postulate that describes the relationship between angles formed by parallel lines and a transversal. Determine whether the statement you create is true.

> A bunch of theorems cover the angles formed by parallel lines and a transversal, but only one of them is a postulate.

Problem 6.2 states the postulate for which you must write the converse: When parallel lines are intersected by a transversal, the corresponding angles formed are congruent. The converse of this statement is, "When lines are intersected by a transversal and the corresponding angles formed are congruent, the lines are parallel." This is a true statement. Like the statement from which it originates, it is a postulate and cannot be verified via deductive reasoning.

Note: Problems 6.19–6.21 refer to the following diagram.

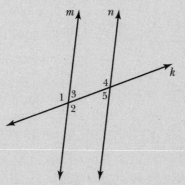

6.19 Given $\angle 1 \cong \angle 4$, prove $m \parallel n$.

> Is this the shortest proof ever? If only they could all be two steps long ....

Statement	Reason
1. $\angle 1 \cong \angle 4$	1. Given
2. $m \parallel n$	2. If two lines intersected by a transversal form congruent corresponding angles, then the lines are parallel

Note: Problems 6.19–6.21 refer to the diagram in Problem 6.19.

6.20 Given $\angle 2 \cong \angle 4$, prove $m \parallel n$.

Statement	Reason
1. $\angle 2 \cong \angle 4$	1. Given
2. $\angle 1 \cong \angle 2$	2. Vertical angles are congruent
3. $\angle 1 \cong \angle 4$	3. Transitive property of equality
4. $m \parallel n$	4. If two lines intersected by a transversal form congruent corresponding angles, the lines are parallel

According to this proof, if alternate interior angles (such as $\angle 2$ and $\angle 4$) are congruent, the lines in question are parallel. The converse is proven in Problem 6.4.

Note: Problems 6.19–6.21 refer to the diagram in Problem 6.19.

6.21 Given $\angle 3$ and $\angle 4$ are supplementary, prove $m \parallel n$.

Statement	Reason
1. $m\angle 3 + m\angle 4 = 180°$	1. Given
2. $m\angle 1 + m\angle 3 = 180°$	2. Angle addition postulate
3. $m\angle 3 + m\angle 4 = m\angle 1 + m\angle 3$	3. Transitive property of equality
4. $m\angle 3 = m\angle 3$	4. Reflexive property of equality
5. $m\angle 4 = m\angle 1$	5. Subtraction property of equality
6. $m \parallel n$	6. If two lines intersected by a transversal form congruent corresponding angles, the lines are parallel

Saying that something equals itself might feel like a waste of time, but Step 4 makes Step 5 easier to understand. In Step 5, you subtracted something from both sides of the equation. What did you subtract? The equal quantities you identified in Step 4.

6.22 According to Problems 16.18–16.21, congruent corresponding angles, congruent alternate interior angles, and supplementary same-side interior angles are sufficient to conclude that lines are parallel. Identify the two remaining theorems used to prove lines parallel that are converses of theorems presented earlier in the chapter.

The first is the converse of the theorem stated in Problem 6.8: If two lines intersected by a transversal form congruent alternate exterior angles, then the lines are parallel. The second is the converse of the theorem stated in Problem 6.10: If two lines intersected by a transversal form supplementary same-side exterior angles, then the lines are parallel.

6.23 Use the diagram below to prove that two lines perpendicular to the same line are parallel to each other.

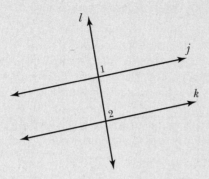

Statement	Reason
1. $j \perp l$; $k \perp l$	1. Given
2. $m\angle 1 = 90°$; $m\angle 2 = 90°$	2. Perpendicular lines form 90° angles
3. $m\angle 1 = m\angle 2$	3. Transitive property of equality
4. $j \parallel k$	4. If two lines intersected by a transversal form congruent corresponding angles, the lines are parallel

6.24 Two lines that are both parallel to a third line are parallel to each other. Prove this using the diagram that follows, in which $p \parallel q$ and $q \parallel r$.

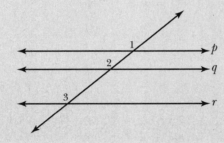

Statement	Reason
1. $p \parallel q$; $q \parallel r$	1. Given
2. $\angle 1 \cong \angle 2$; $\angle 2 \cong \angle 3$	2. Corresponding angles formed by parallel lines and a transversal are congruent
3. $\angle 1 \cong \angle 3$	3. Transitive property of equality
4. $p \parallel r$	4. If two lines intersected by a transversal form congruent corresponding angles, the lines are parallel

This isn't the exact wording of the Fifth Postulate; the original is a lot more complicated. However, this statement is its logical equivalent.

Another fact: exactly one line passes through P and is perpendicular to l.

6.25 Complete the following statement and explain your answer.

According to Euclid's Fifth Postulate, given line l and coplanar point P not on the line, there exists _____ line passing through P that is _____ to l.

The entirety of Euclidean geometry is based upon five postulates. Four of the postulates (including "A straight line may be drawn through any two points" and "All right angles are congruent") are universally accepted as true. However, the Fifth Postulate—commonly called the Parallel Postulate—was more controversial: given line *l* and coplanar point *P* not on the line, there exists <u>exactly one</u> coplanar line passing through *P* that is <u>parallel</u> to *l*. In the diagram below, line *m* is the only coplanar line through *P* that is parallel to *l*.

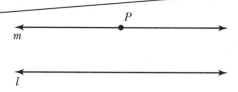

Mathematicians, including Lobachevsky and Riemann, who discard the Fifth Postulate develop logically sound but radically different geometries with theorems that directly contradict those of Euclidean geometry. This book, however, is dedicated to Euclidean geometry and, therefore, must assume that the Fifth Postulate is true.

Note: Problems 6.26–6.28 refer to the diagram that follows.

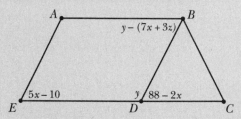

6.26 For what value of *x* is $\overrightarrow{AE} \parallel \overrightarrow{BD}$?

If $\overrightarrow{AE} \parallel \overrightarrow{BD}$, then corresponding angles formed by the lines and the transversal $\overline{EC}$ are congruent. Therefore, $\angle AEC \cong \angle BDC$. Set the measures of the angles equal and solve for *x*.

$$5x - 10 = 88 - 2x$$
$$5x + 2x = 88 + 10$$
$$7x = 98$$
$$x = \frac{98}{7}$$
$$x = 14$$

Note: Problems 6.26–6.28 refer to the diagram in Problem 6.26.

6.27 Given the value of x generated by Problem 6.26, calculate y.

Calculate $m\angle BDC$.

$$
\begin{aligned}
m\angle BDC &= 88 - 2x \\
&= 88 - 2(14) \\
&= 88 - 28 \\
&= 60 \leftarrow
\end{aligned}
$$

Because $\angle BDE$ and $\angle BDC$ are supplementary angles, the sum of their measures is $180°$.

$$
\begin{aligned}
m\angle BDE + m\angle BDC &= 180 \\
y \quad + \quad 60 \quad &= 180 \\
y &= 180 - 60 \\
y &= 120
\end{aligned}
$$

> If you measure $m\angle BDC$ with a protractor, you probably won't get $60°$, but that doesn't mean you got the answer wrong. The lengths of segments and measures of angles should be judged based on algebra and proofs, not the diagram itself.

Note: Problems 6.26–6.28 refer to the diagram in Problem 6.26.

6.28 Given the values of x and y calculated previously, for what value of z are $\overleftrightarrow{AB}$ and $\overleftrightarrow{CE}$ parallel?

If $\overleftrightarrow{AB} \parallel \overleftrightarrow{CE}$, alternate interior angles formed by those lines and the transversal $\overleftrightarrow{BD}$ are congruent: $\angle ABD \cong \angle BDC$. Set the measures of the angles equal and substitute $x = 14$ and $y = 120$ into the equation.

$$
\begin{aligned}
m\angle ABD &= m\angle BDC \\
y - (7x + 3z) &= 88 - 2x \\
120 - (7 \cdot 14 + 3z) &= 88 - 2(14) \\
120 - (98 + 3z) &= 88 - 28 \\
120 - 98 - 3z &= 60
\end{aligned}
$$

Solve for z.

$$
\begin{aligned}
22 - 3z &= 60 \\
-3z &= 60 - 22 \\
-3z &= 38 \\
z &= -\frac{38}{3}
\end{aligned}
$$

Note: Problems 6.29–6.31 refer to the diagram below.

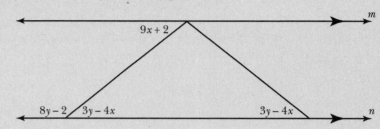

6.29 Calculate x.

Lines m and n are parallel, so alternate interior angles are congruent. Therefore, $3y - 4x = 9x + 2$. Write the equation in standard form.

$$3y - 4x - 9x = 2$$
$$-13x + 3y = 2$$
$$(-1)(-13x + 3y) = (-1)(2)$$
$$13x - 3y = -2$$

You don't HAVE to put the equation in standard form, but it helps when you apply the elimination technique in a few steps. If you don't feel like it, don't worry about making the x-term positive, one requirement of standard form.

Same-side interior angles formed by two parallel lines intersected by a transversal are supplementary. Therefore, $(9x + 2) + (8y - 2) = 180$. Write the equation in standard form.

$$9x + 8y + 2 - 2 = 180$$
$$9x + 8y = 180$$

The two linear equations in standard form constitute the following system of equations.

$$\begin{cases} 13x - 3y = -2 \\ 9x + 8y = 180 \end{cases}$$

The problem directs you to calculate x, so solve the system by eliminating y.

Multiply the top equation by 8, multiply the bottom equation by 3, and then add the equations.

$$
\begin{array}{rrrcr}
104x & - & 24y & = & -16 \\
27x & + & 24y & = & 540 \\
\hline
131x & + & 0 & = & 524
\end{array}
$$

Solve for x.

$$x = \frac{524}{131}$$
$$x = 4$$

Note: Problems 6.29–6.31 refer to the diagram in Problem 6.29.

6.30 Calculate *y*.

Problem 6.29 uses the elimination technique to determine that *x* = 4 for the following system of equations.

$$\begin{cases} 13x - 3y = -2 \\ 9x + 8y = 180 \end{cases}$$

Substitute *x* into either of the equations to calculate the corresponding value of *y*.

$$13x - 3y = -2$$
$$13(4) - 3y = -2$$
$$52 - 3y = -2$$
$$-3y = -2 - 52$$
$$-3y = -54$$
$$y = \frac{-54}{-3}$$
$$y = 18$$

Note: Problems 6.29–6.31 refer to the diagram in Problem 6.29.

6.31 Calculate *z*.

The diagram contains angles with measures $9x + 2$, $8y - 2$, and $3y - 4x$. Substitute *x* = 4 and *y* = 18 into the expressions to evaluate the angles.

$9x + 2$	$8y - 2$	$3y - 4x$
$= 9(4) + 2$	$= 8(18) - 2$	$= 3(18) - 4(4)$
$= 36 + 2$	$= 144 - 2$	$= 54 - 16$
$= 38$	$= 142$	$= 38$

Replace the variable expressions in the diagram with the angle measurements calculated above. Because $m \parallel n$, alternate interior angles are congruent.

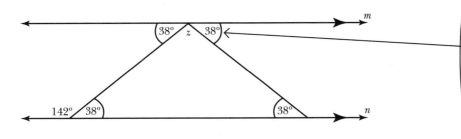

This angle wasn't labeled in the original diagram, but it must be 38° because its alternate interior angle is $3y - 4x = 38°$.

The sum of the measures of the interior angles along m is 180° because the three angles combine to form a straight angle.

$$38 + z + 38 = 180$$
$$76 + z = 180$$
$$z = 180 - 76$$
$$z = 104$$

Note: Problems 6.32–6.33 refer to the diagram below.

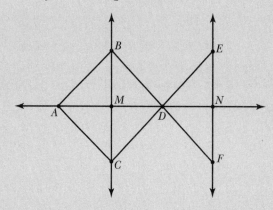

6.32 Given $\angle ABC$ and $\angle DBC$ are complementary and $\overleftrightarrow{CE} \perp \overleftrightarrow{BF}$, prove $\overleftrightarrow{AB} \parallel \overleftrightarrow{CD}$.

Statement	Reason
1. $m\angle ABC + m\angle DBC = 90°$	1. $\angle ABC$ and $\angle DBC$ are complimentary
2. $m\angle ABC + m\angle DBC = m\angle ABD$	2. Angle addition postulate
3. $m\angle ABD = 90°$	3. Transitive property of equality
4. $\overleftrightarrow{AB} \perp \overleftrightarrow{BF}$	4. Lines that intersect at 90° angles are perpendicular
5. $\overleftrightarrow{CD} \perp \overleftrightarrow{BF}$	5. Given
6. $\overleftrightarrow{AB} \parallel \overleftrightarrow{CD}$	6. Two lines perpendicular to the same line are parallel

$\overleftrightarrow{AB}$ and $\overleftrightarrow{CD}$ are both perpendicular to $\overleftrightarrow{BF}$.

Note: Problems 6.32–6.33 refer to the diagram in Problem 6.32.

6.33 Given $\angle AMC$ and $\angle DNE$ are supplementary, prove $\angle BCD \cong \angle FED$.

Statement	Reason
1. $m\angle AMC + m\angle DNE = 180°$	1. Given
2. $m\angle AMC = m\angle BMD$	2. Vertical angles are congruent
3. $m\angle BMD + m\angle DNE = 180°$	3. Substitution property of equality (Steps 1, 2)
4. $\overleftrightarrow{BC} \parallel \overleftrightarrow{EF}$	4. If two lines intersected by a transversal form supplementary same-side interior angles, the lines are parallel
5. $\angle BCD \cong \angle FED$	5. Alternate interior angles formed by parallel lines intersected by a transversal are congruent

In Step 4 the transversal is $\overleftrightarrow{MN}$, but in Step 5 the transversal is $\overleftrightarrow{CE}$.

Perpendicular Lines and Angle Bisectors

Fun facts about things that cut other things in half

Note: Problems 6.34–6.36 refer to line l, the perpendicular bisector of segment $\overline{AB}$.

6.34 Identify the two defining characteristics of *l*.

The perpendicular bisector of a segment divides the segment into two congruent halves. Furthermore, at the point of intersection, the segment and the bisector form 90° angles.

Note: Problems 6.34–6.36 refer to line l, the perpendicular bisector of segment $\overline{AB}$.

6.35 Complete the following statement and provide a diagram that illustrates your answer.

Every point on l is _____ from the _____ of $\overline{AB}$.

Every point on *l* is <u>equidistant</u> from the <u>endpoints</u> of $\overline{AB}$. Consider point *W* in the diagram that follows. The distance between *W* and endpoint *A* is equal to the distance between *W* and endpoint *B*.

> In other words, WA = WB.

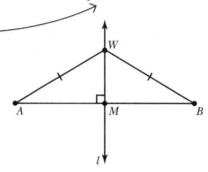

Note: Problems 6.34–6.36 refer to line l, the perpendicular bisector of segment $\overline{AB}$.

6.36 Explain why *l* must pass through *M*, the midpoint of $\overline{AB}$.

The midpoint of a segment is defined as the point on the segment that is equidistant from the endpoints. In this example, if *M* is the midpoint of $\overline{AB}$, then *AM* = *MB*. The perpendicular bisector of $\overline{AB}$ consists of *all* points equidistant from the endpoints of the segment, including *M*, so *M* belongs to the perpendicular bisector.

The following true statements are presented without proof:

♦ A point on the perpendicular bisector of a segment is equidistant from the endpoints of the segment.

♦ Any point equidistant from the endpoints of a segment belongs to the perpendicular bisector of that segment.

Thus, it is both necessary and sufficient for a point to be equidistant from the endpoints of a segment to lie on the perpendicular bisector of that segment.

> If a statement (such as the first bullet) is true, its converse (the second bullet) is not automatically true. In this case, however, both are true.

6.37 Complete the following statement and provide a diagram that illustrates your answer.

The distance between a line and a point not on that line is defined as the length of the _____.

The distance between a line and a point not on that line is defined as the length of the <u>perpendicular segment connecting the point and the line</u>. Consider point P and line m in the diagram below. Let $\overleftrightarrow{PR}$ be the unique line through P that is perpendicular to m. The distance between P and m is equal to PR.

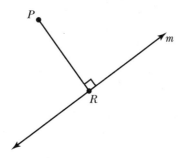

Note: Problems 6.38–6.39 refer to the diagram below, in which $\overline{PM}$ is the perpendicular bisector of $\overline{CD}$.

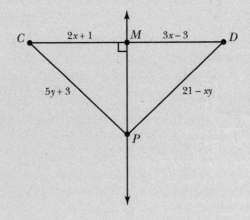

6.38 Calculate x.

A perpendicular bisector intersects a segment at its midpoint, so M is the midpoint of $\overline{CD}$. The midpoint of a segment is equidistant from the endpoints, so $CM = MD$.

$$2x + 1 = 3x - 3$$

Solve for x.

$$2x - 3x = -3 - 1$$
$$-x = -4$$
$$x = 4$$

Note: Problems 6.38–6.39 refer to the diagram in Problem 6.38, in which $\overrightarrow{PM}$ is the perpendicular bisector of $\overline{CD}$.

6.39 Calculate y.

Every point on the perpendicular bisector of a segment is equidistant from the endpoints of that segment, so $PC = PD$.

$$5y + 3 = 21 - xy$$

Substitute $x = 4$ into the equation and solve for y.

The value of x calculated in Problem 6.38

$$5y + 3 = 21 - (4)y$$
$$5y + 3 = 21 - 4y$$
$$5y + 4y = 21 - 3$$
$$9y = 18$$
$$y = \frac{18}{9}$$
$$y = 2$$

6.40 Given $\overleftrightarrow{KM}$ is the perpendicular bisector of $\overline{AB}$, $KA = KC$, and $BN = NC$ in the following diagram, prove $\overleftrightarrow{KN} \perp \overline{BC}$.

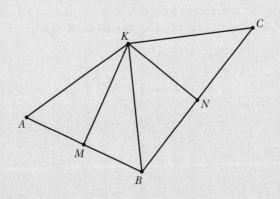

Statement	Reason
1. $\overleftrightarrow{KM}$ is the perpendicular bisector of $\overline{AB}$	1. Given
2. $KA = KB$	2. Points on the perpendicular bisector of a segment are equidistant from the segment's endpoints
3. $KA = KC$	3. Given
4. $KC = KB$	4. Transitive property of equality
5. $BN = NC$	5. Given

6. The perpendicular bisector of $\overline{BC}$ passes through K and N

6. Points equidistant from the endpoints of a segment lie on the perpendicular bisector of that segment

7. $\overleftrightarrow{KN}$ is the perpendicular bisector of $\overline{BC}$

7. Through two points passes exactly one line

8. $\overleftrightarrow{KN} \perp \overline{BC}$

8. Definition of a perpendicular bisector

Step 6 says that K and N are points on the perpendicular bisector. Step 7 says that it takes only two points to define a line, so $\overleftrightarrow{KN}$ actually IS the perpendicular bisector.

This one's easy. If $\overleftrightarrow{KN}$ is the PERPENDICULAR bisector of $\overline{BC}$, then, by definition, they must be perpendicular lines.

6.41 Complete the following statement and provide a diagram that illustrates your answer.

Every point on an angle bisector is _____ from the _____ of the angle it bisects.

According to Problem 6.35, every point on the perpendicular bisector of a segment is equidistant from the endpoints of that segment. Similarly, every point on an angle bisector is <u>equidistant</u> from the <u>sides</u> of the angle it bisects.

In the diagram below, $\overrightarrow{BD}$ bisects $\angle ABC$, so D is equidistant from the sides of the angle, $\overrightarrow{BA}$ and $\overrightarrow{BC}$. Recall that the distance between a point and a line is defined as the length of the perpendicular segment connecting them. Therefore, $XD = YD$.

See Problem 6.37.

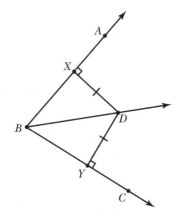

Chapter 7
TRIANGLES AND POLYGONS

Three-sided shapes and beyond

This chapter is the first in a lengthy progression of chapters dedicated to the study of polygons. It begins by introducing the triangle, including the means by which triangles are classified, the identification of special line segments used in the study of triangles, and a collection of theorems relating the angles of a triangle. The chapter concludes by investigating generic polygons, including theorems that describe the diagonals and angles of a polygon.

Polygons are shapes made of line segments. You already know the names of many polygons, like the pentagon and octagon, but you'll start out with the simplest polygon, the triangle. Triangles are made up of three line segments (sides) and three vertices (a vertex is the point where the sides of a polygon meet, and the term "vertices" is the plural of "vertex").

Next, you move on to medians, altitudes, angle bisectors, and perpendicular bisectors, which are line segments you can draw on any triangle. The most interesting things about them are the points where they intersect. After that, you push through a bunch of triangle theorems, including the most famous one: the angles of every triangle add up to 180°.

The chapter ends by discussing polygons generically; in other words, these theorems and properties apply no matter how many sides the shape has.

Classifying Triangles

Naming triangles and their parts

7.1   Name the triangle below and identify its sides and vertices.

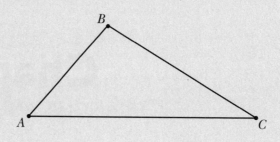

The sides of the triangle are segments: $\overline{AB}$, $\overline{BC}$, and $\overline{AC}$. The vertices of the triangle are the points at which the sides meet: A, B, and C. Name the triangle by listing the vertices clockwise or counterclockwise; the vertex where you begin is irrelevant. Therefore, $\triangle ABC$, $\triangle BCA$, $\triangle CAB$, $\triangle ACB$, $\triangle CBA$, and $\triangle BAC$ are all equally valid names for the triangle.

Note that the vertices are explicitly drawn as points in this diagram, but mathematical convention omits points in the diagrams of polygons. Vertices are still labeled where sides intersect, but actual points are rarely drawn.

Note: Problems 7.2–7.3 refer to the triangle below.

> Notice the words "at least." This means that a triangle with two OR three congruent sides is an isosceles triangle.

7.2   Classify $\triangle RST$ according to the length of its sides.

The diagram indicates that two sides of the triangle are congruent: $\overline{RS} \cong \overline{ST}$. A triangle with at least two congruent sides is described as isosceles.

Note: Problems 7.2–7.3 refer to the diagram in Problem 7.2.

7.3   Classify $\triangle RST$ according to the measures of its angles.

A triangle that contains a right angle is described as a right triangle. Therefore, $\triangle RST$ is a right triangle because $\angle S$ is a right angle.

Note: Problems 7.4–7.5 refer to the triangle below.

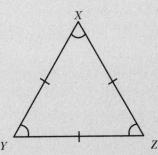

7.4 Classify △*XYZ* according to the lengths of its sides.

All three sides of the triangle are congruent: $\overline{XY} \cong \overline{YZ} \cong \overline{XZ}$. According to Problem 7.2, a triangle with at least two congruent sides is isosceles, so △*XYZ* is an isosceles triangle. However, it is more descriptive to classify △*XYZ* as an equilateral triangle, which indicates that all *three* sides of the triangle are congruent. ←

All equilateral triangles are automatically isosceles—if a triangle has three congruent sides, then it definitely has at least two congruent sides.

Note: Problems 7.4–7.5 refer to the diagram in Problem 7.4.

7.5 Classify △*XYZ* according to the measures of its angles.

Problem 7.4 states that a triangle with three congruent sides is described as equilateral. Similarly, a triangle with three congruent angles is described as equiangular. The diagram indicates that $\angle X \cong \angle Y \cong \angle Z$, so △*XYZ* is an equiangular triangle.

Note: Problems 7.6–7.7 refer to the triangle below. Assume that the diagram is drawn to scale. ←

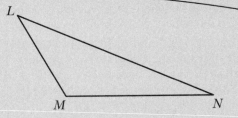

7.6 Classify △*LMN* according to the lengths of its sides.

Unlike the triangles referenced in Problems 7.2–7.5, none of the sides of △*LMN* are congruent. Triangles with three sides of different lengths are described as scalene.

Normally, you can't assume that a diagram is precise. For example, an angle may actually be bigger than it appears. Otherwise, geometry would be all about measuring things with rulers and protractors. In this case, however, the diagram is precise.

Note: Problems 7.6–7.7 refer to the diagram in Problem 7.6. Assume that the diagram is drawn to scale.

7.7   Classify △*LMN* according to the measures of its angles.

If a triangle contains an obtuse angle, the triangle is also described as obtuse. Because *m∠LMN* > 90°, ∠*LMN* is an obtuse angle and △*LMN* is an obtuse triangle. Similarly, if all the angles of a triangle are acute (i.e., each angle measures less than 90°), the triangle is acute.

7.8   Explain the circumstances under which the side of a triangle may also be described as a hypotenuse. Illustrate one such example with an appropriately labeled diagram.

> In Chapter 10, you'll prove that the longest side of a triangle is the side opposite the biggest angle, and vice versa. The bigger the angle, the bigger the opposite side.

The longest side of a right triangle—the side opposite the right angle—is called the hypotenuse of the triangle. Therefore, only a right triangle may contain a hypotenuse. The remaining two sides of a right triangle are called the legs.

In the diagram that follows, △*POQ* is a right triangle because *m∠P* = 90°. The side opposite the right angle—and, thus, the longest side of the triangle—is $\overline{OQ}$, the hypotenuse of the triangle.

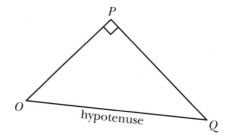

7.9   Explain how to determine which side of a triangle is its base.

Though you may infer that the word "base" indicates the "bottom" of the triangle, any side of the triangle may serve as the base. There are a few situations in which the base of the triangle serves a particular purpose (see Problem 7.10), but barring those exceptions, any of the triangle's sides may be designated the base.

7.10 Construct △*XYZ* given $\overline{XY} \cong \overline{YZ}$ and $\overline{XZ} \not\cong \overline{XY}$. Label the base, legs, base angles, and vertex angle of the isosceles triangle.

Exactly two sides of △*XYZ* are congruent, so △*XYZ* is isosceles. The legs of an isosceles triangle are the congruent sides, so the legs of △*XYZ* are $\overline{XY}$ and $\overline{YZ}$. The remaining side of the triangle is the base. In the diagram below, $\overline{XZ}$ is the base of △*XYZ*.

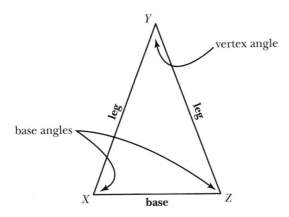

The base angles are opposite the legs of an isosceles triangle; the vertex angle is opposite the base. The base angles of △*XYZ* are ∠*X* and ∠*Z*; the vertex angle is ∠*Y*.

Note: Problems 7.11–7.12 refer to the diagram that follows.

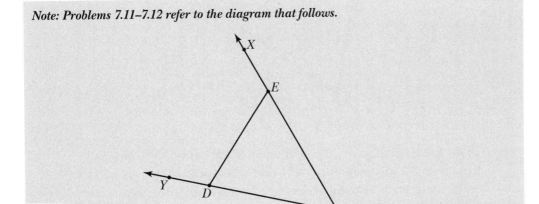

7.11 Identify the interior angles of △*DEF*.

The interior angles of a triangle are located within the two-dimensional region bounded by the triangle's sides. The interior angles of △*DEF* are ∠*EDF*, ∠*EFD*, and ∠*FED*.

In shockingly unsurprising news, the interior angles of a triangle are located INSIDE the triangle.

Note: Problems 7.11–7.12 refer to the diagram in Problem 7.11.

7.12 Identify the exterior angles of △*DEF*.

A triangle has three exterior angles, each the supplement of an interior angle. For instance, the supplement of interior angle ∠*EDF* is the exterior angle ∠*EDY*. The other exterior angles of △*DEF* are ∠*DEX* and ∠*ZFE*.

Special Triangle Segments
Medians, altitudes, bisectors

7.13 Define the median of a triangle.

The median of a triangle is a segment that extends from a vertex to the midpoint of the opposite side. Because triangles have three vertices, triangles have three medians as well.

7.14 Complete the following statement and illustrate your answer using △*ABC*.

The medians of a triangle intersect at a single point called the _____, which serves as the triangle's "balance point" or center of gravity.

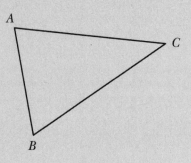

In the following diagram, the midpoint of each side of △*ABC* has been identified: *X* is the midpoint of $\overline{AB}$, *Y* is the midpoint of $\overline{AC}$, and *Z* is the midpoint of $\overline{BC}$. The triangle has three medians, each connecting a vertex to the midpoint of the opposite side. The medians of △*ABC* are $\overline{AZ}$, $\overline{BY}$, and $\overline{CX}$.

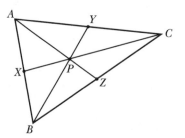

In other words, if △*ABC* were made out of wood, you could balance the shape by placing your finger underneath at point P.

Notice that the medians intersect at point *P*. The medians of a triangle intersect at a single point called the <u>centroid</u>, which serves as the triangle's "balance point" or center of gravity.

Note: Problems 7.15–7.16 refer to the figure below, △DEF and its medians $\overline{DN}$, $\overline{EM}$, and $\overline{FL}$.

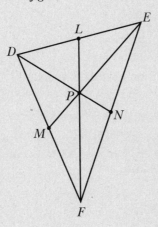

7.15 Given $ME = 24$, calculate MP and PE.

According to the segment addition postulate, the length of $\overline{ME}$ is equal to the sum of the lengths of $\overline{MP}$ and $\overline{PE}$.

$$MP + PE = ME$$

The segment connecting a vertex to a centroid is twice as long as the segment connecting the centroid to the midpoint of the side opposite that vertex. Therefore, $PE = 2(MP)$. Replace PE in the above segment addition equation with $2(MP)$.

$$MP + 2(MP) = ME$$
$$3(MP) = ME$$

The problems state that $ME = 24$. Use this information to solve for MP.

$$3(MP) = 24$$
$$MP = \frac{24}{3}$$
$$MP = 8$$

Recall that $PE = 2(MP)$.

$$PE = 2(MP)$$
$$PE = 2(8)$$
$$PE = 16$$

Therefore, $MP = 8$ and $PE = 16$. The length of the segment connecting centroid P to vertex E is 16, twice as long as the segment connecting P to M, the midpoint of the side opposite vertex P.

Note: Problems 7.15–7.16 refer to △DEF in Problem 7.15, which has medians $\overline{DN}$, $\overline{EM}$, and $\overline{FL}$.

7.16 Given $LP = 7$, calculate LF.

Apply the segment addition postulate to define LF.

$$LP + PF = LF$$

According to the theorem introduced in Problem 7.15, the segment connecting centroid P to vertex F is twice as long as the segment connecting P to L, the midpoint of the side opposite F.

$$PF = 2(LP)$$

The problem states that $LP = 7$. Use this information to calculate PF.

$$PF = 2(7)$$
$$PF = 14$$

Substitute $LP = 7$ and $PF = 14$ into the segment addition equation from the beginning of the problem to calculate LF.

$$LP + PF = LF$$
$$7 + 14 = LF$$
$$21 = LF$$

Note: Problems 7.17–7.18 refer to △GHI that follows.

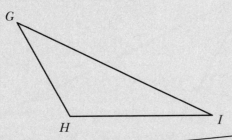

The altitude describes a very important triangle characteristic—its height. In Chapter 17, when you calculate the area of a triangle, you'll need to know the base and the height—one side of the triangle and the altitude that extends to that side.

7.17 Define the altitude of a triangle and draw the altitude that extends from vertex *G*.

The altitude of a triangle is a segment extending from the vertex that is perpendicular to a line containing the opposite side. In some cases, the altitude of a triangle does not actually intersect the opposite side; this is one such case. In the diagram below, side $\overline{HI}$ is extended beyond its endpoint *H* as a visual aid to facilitate the construction of altitude $\overline{GX}$.

The side of the triangle doesn't magically grow. Some people extend the sides using dotted lines so that it's easier to draw a perpendicular segment.

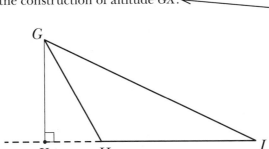

Note: Problems 7.17–7.18 refer to the diagram in Problem 7.17.

7.18 Complete the following statement and illustrate your answer using △GHI.

The altitudes of a triangle intersect at a single point called the _____.

Construct the altitudes of △GHI, as illustrated by the following diagram. You may find it necessary to extend side $\overline{GH}$ beyond endpoint *H* to construct altitude $\overline{IZ}$.

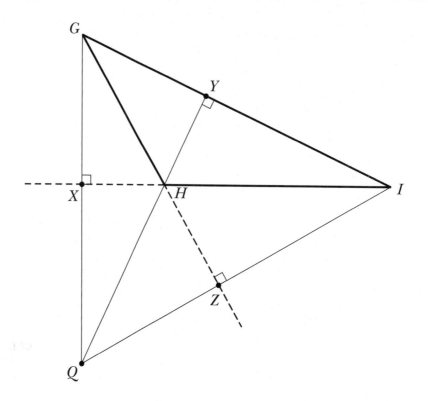

The altitudes of a triangle intersect at a single point called the <u>orthocenter</u>. In the above diagram, altitudes $\overline{GX}$, $\overline{HY}$, and $\overline{IZ}$ intersect at orthocenter Q.

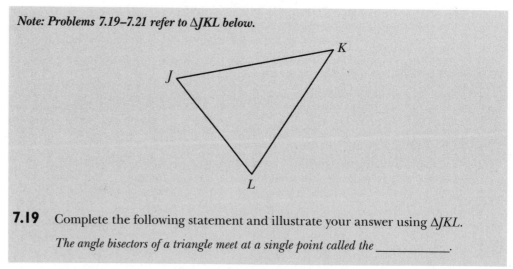

Note: Problems 7.19–7.21 refer to △JKL below.

7.19 Complete the following statement and illustrate your answer using △*JKL*.

The angle bisectors of a triangle meet at a single point called the _____.

The angle bisectors of a triangle meet at a single point called the <u>incenter</u>. In the following diagram, $\overline{JR}$ bisects ∠*KJL*, $\overline{LR}$ bisects ∠*JLK*, and $\overline{KR}$ bisects ∠*JKL*. The angle bisectors meet at point *R*, the incenter of the triangle.

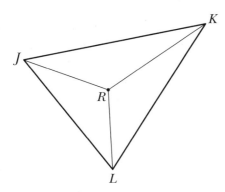

Note: *Problems 7.19–7.21 refer to △JKL and point R, as defined by Problem 7.19.*

7.20 Complete the following statement and illustrate your answer using △JKL.

R is equidistant from the _____, and a circle can be drawn with center R that is _____ the triangle.

Because it is the incenter of △JKL, R is equidistant from the <u>sides of △JKL</u>, and a circle can be drawn with center R that is <u>inscribed in</u> the triangle. The common distance between each side of the triangle and the incenter serves as the radius of the circle.

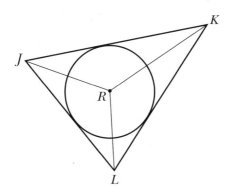

Circles are discussed in Chapter 13, but for now, a simple definition suffices: a circle is the set of points that are the same distance from the center. R is the same distance from the sides of △JKL, so you can draw a circle inside the triangle that touches all three sides exactly once.

Note: Problems 7.19–7.21 refer to △JKL and point R, as defined by Problem 7.19.

7.21 Using the following diagram, prove that the incenter of a triangle is equidistant from the sides of the triangle.

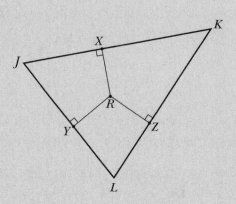

Begin by identifying the information that is given and what you must prove. You are given △JKL with incenter R, and you must prove that RX = RY = RZ.

R is on the angle bisector of ∠J, so it's the same distance from the sides of angle J, $\overline{JK}$ and $\overline{JL}$. Those distances are RX and RY, respectively. Same thing goes for R bisecting ∠L.

Statement	Reason
1. R is the incenter of △JKL	1. Given
2. R lies on the angle bisectors of ∠J, ∠K, and ∠L	2. The incenter of a triangle is the point at which the angle bisectors intersect
3. RX is the distance between R and $\overline{JK}$; RY is the distance between R and $\overline{JL}$; RZ is the distance between R and $\overline{LK}$	3. The distance between a point and a line is the perpendicular segment connecting them
4. RX = RY; RY = RZ	4. Every point on an angle bisector is equidistant from the sides of the angle
5. RX = RY = RZ	5. Transitive property of equality

Note: *Problems 7.22–7.24 refer to △XYZ below.*

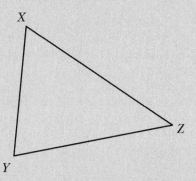

7.22 Complete the following statement and illustrate your answer using △*XYZ*.

The perpendicular bisectors of a triangle's sides meet at a single point P called the _____.

The perpendicular bisectors of a triangle's sides meet at a single point *P* called the <u>circumcenter</u>. In the diagram below, $\overline{AP}$ is the perpendicular bisector of $\overline{XY}$, $\overline{BP}$ is the perpendicular bisector of $\overline{XZ}$, and $\overline{CP}$ is the perpendicular bisector of $\overline{YZ}$. The perpendicular bisectors intersect at the circumcenter of △*XYZ*, point *P*.

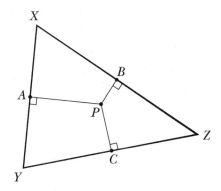

Note: Problems 7.22–7.24 refer to △XYZ and point P, as defined by Problem 7.22.

7.23 Complete the following statement and illustrate your answer using △XYZ.

P is equidistant from the _____, and a circle can be drawn with center P that is _____ the triangle.

Because it is the circumcenter of △XYZ, P is equidistant from the <u>vertices of</u> <u>△XYZ</u>, and a circle can be drawn with center P that is <u>circumscribed about</u> the triangle. The common distance between each vertex of the triangle and the circumcenter serves as the radius of the circle.

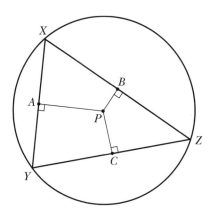

Note: Problems 7.22–7.24 refer to △XYZ and point P, as defined by Problem 7.22.

7.24 Using the diagram below, prove that the circumcenter of a triangle is equidistant from the vertices of the triangle.

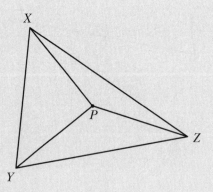

Begin by identifying the information that is given and what you must prove. You are given △XYZ with circumcenter P, and you must prove that PX = PY = PZ.

Statement	Reason
1. *P* is the circumcenter of ΔXYZ	1. Given
2. *P* lies on the perpendicular bisectors of $\overline{XY}$, $\overline{YZ}$, and $\overline{XZ}$	2. The circumcenter of a triangle is the point at which the perpendicular bisectors of the sides intersect
3. *PX* = *PY*; *PY* = *PZ*	3. Every point on the perpendicular bisector of a segment is equidistant from the endpoints of the segment ←
4. *PX* = *PY* = *PZ*	4. Transitive property of equality

This is very similar to the proof in Problem 7.21. The theorem about points on angle bisectors is found in Problem 6.41, and the theorem about points on perpendicular bisectors debuts in Problem 6.35.

7.25 According to Problem 7.18, the orthocenter of a triangle may be located outside the triangle. Which one of the other "centers" may be located outside a triangle? Provide an illustration that justifies your answer.

The orthocenter in Problem 7.18 is located outside of the triangle because that triangle is obtuse. Similarly, point *P*, the circumcenter of obtuse triangle *QRS* in the diagram below, is outside the triangle.

It's either the incenter, the circumcenter, or the centroid. Two of them are ALWAYS located inside the triangle.

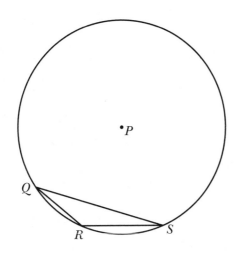

Triangle Theorems
Lots of different ways the angles are related

7.26 Use the diagram below to prove that the sum of the interior angles of a triangle is 180°.

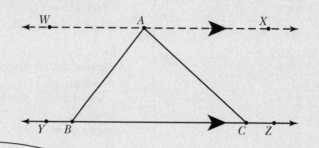

See Problem 6.25.

According to the Parallel Postulate, there exists exactly one line through A that is parallel to $\overline{BC}$. The diagram identifies this line as $\overleftrightarrow{WX}$, and although it is not part of the triangle, the line is needed to prove that $m\angle BAC + m\angle ABC + m\angle ACB = 180°$.

Statement	Reason
1. $\overleftrightarrow{WX} \parallel \overleftrightarrow{BC}$	1. Given
2. $m\angle WAB + m\angle BAC + m\angle XAC = 180°$	2. Angle addition postulate
3. $\angle WAB \cong \angle ABC$; $\angle XAC \cong \angle ACB$	3. Parallel lines intersected by a transversal form congruent alternate interior angles
4. $m\angle ABC + m\angle BAC + m\angle ACB = 180°$	4. Substitution property of equality (Steps 2, 3)

7.27 Use indirect reasoning and the following diagram to prove that a triangle may have, at most, only one right angle.

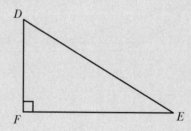

Indirect reasoning is often used to prove that something is untrue or cannot occur. In this problem, you are asked to demonstrate that a triangle may contain only one right angle. To do so, you must prove that a triangle *cannot* contain two or three right angles.

The most common method of indirect proof is called proof by contradiction. This method proves the statement "if p, then q" by demonstrating that the related statement, "if p, then not q" must be false. Therefore, the first step in a proof by contradiction problem is to negate the conclusion—the q phrase of the "if p, then q" statement.

Begin by proving that a right triangle cannot contain two right angles. To do so, assume the opposite of what you set out to prove. In other words, assume that a triangle *can* contain two right angles by setting $m\angle D = m\angle F = 90°$. Demonstrate that the assumption leads to a logical contradiction.

Statement	Reason
1. $m\angle F = 90°$; $m\angle D = 90°$	1. Given
2. $m\angle D + m\angle E + m\angle F = 180°$	2. The sum of the interior angles of a triangle is 180°
3. $90° + m\angle E + 90° = 180°$	3. Substitution property of equality (Steps 1, 2)
4. $180° + m\angle E = 180°$	4. Combination of like terms
5. $180° = 180°$	5. Reflexive property of equality
6. $m\angle E = 0°$	6. Subtraction property of equality

The proof has reached a contradiction; a triangle cannot contain an angle that measures 0°. Therefore, a triangle cannot contain two right angles. However, you have proven only one case. If a right triangle can contain only one right angle, that means it cannot contain two right angles (as the previous proof guarantees) *and* it cannot contain three right angles.

Assume that ΔDEF contains three right triangles and attempt to reach a logical contradiction.

Statement	Reason
1. $m\angle D = 90°$; $m\angle E = 90°$; $m\angle F = 90°$	1. Given
2. $m\angle D + m\angle E + m\angle F = 180°$	2. The sum of the interior angles of a triangle is 180°
3. $90° + 90° + 90° = 180°$	3. Substitution property of equality (Steps 1, 2)
4. $270° = 180°$	4. Combination of like terms

Clearly, $180° \neq 270°$, so a right triangle cannot contain three right angles. Therefore, a right triangle cannot contain two right angles, nor can it contain three right angles, so it must contain, at most, one right angle.

7.28 Use the diagram below to prove that the acute angles of a right triangle are complementary.

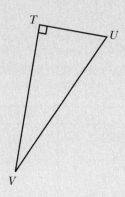

Given $m\angle T = 90°$, prove $\angle U$ and $\angle V$ are complementary.

Statement	Reason
1. $m\angle T = 90°$	1. Given
2. $m\angle T + m\angle U + m\angle V = 180°$	2. The sum of the interior angles of a triangle is 180°
3. $90° + m\angle U + m\angle V = 180°$	3. Substitution property of equality (Steps 1, 2)
4. $m\angle U + m\angle V = 90°$	4. Subtraction property of equality
5. $\angle U$ and $\angle V$ are complementary	5. Definition of complementary angles

7.29 Use the following diagram to prove that the measure of an exterior angle is equal to the sum of the measures of the remote interior angles.

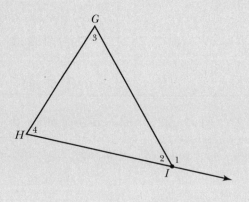

Given the angles labeled in the diagram, prove $m\angle 1 = m\angle 3 + m\angle 4$.

Statement	Reason
1. $\angle 1$ is an exterior angle	1. Definition of exterior angle of $\triangle GHI$
2. $\angle 3$ and $\angle 4$ are remote with respect to $\angle 1$	2. Definition of remote ← interior angles
3. $m\angle 1 + m\angle 2 = 180°$	3. Angle addition postulate
4. $m\angle 2 + m\angle 3 + m\angle 4 = 180°$	4. The sum of the interior angles of a triangle is 180°
5. $m\angle 1 + m\angle 2 =$ $m\angle 2 + m\angle 3 + m\angle 4$	5. Transitive property of equality
6. $m\angle 2 = m\angle 2$	6. Reflexive property of equality
7. $m\angle 1 = m\angle 3 + m\angle 4$	7. Subtraction property of equality

> $\angle 3$ and $\angle 4$ are the two angles inside the triangle that are farthest away from $\angle 1$ ($\angle 2$ is touching $\angle 1$, and angles that touch can't be far away from each other), which makes them remote angles.

> Steps 4 and 5 contain two expressions that equal 180°, so set those expressions equal to each other.

7.30 Calculate $m\angle C$ in the diagram below.

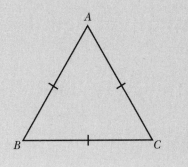

The sum of the measures of the interior angles of any triangle is 180°.

$$m\angle A + m\angle B + m\angle C = 180°$$

Because its sides are congruent, $\triangle ABC$ is equilateral. All equilateral triangles are also equiangular. Therefore, $\angle A \cong \angle B \cong \angle C$ and $m\angle A = m\angle B = m\angle C$. All three angles have the same measure, so you can substitute $m\angle C$ for both $m\angle A$ and $m\angle B$.

$$m\angle C + m\angle C + m\angle C = 180°$$
$$3(m\angle C) = 180°$$
$$m\angle C = \frac{180°}{3}$$
$$m\angle C = 60°$$

All equiangular (and equilateral) triangles have three congruent interior angles that measure 60°.

Note: Problems 7.31–7.33 refer to the following diagram, in which m∠DEF = 75˚.

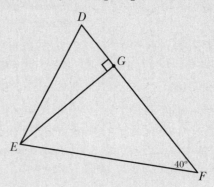

7.31 Calculate *m∠D*.

The sum of the interior angles of Δ*DEF* is 180˚.

$$m\angle D + m\angle DEF + m\angle F = 180°$$

The problem provides angle measurements for two of these interior angles: *m∠F* = 40˚ and *m∠DEF* = 75˚. Substitute them into the equation and solve for *m∠D*.

$$m\angle D + 75° + 40° = 180°$$
$$m\angle D + 115° = 180°$$
$$m\angle D = 180° - 115°$$
$$m\angle D = 65°$$

Note: Problems 7.31–7.33 refer to the diagram in Problem 7.31, in which m∠DEF = 75˚.

7.32 Calculate *m∠DEG*.

The diagram contains three distinct triangles: Δ*DEF*, Δ*DEG*, and Δ*EFG*. Problem 7.31 referenced Δ*DEF* to calculate ∠*D*. Similarly, you can reference Δ*DEG* in this problem. The sum of the measures of the interior angles of Δ*DEG* also equals 180˚.

$$m\angle D + m\angle DEG + m\angle DGE = 180°$$

And m∠EGF = 90˚.

Because $\overline{DG} \perp \overline{GE}$, *m∠DGE* = 90˚. According to Problem 7.31, *m∠D* = 65˚. Substitute these values into the equation and solve for *m∠DEG*.

$$65° + m\angle DEG + 90° = 180°$$
$$m\angle DEG + 155° = 180°$$
$$m\angle DEG = 180° - 155°$$
$$m\angle DEG = 25°$$

Note: Problems 7.31–7.33 refer to the diagram in Problem 7.31, in which m∠DEF = 75˚.

7.33 Calculate *m∠GEF*.

According to the angle addition postulate, *m∠DEF* is equal to the sum of the measures of the smaller angles that comprise it.

$$m\angle DEF = m\angle DEG + m\angle GEF$$

You are given *m∠DEF* = 75˚, and Problem 7.32 states that *m∠DEG* = 25˚. Substitute these values into the equation to calculate *m∠GEF*.

$$75° = 25° + m\angle GEF$$
$$75° - 25° = m\angle GEF$$
$$50° = m\angle GEF$$

Note: Problems 7.34–7.36 refer to the following diagram, in which m∠BCD = 40˚.

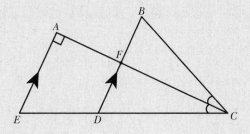

7.34 Calculate *m∠E*.

According to the diagram, ∠BCF ≅ ∠DCF. Apply the angle addition postulate to express *m∠BCD* as the sum of the measures of the congruent angles.

$$m\angle BCF + m\angle DCF = m\angle BCD$$

Because ∠BCF ≅ ∠DCF, *m∠BCF* = *m∠DCF*.

$$m\angle DCF + m\angle DCF = m\angle BCD$$
$$2(m\angle DCF) = m\angle BCD$$

You are given *m∠BCD* = 40˚. Solve for *m∠DCF*.

$$2(m\angle DCF) = 40°$$
$$m\angle DCF = \frac{40°}{2}$$
$$m\angle DCF = 20°$$

> You can substitute m∠DCF for m∠BCF in the angle addition equation because congruent angles have the same measure.

> ∠BCD is bisected by $\overrightarrow{CF}$, so each of the smaller angles is half as big. Half of 40 is 20.

The sum of the measures of the interior angles of $\triangle ACE$ equals 180°. Substitute the two known angle measurements to determine the third.

$$m\angle A + m\angle ACE + m\angle E = 180°$$
$$90° + 20° + m\angle E = 180°$$
$$110° + m\angle E = 180°$$
$$m\angle E = 180° - 110°$$
$$m\angle E = 70°$$

Note: Problems 7.34–7.36 refer to the diagram in Problem 7.34, in which m∠BCD = 40°.

7.35 Calculate $m\angle BDC$.

When parallel lines $\overrightarrow{EA}$ and $\overrightarrow{DB}$ are intersected by transversal $\overleftrightarrow{EC}$, the corresponding angles formed are congruent. Therefore, $\angle E \cong \angle BDC$. According to Problem 7.34, $m\angle E = 70°$; therefore, $m\angle BDC = 70°$.

Note: Problems 7.34–7.36 refer to the diagram in Problem 7.34, in which m∠BCD = 40°.

7.36 Calculate $m\angle B$.

The sum of the measures of the interior angles of $\triangle BDC$ is equal to 180°.

$$m\angle B + m\angle BDC + m\angle BCD = 180°$$

The problem states that $m\angle BCD = 40°$; according to Problem 7.35, $m\angle BDC = 70°$. Substitute these values into the equation to determine $m\angle B$.

$$m\angle B + 70° + 40° = 180°$$
$$m\angle B + 110° = 180°$$
$$m\angle B = 180° - 110°$$
$$m\angle B = 70°$$

You can determine that $m\angle B = 70°$ without performing any calculations if you notice that two angles of $\triangle ACE$ are congruent to two angles of $\triangle BCF$: $m\angle A = m\angle BFC = 90°$ and $m\angle ACE = m\angle BCF = 20°$. When two angles of two different triangles are congruent, then the remaining angles of each triangle are also congruent: $m\angle E = m\angle B = 70°$.

7.37 Calculate $m\angle 1$ in the diagram below.

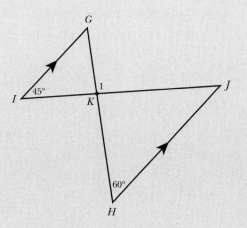

The parallel lines in the diagram are intersected by two different transversals, each creating a congruent pair of alternate interior angles: $\angle G \cong \angle H$ and $\angle I \cong \angle J$. Therefore, $m\angle G = 60°$ and $m\angle J = 45°$.

Because $\angle 1$ is an exterior angle of $\triangle KJH$, its measure is equal to the sum of the measures of the remote interior angles.

$$m\angle 1 = m\angle H + m\angle J$$

Substitute known values into the equation and solve for $m\angle 1$.

$$m\angle 1 = 60° + 45°$$
$$m\angle 1 = 105°$$

> You can also figure out that m∠HKJ = 75° (because the angles of ΔKJH have to add up to 180°) and then subtract 75° from 180° to get m∠1 (because ∠HKJ and ∠1 are supplementary).

7.38 Calculate x and y in the diagram below.

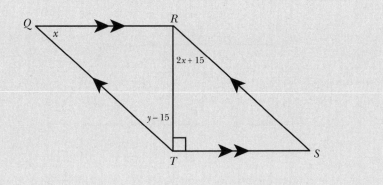

Parallel lines $\overleftrightarrow{QR}$ and $\overleftrightarrow{TS}$ are intersected by transversal $\overleftrightarrow{RT}$, creating congruent alternate interior angles: $\angle QRT \cong \angle STR$. Therefore, $m\angle QRT = m\angle STR = 90°$.

Similarly, parallel lines $\overleftrightarrow{QT}$ and $\overleftrightarrow{RS}$ are intersected by transversal $\overleftrightarrow{RT}$, creating congruent alternate interior angles: $\angle QTR \cong \angle SRT$. Therefore, $y - 15 = 2x + 15$. Add 15 to both sides of the equation to solve for y: $y = 2x + 30$.

The interior angles of $\triangle QRT$ have a sum of $180°$. Substitute known values into the equation.

$$m\angle Q + m\angle QRT + m\angle RTQ = 180$$
$$x + 90 + \left[y - 15\right] = 180$$

Substitute $y = 2x + 30$ into the equation and solve for x.

$$x + 90 + \left[(2x + 30) - 15\right] = 180$$
$$x + 90 + \left[2x + 15\right] = 180$$
$$3x + 105 = 180$$
$$3x = 180 - 105$$
$$3x = 75$$
$$x = \frac{75}{3}$$
$$x = 25$$

Calculate y by substituting $x = 25$ into the equation previously solved for y.

$$y = 2x + 30$$
$$= 2(25) + 30$$
$$= 50 + 30$$
$$= 80$$

> That means D is a point on side $\overline{CE}$. Each vertex of a polygon represents the intersection of two sides, and those sides can't lie on the same line.

Classifying Polygons

Shapes that may have more than three sides

7.39 Explain why *ABCDE* is an inaccurate name for the polygon in the figure below.

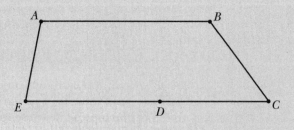

> Those aren't the only correct names. As long as you start at one vertex and go around the shape (whether you go clockwise doesn't matter), you're fine. Just make sure you don't include the point where you started! That means the name *ABCEA* is wrong.

A polygon is identified by its vertices as you travel clockwise or counterclockwise about the figure. *D* is not a vertex of the polygon because consecutive sides of a polygon cannot be collinear. The correct name of the polygon should include four vertices, not five, because the polygon has four sides. Correct names for this polygon include *ABCE*, *EABC*, and *BAEC*.

7.40 Explain why *STUVWXYZ* in the following diagram is not a polygon.

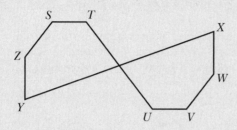

Each side of a polygon must intersect exactly two other sides of the polygon, and those points of intersection must occur at the vertices. Simply put, the sides of a polygon may not "overlap." In this diagram, sides $\overline{TU}$ and $\overline{XY}$ intersect at a point that is not a vertex of the figure, so *STUVWXYZ* is not a polygon.

Note: Problems 7.41–7.42 refer to the diagram below.

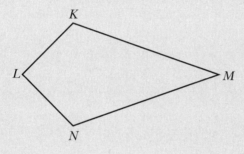

7.41 Name the polygon.

A vertex is an endpoint of a side, a "corner" of the polygon. The plural form of vertex is vertices.

The polygon has four vertices, so the name of the polygon should include exactly four letters. Begin at any vertex, and list the remaining vertices by traveling clockwise or counterclockwise about the polygon. The polygon has eight possible correct names: *KMNL, MNLK, NLKM, LKMN, KLNM, LNMK, NMKL,* and *MKLN*.

Note: Problems 7.41–7.42 refer to the diagram in Problem 7.41.

7.42   Explain why the polygon is convex.

When you extend the sides of a convex polygon, the resulting lines do not intersect the interior of the figure. Consider the diagram below, which extends the sides of the polygon infinitely in each direction.

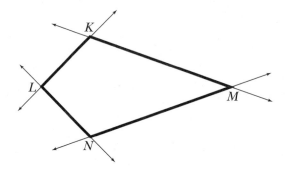

None of the lines that are generated by extending the sides of the polygon pass through the interior of the polygon, so *KMNL* is convex.

7.43   Explain why the polygon below is not convex.

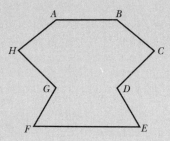

When you extend the sides of a convex polygon, the resulting lines do not intersect the interior of the figure. However, when you extend sides $\overline{HG}$ and $\overline{CD}$, the resulting lines pass through the inside of the figure. In the diagram below, the extended sides that prevent *ABCDEFGH* from meeting the requirements of a convex polygon are illustrated as dotted lines.

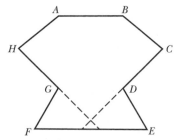

7.44 Draw two hexagons, one that is regular and one that is not. Explain your answer.

A hexagon is a six-sided polygon.

All the sides of a regular polygon are congruent, as are all the interior angles. In the diagram below, hexagon *ABCDEF* is regular, but hexagon *LMNOPQ* is not.

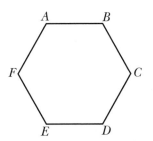 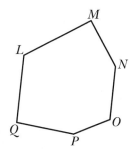

Note: Problems 7.45–7.47 investigate the relationship between the diagonals and the interior angles of a polygon.

7.45 Complete the following statement and illustrate your answer.

A convex octagon has _____ diagonals that divide the interior of the octagon into _____ nonoverlapping triangles.

An octagon is an eight-sided polygon.

Segments connecting consecutive (adjacent) vertices of a polygon are the sides of the polygon, whereas segments connecting nonconsecutive (nonadjacent) vertices are diagonals of the polygon. Consider vertex *A* in the regular octagon below; vertices *B* and *H* are adjacent to *A*, whereas *C, D, E, F,* and *G* are nonadjacent. To construct the diagonals extending from *A*, draw segments connecting *A* to each of the nonadjacent vertices.

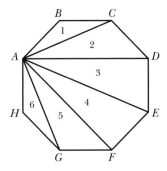

As the above diagram demonstrates, a convex octagon has <u>five</u> diagonals that divide the interior of the octagon into <u>six</u> nonoverlapping triangles.

Note: Problems 7.45–7.47 investigate the relationship between the diagonals and the interior angles of a polygon.

7.46 Calculate the sum of the measures of the interior angles of a convex octagon.

According to Problem 7.45, the diagonals of a convex octagon divide the figure into six nonoverlapping triangles. The sum of the measures of the interior angles of each triangle is 180°, so the sum of the measures of the interior angles of six triangles is 6(180°) = 1,080°. Therefore, the sum of the interior angles of a convex hexagon is 1,080°.

Note: Problems 7.45–7.47 investigate the relationship between the diagonals and the interior angles of a polygon.

7.47 Generalize the techniques used in Problems 7.45–7.46 to calculate the sum of the measures of the interior angles of any n-sided convex polygon.

Problem 7.45 states that an eight-sided polygon has five diagonals that split the figure into six nonoverlapping triangles. Note that there are two fewer triangles than the number of sides. The same numerical relationship holds true for any convex polygon. Therefore, a polygon with n sides may be split into $n-2$ nonoverlapping triangles.

In Problem 7.46, the sum of the measures of the interior angles is calculated by multiplying the total number of triangles in the polygon by 180°. When the polygon has n sides, the sum is equal to 180°$(n-2)$.

> There are $n - 2$ triangles, so multiply $(n - 2)$ by 180°.

7.48 Calculate the sum of the interior and exterior angles of a convex pentagon.

According to Problem 7.47, the sum of the measures of the interior angles of an n-sided polygon is 180°$(n-2)$. Evaluate the formula given a pentagon has $n = 5$ sides.

$$180°(5-2) = 180°(3) = 540°$$

Therefore, the sum of the measures of the interior angles is 540°. The sum of the measures of the exterior angles is constant for all convex polygons. Regardless of the number of sides, the sum is 360°.

> ALL convex polygons, (triangles, rectangles, pentagons, hexagons, and so on) have the exact same exterior angle sum: 360°.

7.49 Calculate the measure of one interior angle and one exterior angle of a regular decagon.

Regular polygons are convex, so the sum of the measures of the exterior angles is equal to 360°. A decagon has 10 sides, so it also has 10 interior and 10 exterior angles. To determine the measure of a single exterior angle, divide the total sum (360°) by the number of angles (10). ←

$$\text{measure of one exterior angle} = \frac{360°}{10} = 36°$$

Apply the formula $180°(n-2)$ to calculate the sum of the measures of the interior angles.

$$180°(n-2) = 180°(10-2)$$
$$= 180°(8)$$
$$= 1{,}440°$$

Divide the sum by the number of angles to calculate the measure of one interior angle.

$$\text{measure of one interior angle} = \frac{1{,}440°}{10} = 144°$$

In a regular polygon, all the angles are congruent, so you can divide the total angle sum by the number of sides. If the angles weren't congruent, you'd have no way of knowing how big each angle was.

7.50 Calculate the measure of one interior angle and one exterior angle of a regular nonagon.

A nonagon is a nine-sided polygon, so set $n = 9$ and evaluate the formula for the sum of the measures of the interior angles.

$$180°(n-2) = 180°(9-2)$$
$$= 180°(7)$$
$$= 1{,}260°$$

Divide the sum by the number of interior angles to calculate the measure of a single interior angle.

$$\text{measure of one interior angle} = \frac{1{,}260°}{9} = 140°$$

The sum of the measures of the exterior angles is 360°, a value shared by all convex polygons regardless of the number of sides present. Divide 360° by the number of exterior angles to calculate the measure of a single exterior angle.

$$\text{measure of one exterior angle} = \frac{360°}{9} = 40°$$

Note that, like Problem 7.49, the single interior angle and exterior angle are supplementary: 140° + 40° = 180°.

Chapter 8
CONGRUENT TRIANGLES

With a brief detour for isosceles triangles

Investigating congruent triangles is necessary to expand beyond the study of triangles to polygons with a higher number of vertices. Corresponding parts of congruent figures are, in turn, congruent; this property provides the foundation for many of the theorems in the proceeding chapters. This chapter also details the isosceles triangle theorem—because its proof requires congruent triangles—and its corollaries.

Congruent triangles are nothing fancy. They're just identical twin triangles that would fit perfectly on top of each other if you stacked them. In other words, each angle of one triangle is congruent to the corresponding angle on the other triangle, and the same goes for the sides.

You can prove that triangles are congruent in a lot of ways, all based on the angles and the sides of the triangles. You'll see postulates like SSS (side-side-side), which means two triangles are congruent because the lengths of the sides of one triangle match the lengths of the sides of another triangle.

You'll see the abbreviation CPCTC a lot. It stands for "corresponding parts of congruent triangles are congruent." In other words, if two triangles are congruent, then the pairs of sides and angles that match up are congruent also.

Identifying Congruent Figures

Shapes with congruent corresponding parts

Note: Problems 8.1–8.3 refer to the congruent figures in the following diagram.

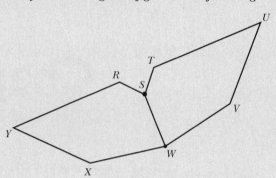

8.1 Complete the following statement: $RSWXY \cong$ _____.

Congruent figures should be named so that corresponding vertices are listed in the same order. The first two vertices of $RSWXY$ identify the shortest segment in the left pentagon, $\overline{RS}$. The second and third vertices name the side that is shared between the congruent pentagons, $\overline{SW}$. Thus, beginning at R, the vertices of $RSWXV$ are listed clockwise about the figure.

The first two vertices in the name of the right pentagon should identify its shortest side as well: $\overline{TS}$. The second and third vertices should name the shared side, $\overline{SW}$. The remaining vertices are named counterclockwise about the figure. Therefore, $RSWXY \cong \underline{TSWVU}$.

Note: Problems 8.1–8.3 refer to the congruent figures in Problem 9.1.

8.2 Complete the following statement: $\overline{UV} \cong$ _____.

If $RSWXY \cong TSWVU$, then $\overline{UV} \cong \overline{YX}$. Corresponding sides of congruent polygons are congruent. The segments $\overline{UV}$ and $\overline{YX}$ correspond because the vertices U and V appear in the same relative position in the name $TSWVU$ as the vertices Y and X appear in the name $RSWXY$.

Note: Problems 8.1–8.3 refer to the congruent figures in Problem 9.1.

8.3 Complete the following statement: $\angle XWS \cong$ _____.

Because $RSWXY \cong TSWVU$, the corresponding angles of the polygons are congruent. The angle in $TSWVU$ that corresponds to $\angle XWS$ is $\angle VWS$. Therefore, $\angle XWS \cong \underline{\angle VWS}$.

Name the points in the same order on both figures. If you had named the left pentagon YXWSR, the right one would be named UVWST.

$\overline{XY}$ is also correct, even though it doesn't name the vertices in the same order. Changing the order of the endpoints doesn't affect the segment.

Just as corresponding angles made by parallel lines and a transversal are in the same relative location, so are corresponding parts of congruent figures.

Note: Problems 8.4–8.6 refer to the figure below.

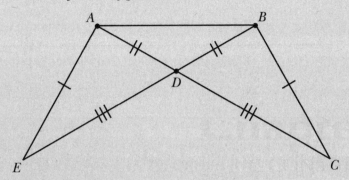

8.4 Identify the triangles that are congruent according to the tick marks on the diagram, ensuring that the vertices in your statement correspond correctly.

According to the diagram, $\overline{AE} \cong \overline{BC}$, $\overline{AD} \cong \overline{BD}$, and $\overline{ED} \cong \overline{CD}$. According to the side-side-side postulate (hereafter referred to as the SSS postulate), if three sides of one triangle are congruent to three sides of another triangle, then those triangles are congruent. Therefore, $\triangle ADE \cong \triangle BDC$. ←

> You can state the answer in other correct ways, as in $\triangle EAD \cong \triangle CBD$ or $\triangle DAE \cong \triangle DBC$. Just make sure you list the vertices in the correct corresponding order.

Note: Problems 8.4–8.6 refer to the figure in Problem 8.4.

8.5 Given $EB = CA$, identify another pair of congruent triangles.

Consider the large triangles: $\triangle EAB$ and $\triangle CBA$. This problem states that $\overline{EB} \cong \overline{CA}$; according to the diagram, $\overline{AE} \cong \overline{BC}$. Notice that the triangles share the third side, $\overline{AB}$, which is congruent to itself according to the reflexive property of congruency: $\overline{AB} \cong \overline{BA}$. According to the SSS postulate, $\triangle EAB \cong \triangle CBA$.

Note: Problems 8.4–8.6 refer to the figure in Problem 8.4.

8.6 Given $EB = CA$, explain why $\angle E \cong \angle C$.

According to Problem 8.5, if $EB = CA$, then $\triangle EAB \cong \triangle CBA$. Corresponding parts of congruent triangles are congruent, a statement used so often in geometry that it is represented by the acronym CPCTC. Because $\angle E$ and $\angle C$ are corresponding angles, $\angle E \cong \angle C$. ←

> "Parts" = angles and sides

Note: Problems 8.7–8.8 refer to congruent polygons: ABCD ≅ KLMN.

8.7 Complete the statement and justify your answer: $\overline{AB} \cong$ _____.

If *ABCD* and *KLMN* are congruent quadrilaterals, then they are named so that their vertices correspond. Therefore, the segment connecting the first two vertices in the name $ABCD$ $\left(\overline{AB}\right)$ corresponds to the segment connecting the first two vertices of $KLMN$ $\left(\overline{KL}\right)$. Corresponding parts of congruent polygons are congruent, so $\overline{AB} \cong \overline{KL}$.

A quadrilateral is a generic four-sided polygon.

If you are not satisfied answering this question by means of the order in which the vertices are named, draw a diagram that illustrates the quadrilaterals, making sure to label the vertices correctly. One such diagram appears here, and it may help you visualize the quadrilaterals.

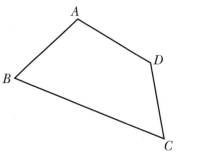 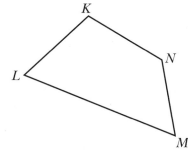

Note: Problems 8.7–8.8 refer to congruent polygons: ABCD ≅ KLMN.

8.8   Complete the statement and justify your answer: $\angle ADC \cong$ _____.

The vertices of the polygon that constitute $\angle ADC$ appear first, fourth, and third in the name *ABCD*. Therefore, the first, fourth, and third vertices in the name *KLMN* constitute the corresponding angle: $\angle ADC \cong \angle KNM$. If an answer based solely on the names of the polygons is unsatisfying, use the diagram generated in Problem 8.7 to verify the answer visually.

Note: Problems 8.9–8.10 refer to the following diagram, in which $\triangle FGH \cong \triangle IJH$.

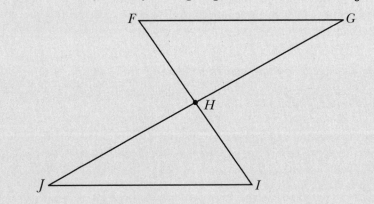

8.9   Prove that H is the midpoint of $\overline{JG}$.

CPCTC stands for "corresponding parts of congruent triangles are congruent."

Statement	Reason
1. $\triangle FGH \cong \triangle IJH$	1. Given
2. $\overline{GH} \cong \overline{JH}$	2. CPCTC
3. H is the midpoint of $\overline{JG}$	3. The midpoint of a segment is the unique point that bisects the segment into two congruent halves

Note: Problems 8.9–8.10 refer to the diagram in Problem 8.9, in which ΔFGH ≅ ΔIJH.

8.10 Prove that $\overline{FG} \parallel \overline{IJ}$.

Statement	Reason
1. ΔFGH ≅ ΔIJH	1. Given
2. ∠F ≅ ∠I	2. CPCTC
3. $\overline{FG} \parallel \overline{IJ}$	3. If two lines are intersected by a transversal and the alternate interior angles formed are congruent, then the lines are parallel

Triangle Congruence Postulates and Theorems

SSS, SAS, ASA, AAS, and HL

8.11 Identify the congruence postulate that asserts ΔLMN ≅ ΔONM in the following diagram.

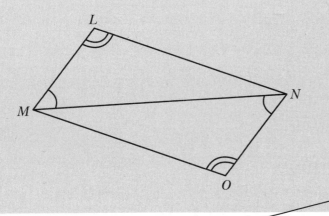

This side is shared by both triangles, so the triangles have one congruent side by default.

Because ∠L ≅ ∠O, ∠LMN ≅ ∠ONM, and $\overline{MN} \cong \overline{MN}$, ΔLMN ≅ ΔONM by the angle-angle-side (AAS) congruence theorem.

The order of the angles and sides is essential when identifying congruent triangles. In the case of the AAS theorem, if two angles *and a nonincluded side* of one triangle are congruent to two angles *and a nonincluded side* of another triangle, then the triangles are congruent.

Just as the "S" in AAS does not appear between the As, the congruent side does not appear between the angles. Because the endpoints *M* and *N* do not connect the vertices of the congruent angles in either of the triangles, $\overline{MN}$ is a nonincluded side.

The included sides in this example are $\overline{LM}$ and $\overline{ON}$ because they connect the vertices of the congruent angles. If the angles were the same and $\overline{LM} \cong \overline{ON}$, then the triangles are congruent by the ASA postulate.

Note: Problems 8.12–8.15 refer to the following diagram, in which F is the midpoint of $\overline{BE}$ and $\overline{CD}$.

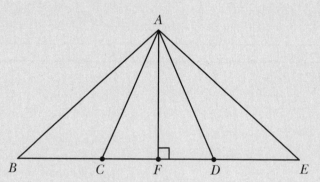

8.12 Justify the following statement: $\triangle ACF \cong \triangle ADF$, according to the side-angle-side (SAS) postulate.

> Included angles are formed by the congruent sides. For example, sides $\overline{CF}$ and $\overline{AF}$ form $\angle CFA$, so $\angle CFA$ is the included angle. Both of these are right angles.

According to the SAS postulate, if two sides and the included angle of one triangle are congruent to two sides and the included angle of another triangle, then the triangles are congruent.

This problem states that F is the midpoint of $\overline{CD}$, so F bisects $\overline{CD}$ into two congruent segments: $\overline{CF} \cong \overline{DF}$. The triangles share common side $\overline{AF}$ and therefore share another pair of congruent sides by default. Finally, the included angles are congruent: $\angle CFA \cong \angle DFA$. Therefore $\triangle ACF \cong \triangle ADF$ according to the SAS postulate.

Note: Problems 8.12–8.15 refer to the diagram in Problem 8.12, in which F is the midpoint of $\overline{BE}$ and $\overline{CD}$.

8.13 Prove $BC = ED$.

> Start with the equation BF = EF in Step 2. Step 3 states that you can substitute BC + CF for BF and can substitute ED + DF for EF.

> Step 2 states that CF = DF, so you can subtract CF from the left side of the equation in Step 4 and subtract DF from the right side.

Statement	Reason
1. F is the midpoint of $\overline{BE}$; F is the midpoint of $\overline{CD}$	1. Given
2. $BF = EF$; $CF = DF$	2. A midpoint bisects a segment into two congruent segments
3. $BC + CF = BF$; $ED + DF = EF$	3. Segment addition postulate
4. $BC + CF = ED + DF$	4. Substitution property of equality (Steps 2, 3)
5. $BC = ED$	5. Subtraction property of equality

Note: Problems 8.12–8.15 refer to the diagram in Problem 8.12, in which F is the midpoint of BE and CD.

8.14 Justify the following statement: $\angle BCA \cong \angle EDA$.

Note that $\angle BCA$ and $\angle ACF$ are supplementary, as are $\angle EDA$ and $\angle ADF$. According to Problem 8.12, $\triangle ACF \cong \triangle ADF$. Because corresponding parts of congruent triangles are congruent, $\angle ACF \cong \angle ADF$. Therefore, $\angle BCA \cong \angle EDA$ because supplements of congruent angles are congruent.

> See Problem 5.38.

Note: Problems 8.12–8.15 refer to the diagram in Problem 8.12, in which F is the midpoint of BE and CD.

8.15 Explain why $\triangle ABC \cong \triangle AED$, according to the SAS postulate.

Two sides of $\triangle ABC$ and the included angle are congruent to two sides of $\triangle AED$ and the included angle: $\overline{AC} \cong \overline{AD}$ because $\triangle ACF \cong \triangle ADF$ and corresponding parts of congruent triangles are congruent; $\overline{BC} \cong \overline{ED}$, according to Problem 8.13; and $\angle BCA \cong \angle EDA$, according to Problem 8.14.

Note: Problems 8.16–8.18 refer to the following diagram.

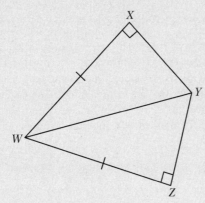

8.16 Explain why the given information alone is not enough to prove that $\triangle WXY \cong \triangle WZY$, according to the SAS postulate.

The diagram states that $\overline{WX} \cong \overline{WZ}$, and the triangles share a side, so $\overline{WY} \cong \overline{WY}$. However, the congruent angles ($\angle X \cong \angle Z$) are not included between the congruent sides, and there is no SSA postulate of triangle congruence.

> So no congruence postulate applies when an angle is not included between two sides. No angle-side-side postulate exists.

Although the given information is not enough on its own to prove the triangles congruent according to the SAS postulate, you cannot assume that the triangles are not congruent.

A theorem can be proven, but a postulate (also called an axiom or property) has to be taken at face value—it can't be proven. The only other theorem is AAS, and you're given information about only one angle in this example, so that can't be the answer.

Note: Problems 8.16–8.18 refer to the diagram in Problem 8.16.

8.17 Complete the following statement and justify your answer.

$\triangle WXY \cong \triangle WZY$ according to the _____ theorem.

Notice that the statement contains the word "theorem," and SSS, SAS, and ASA are all postulates. Also notice that the diagram contains right triangles. The hypotenuse-leg (or HL) theorem applies only to right triangles (because only right triangles contain a hypotenuse). According to this theorem, if the hypotenuse and one leg of a right triangle are congruent to the hypotenuse and one leg of another right triangle, then the triangles are congruent.

In this diagram, the right triangles share the same hypotenuse, $\overline{WY}$. Therefore, the hypotenuses are congruent. Furthermore, side $\overline{WX}$ in $\triangle WXY$ is congruent to $\overline{WZ}$ in $\triangle WZY$, so one leg of each triangle is congruent to one leg of the other triangle. Hence, $\triangle WXY \cong \triangle WZY$ according to the <u>hypotenuse-leg</u> theorem.

The hypotenuse is the longest side of a right triangle, and it's always opposite the right angle.

Note: Problems 8.16–8.18 refer to the diagram in Problem 8.16.

8.18 Identify two reasons why $\angle XWY \cong \angle ZWY$.

As Problem 8.17 states, $\triangle WXY \cong \triangle WZY$ according to the hypotenuse-leg theorem. Therefore, $\angle XWY \cong \angle ZWY$ because corresponding parts of congruent triangles are congruent.

Problem 6.37 states that the distance between a line and a point not on that line is the length of the perpendicular segment that connects them.

Notice that XY represents the distance between point Y and line $\overrightarrow{XW}$. Similarly, ZY represents the distance between point Y and line $\overrightarrow{ZW}$. Because $\triangle WXY \cong \triangle WZY$, those lengths are equal. According to Problem 7.41, a point equidistant from the sides of an angle lies on the angle bisector. Therefore, $\overrightarrow{WY}$ bisects $\angle XWZ$, dividing it into two congruent angles, $\angle XWY$ and $\angle ZWY$.

Note: Problems 8.19–8.22 refer to the following diagram, in which BC = DE.

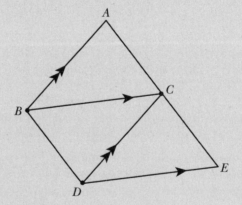

8.19 Justify the following statement: $\angle ACB \cong \angle CED$.

Parallel lines $\overleftrightarrow{BC}$ and $\overleftrightarrow{DE}$ are intersected by transversal $\overleftrightarrow{AE}$, forming congruent corresponding angles $\angle ACB$ and $\angle CED$.

Note: Problems 8.19–8.22 refer to the diagram in Problem 8.19, in which BC = DE.

8.20 Justify the following statement: $\angle ABC \cong \angle CDE$.

Parallel lines $\overrightarrow{BA}$ and $\overrightarrow{DC}$ are intersected by transversal $\overrightarrow{BC}$, forming congruent alternate interior angles $\angle ABC$ and $\angle BCD$. Now consider parallel lines $\overrightarrow{BC}$ and $\overrightarrow{DE}$, intersected by transversal $\overrightarrow{DC}$. These lines form congruent alternate interior angles as well: $\angle BCD \cong \angle CDE$.

Therefore, $\angle ABC \cong \angle BCD$ and $\angle BCD \cong \angle CDE$. According to the transitive property of congruency, $\angle ABC \cong \angle CDE$.

Note: Problems 8.19–8.22 refer to the diagram in Problem 8.19, in which BC = DE.

8.21 Identify the triangle congruence postulate that confirms $\triangle ABC \cong \triangle CDE$.

If two angles and the included side of one triangle are congruent to two angles and the included side of another triangle, then the triangles are congruent by the angle-side-angle (ASA) postulate. According to Problem 8.19, $\angle ACB \cong \angle CED$; Problem 8.20 states that $\angle ABC \cong \angle CDE$; and you are given $\overline{BC} \cong \overline{DE}$, the included sides of each triangle, are congruent. Therefore, $\triangle ABC \cong \triangle CDE$ by the ASA postulate.

> The included side connects the vertices of the congruent angles.

Note: Problems 8.19–8.22 refer to the diagram in Problem 8.19, in which BC = DE.

8.22 Identify the triangle congruence postulate that confirms $\triangle ABC \cong \triangle DCB$.

Because $\triangle ABC \cong \triangle CDE$ (as explained in Problem 8.21), corresponding parts of those triangles are congruent. Therefore, $\overline{AB} \cong \overline{DC}$. The triangles also contain shared side $\overline{BC}$. Recall that $\angle ABC \cong \angle DCB$. Therefore, the triangles are congruent according to the SAS postulate, as illustrated below.

> See Problem 8.20.

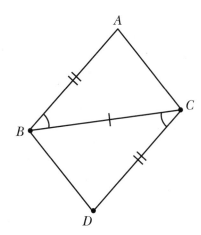

These triangles aren't congruent, but they are "similar," a topic covered in Chapter 11.

8.23 Given △QRS and △TUV such that ∠Q ≅ ∠T, ∠R ≅ ∠U, and ∠S ≅ ∠V, is △QRS ≅ △TUV? Justify your response using an illustration.

This problem asks the following question: If three angles of one triangle are congruent to three angles of another triangle, are the triangles congruent? In other words, is there an angle-angle-angle (AAA) congruence postulate?

Two triangles constructed of the same angles have the same shape but not necessarily the same size, so △QRS and △TUV are not necessarily congruent. Consider the following diagram, in which corresponding angles are congruent but the sides of the triangles clearly differ.

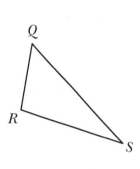
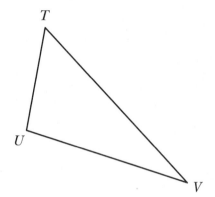

Note: Problems 8.24–8.25 refer to the following diagram, in which WY = XZ.

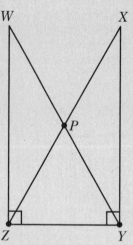

8.24 Explain why △WZY ≅ △XYZ, according to the hypotenuse-leg theorem.

The problem states that the hypotenuses of the right triangles are congruent—if WY = XZ, then $\overline{WY} ≅ \overline{XZ}$. The triangles share leg $\overline{ZY}$, so those legs are congruent. Because the hypotenuse and one leg of △WZY are congruent to the hypotenuse and one leg of △XYZ, the triangles are congruent according to the hypotenuse-leg theorem.

Note: Problems 8.24–8.25 refer to the diagram in Problem 8.24, in which WY = XZ.

8.25 Explain why $\triangle WPZ \cong \triangle XPY$, according to the AAS theorem.

For more information about vertical angles, see Problem 4.41. Remember, vertical angles are always congruent.

Notice that $\angle WPZ$ and $\angle XPY$ are vertical angles, so $\angle WPZ \cong \angle XPY$. As Problem 8.24 states, $\triangle WZY \cong \triangle XYZ$, so the corresponding parts of the triangles are congruent. Of specific interest are one angle and one side: $\angle W \cong \angle X$ and $\overline{WZ} \cong \overline{XY}$. Because two angles and a nonincluded side of $\triangle WPZ$ are congruent to two angles and a nonincluded side of $\triangle XPY$, the triangles are congruent according to the AAS theorem.

It may be helpful to eliminate extraneous information from the diagram to better visualize the congruent sides and angles of the congruent triangles. Note that the congruent sides in this example are not included between the included angles, as evidenced in the diagram below.

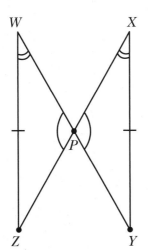

8.26 According to the hypotenuse-angle (HA) theorem, if the hypotenuse and one acute angle of a right triangle are congruent to the hypotenuse and one acute angle of another right triangle, then the right triangles are congruent. Prove the HA theorem using the following diagram.

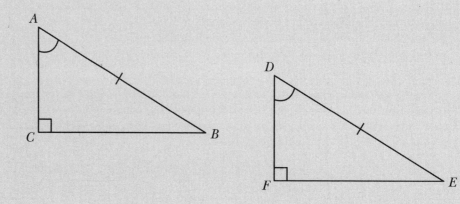

You could prove that ∠B ≅ ∠E (if two angles of one triangle are congruent to two angles of another triangle, then the third angles of the triangles are congruent) and prove that the triangles are congruent by ASA. By the way, this is also how you prove the AAS theorem, in case you're curious.

Statement	Reason
1. $\angle A \cong \angle D$; $\overline{AB} \cong \overline{DE}$	1. Given
2. $m\angle C = 90°$; $m\angle F = 90°$	2. Right angles measure 90°
3. $\angle C \cong \angle F$	3. Angles with equal measures are congruent
4. $\triangle ABC \cong \triangle DEF$	4. AAS theorem

Besides the HA theorem in Problem 8.26 and the LL theorem in this problem, there's also a leg-angle (LA) theorem. It states that two right triangles are congruent if they've got one acute angle and one leg that are congruent.

8.27 According to the leg-leg (LL) theorem, if the legs of one right triangle are congruent to the legs of another right triangle, then the right triangles are congruent. Prove the LL theorem using the following diagram.

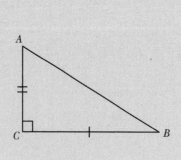
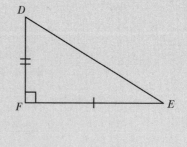

Statement	Reason
1. $\overline{AC} \cong \overline{DF}$; $\overline{CB} \cong \overline{FE}$	1. Given
2. $m\angle C = 90°$; $m\angle F = 90°$	2. Right angles measure 90°
3. $\angle C \cong \angle F$	3. Angles with equal measures are congruent
4. $\triangle ABC \cong \triangle DEF$	4. SAS postulate

Isosceles Triangle Theorem

Base angles of an isosceles triangle are congruent

8.28 The isosceles triangle theorem states that the base angles (the angles opposite the congruent sides of the triangle) are congruent. Prove the isosceles triangle theorem using the following diagram, in which $\overrightarrow{XP}$ bisects the vertex angle, $\angle X$.

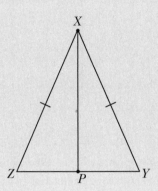

To prove the isosceles triangle theorem using this diagram, you must prove $\angle Z \cong \angle Y$ given $\overline{XZ} \cong \overline{XY}$.

Statement	Reason
1. $\overline{XZ} \cong \overline{XY}$	1. Given
2. $\angle ZXP \cong \angle YXP$	2. An angle bisector divides an angle into two congruent angles
3. $\overline{XP} \cong \overline{XP}$	3. Reflexive property of congruency
4. $\triangle XZP \cong \triangle XYP$	4. SAS postulate
5. $\angle Z \cong \angle Y$	5. CPCTC

8.29 Prove the converse of the isosceles triangle theorem using the following diagram, in which $\overline{XP}$ bisects the vertex angle, $\angle X$.

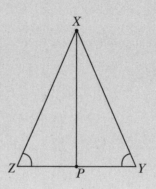

The isosceles triangle theorem states that a triangle with two congruent sides has two congruent angles opposite those sides. The converse, therefore, states that a triangle with two congruent angles has two congruent sides opposite those angles. In this diagram, given $\angle Z \cong \angle Y$, you must prove $\overline{XZ} \cong \overline{XY}$.

Statement	Reason
1. $\angle Z \cong \angle Y$	1. Given
2. $\angle ZXP \cong \angle YXP$	2. An angle bisector divides an angle into two congruent angles
3. $\overline{XP} \cong \overline{XP}$	3. Reflexive property of congruency
4. $\triangle XZP \cong \triangle XYP$	4. AAS postulate
5. $\overline{XZ} \cong \overline{XY}$	5. CPCTC

The vertex angle is opposite the base.

8.30 Use the following diagram to prove that the angle bisector of the vertex angle of an isosceles triangle is also the perpendicular bisector of the base.

The base is the odd man out in the isosceles triangle. The triangle has two congruent sides and then the lonely old base that's left over.

Statement	Reason
1. $\overline{XZ} \cong \overline{XY}$; $\angle ZXP \cong \angle YXP$	1. Given
2. $\overline{XP} \cong \overline{XP}$	2. Reflexive property of equality
3. $\triangle XZP \cong \triangle XYP$	3. SAS postulate
4. $\overline{ZP} \cong \overline{YP}$	4. CPCTC
5. P is the midpoint of $\overline{ZY}$	5. A midpoint is equidistant from the endpoints of the segment
6. $\overline{XP}$ bisects $\overline{ZY}$	6. Segment bisectors intersect a segment at its midpoint
7. $\angle ZPX \cong \angle YPX$	7. CPCTC
8. $m\angle ZPX + m\angle YPX = 180°$	8. Angle addition postulate
9. $m\angle ZPX + m\angle ZPX = 180°$	9. Substitution property of equality (Steps 7, 8)
10. $2(m\angle ZPX) = 180°$	10. Combination of like terms
11. $m\angle ZPX = 90°$	11. Division property of equality
12. $\overline{XP} \perp \overline{ZY}$	12. Perpendicular lines intersect at right angles
13. $\overline{XP}$ is the perpendicular bisector of $\overline{ZY}$	13. Steps 6, 12 ←

> Step 6 proves the bisector part and Step 12 proves the perpendicular part.

8.31 Use the following diagram to prove that an equiangular triangle is also equilateral.

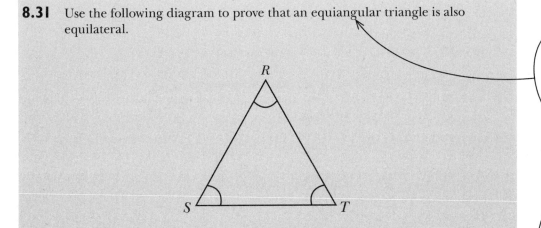

> All three interior angles of an equiangular triangle are congruent.

Statement	Reason
1. $\angle S \cong \angle T$	1. Given
2. $\overline{RS} \cong \overline{RT}$	2. Converse of the isosceles triangle theorem ←
3. $\angle R \cong \angle S$	3. Given
4. $\overline{ST} \cong \overline{RT}$	4. Converse of the isosceles triangle theorem

> The isosceles triangle theorem proves that base angles opposite congruent sides are congruent. Its converse proves that the sides opposite congruent base angles are congruent.

continues

continued

Statement	Reason
5. $\overline{ST} \cong \overline{RS} \cong \overline{RT}$	5. Transitive property of congruency
6. $\triangle RST$ is equilateral	6. All three sides of an equilateral triangle are congruent (Steps 2, 4, 5)

8.32 Given that the vertex angle of an isosceles triangle measures 112°, calculate the measure of one base angle of the triangle.

The sum of the measures of the interior angles of any triangle is 180°. Let x represent the measure of the vertex angle, and let y represent the measure of one base angle. Recall that an isosceles triangle contains one vertex angle and two congruent base angles.

$$x + y + y = 180$$
$$x + 2y = 180$$

You are given the measure of the vertex angle: $x = 112$. Substitute this value into the equation and solve for y.

$$112 + 2y = 180$$
$$2y = 180 - 112$$
$$2y = 68$$
$$y = \frac{68}{2}$$
$$y = 34$$

Each base angle of the isosceles triangle measures 34°.

Note: Problems 8.33–8.34 refer to the following diagram, in which AC = AB and $\overline{AM}$ is a median.

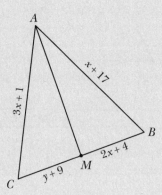

8.33 Calculate x.

Note that $\triangle ABC$ is isosceles, because the problem states that $AC = AB$. Set the lengths of the legs equal and solve for x.

$$AC = AB$$
$$3x + 1 = x + 17$$
$$3x - x = 17 - 1$$
$$2x = 16$$
$$x = \frac{16}{2}$$
$$x = 8$$

Note: Problems 8.33–8.34 refer to the diagram in Problem 8.33, in which AC = AB and $\overline{AM}$ is a median.

8.34 Calculate *y*.

A median of a triangle extends from a vertex to the midpoint of the opposite side. The problem states that $\overline{AM}$ is a median, so *M* must be the midpoint of $\overline{CB}$. Therefore, *M* bisects $\overline{CB}$ into two congruent segments.

$$CM = BM$$
$$y + 9 = 2x + 4$$

According to Problem 8.33, *x* = 8. Substitute this value into the equation and solve for *y*.

$$y + 9 = 2(8) + 4$$
$$y + 9 = 16 + 4$$
$$y + 9 = 20$$
$$y = 20 - 9$$
$$y = 11$$

8.35 Demonstrate that a triangle with interior angles that measure $4x + 10$, $7x - 5$, and $24x$ degrees is isosceles; identify the measures of the interior angles.

The sum of the measures of the interior angles of a triangle is 180°. Add the given angle measurements and solve for *x*.

$$(4x + 10) + (7x - 5) + (24x) = 180$$
$$(4x + 7x + 24x) + (10 - 5) = 180$$
$$35x + 5 = 180$$
$$35x = 180 - 5$$
$$35x = 175$$
$$x = \frac{175}{35}$$
$$x = 5$$

Evaluate the angle measurements by substituting *x* = 5 into each.

$$4x + 10 = 4(5) + 10 \qquad 7x - 5 = 7(5) - 5 \qquad 24x = 24(5)$$
$$= 20 + 10 \qquad\qquad\quad = 35 - 5 \qquad\qquad = 120°$$
$$= 30° \qquad\qquad\qquad = 30°$$

The triangle contains two congruent angles measuring 30°. According to the converse of the isosceles triangle theorem, a triangle with two congruent angles has two congruent sides opposite those angles. Therefore, this triangle is isosceles.

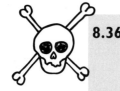

8.36 Given $m\angle 4 = 80°$ in the following diagram, calculate the measures of the eight other numbered angles.

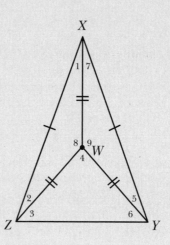

The sum of the measures of the interior angles of any triangle is 180°. Apply this principle to $\triangle WYZ$.

$$m\angle 4 + m\angle 6 + m\angle 3 = 180°$$

Substitute the known value of $m\angle 4$ into the equation and subtract it from both sides.

$$80° + m\angle 6 + m\angle 3 = 180°$$
$$m\angle 6 + m\angle 3 = 180° - 80°$$
$$m\angle 6 + m\angle 3 = 100°$$

The diagram indicates that $WZ = WY$, so $\triangle WZY$ is isosceles. Therefore, the sum of the measures of the two congruent base angles is 100°. To calculate the measure of each angle, divide 100° by 2.

$$m\angle 6 = m\angle 3 = \frac{100°}{2} = 50°$$

The large triangle, $\triangle XYZ$, is isosceles as well because $XZ = XY$. Therefore, $m\angle XZY = m\angle XYZ$.

$$m\angle XZY = m\angle XYZ$$
$$m\angle 2 + m\angle 3 = m\angle 5 + m\angle 6$$

You know that $m\angle 3 = m\angle 6$, so subtract the equivalent values from both sides of the equation.

$$m\angle 2 = m\angle 5$$

This is sufficient information to conclude that $\triangle XZW \cong \triangle XYW$ according to the SAS postulate: $\overline{XZ} \cong \overline{XY}$, $\angle 2 \cong \angle 5$, and $\overline{ZW} \cong \overline{YW}$. This allows you to conclude that $m\angle 8 = m\angle 9$, because corresponding angles of congruent triangles are congruent.

Notice that $m\angle 8 + m\angle 9 + m\angle 4 = 360°$. Substitute $m\angle 4 = 80°$ into the equation and subtract it from both sides.

$$m\angle 8 + m\angle 9 + 80° = 360°$$
$$m\angle 8 + m\angle 9 = 360° - 80°$$
$$m\angle 8 + m\angle 9 = 280°$$

> If you add these three angles, you get a full circle. If you start at $\overline{ZW}$ and go clockwise through angles 8, 9, and finally 4, you end up back at $\overline{ZW}$.

If the sum of two equal angle measurements is 280°, then each angle measures 140°. Therefore, $m\angle 8 = m\angle 9 = 140°$.

The sum of the measures of the interior angles of $\triangle XWZ$ is 180°.

$$m\angle 1 + m\angle 8 + m\angle 2 = 180°$$

> If you add $m\angle 8$ and $m\angle 9$, you get 280°, but you know that $m\angle 8 = m\angle 9$, so each of the measurements is $280° \div 2 = 140°$.

Substitute the known value $m\angle 8 = 140°$ into the equation and subtract it from both sides.

$$m\angle 1 + 140° + m\angle 2 = 180°$$
$$m\angle 1 + m\angle 2 = 180° - 140°$$
$$m\angle 1 + m\angle 2 = 40°$$

Because $XW = ZW$, $\triangle XWZ$ is isosceles, and according to the isosceles triangle theorem, $m\angle 1 = m\angle 2$. If the sum of the two congruent angles is 40°, then each angle measures 20°: $m\angle 1 = m\angle 2 = 20°$.

Recall that $\triangle XWZ \cong \triangle XWY$, so $m\angle 1 = m\angle 7 = 20°$ and $m\angle 2 = m\angle 5 = 20°$, because corresponding angles of congruent triangles are congruent.

In conclusion, here are the measures of the numbered angles, presented sequentially: $m\angle 1 = 20°$, $m\angle 2 = 20°$, $m\angle 3 = 50°$, $m\angle 4 = 80°$, $m\angle 5 = 20°$, $m\angle 6 = 50°$, $m\angle 7 = 20°$, $m\angle 8 = 140°$, and $m\angle 9 = 140°$.

8.37 Given $m\angle B = 55°$ and $m\angle D = 35°$ in the following diagram, prove $\overline{AC} \perp \overline{CE}$.

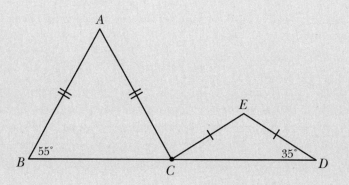

So technically, an equilateral triangle (which has three congruent sides) is also isosceles, because AT LEAST two of its sides are congruent.

Statement	Reason
1. $m\angle B = 55°$; $m\angle D = 35°$	1. Given
2. $AB = AC$; $EC = ED$	2. Given
3. $\triangle ACB$ is isosceles; $\triangle EDC$ is isosceles	3. An isosceles triangle contains at least two congruent sides
4. $m\angle ACB = m\angle B$; $m\angle ECD = m\angle D$	4. Isosceles triangle theorem
5. $m\angle ACB = 55°$; $m\angle ECD = 35°$	5. Substitution property of equality (Steps 1, 4)
6. $m\angle ACB + m\angle ACE + m\angle ECD = 180°$	6. Angle addition postulate
7. $55° + m\angle ACE + 35° = 180°$	7. Substitution property of equality (Steps 5, 6)
8. $m\angle ACE + 90° = 180°$	8. Combination of like terms
9. $m\angle ACE = 90°$	9. Subtraction property of equality
10. $\overline{AC} \perp \overline{CE}$	10. Perpendicular lines intersect at 90° angles

Proving Triangles Congruent
Featuring CPCTC

8.38 Given B is the midpoint of $\overline{AE}$ and $\overline{CD}$ in the following diagram, prove $\triangle ABC \cong \triangle EBD$.

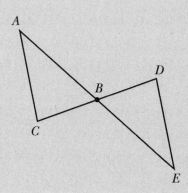

Statement	Reason
1. B is the midpoint of $\overline{AE}$; B is the midpoint of $\overline{CD}$	1. Given
2. $\overline{AB} \cong \overline{EB}$; $\overline{CB} \cong \overline{DB}$	2. A midpoint bisects a segment into two congruent segments
3. $\angle ABC \cong \angle EBD$	3. Vertical angles are congruent
4. $\triangle ABC \cong \triangle EBD$	4. SAS postulate

8.39 Given X is the midpoint of $\overline{VW}$ in the following diagram and $\angle XZY \cong \angle XYZ$, prove $\triangle XVZ \cong \triangle XWY$.

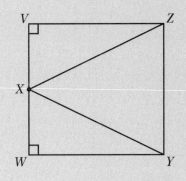

Statement	Reason
1. $\angle XZY \cong \angle XYZ$	1. Given
2. $\overline{XZ} \cong \overline{XY}$	2. Converse of the isosceles triangle theorem

The sides of $\triangle XYZ$ opposite the congruent angles are congruent.

continues

continued

Statement	Reason
3. X is the midpoint of $\overline{VW}$	3. Given
4. $\overline{VX} \cong \overline{WX}$	4. A midpoint bisects a segment into two congruent segments
5. $\triangle XVZ \cong \triangle XWY$	5. HL theorem

8.40   Given $\overline{LM} \parallel \overline{ON}$ and $\overline{LO} \parallel \overline{MN}$ in the following diagram, prove $\triangle OML \cong \triangle MON$.

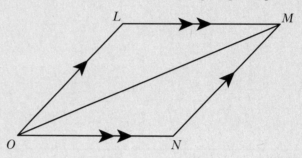

Statement	Reason
1. $\overline{LM} \parallel \overline{ON}$; $\overline{LO} \parallel \overline{MN}$	1. Given
2. $\angle LMO \cong \angle NOM$; $\angle LOM \cong \angle NMO$	2. Parallel lines intersected by a transversal form congruent alternate interior angles
3. $\overline{MO} \cong \overline{OM}$	3. Reflexive property of congruency
4. $\triangle OML \cong \triangle MON$	4. ASA postulate

Note: Problems 8.41–8.42 refer to equilateral triangle ABC, which has centroid P and medians $\overline{CD}$, $\overline{BF}$, and $\overline{AE}$.

8.41   Prove $\triangle ACE \cong \triangle ABE$.

An equilateral triangle consists of three congruent sides, so $\overline{AB} \cong \overline{BC} \cong \overline{CA}$. The medians of a triangle extend from the vertices of a triangle to the midpoint of the opposite side. Construct a diagram that illustrates the equilateral triangle and its medians.

$\overline{CD}$ is a median, so it extends from C to the midpoint of the opposite side of the triangle. This means that D is the midpoint of $\overline{AB}$.

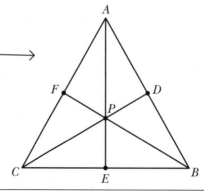

Note that the medians intersect at the centroid of the triangle, point P. Complete the proof using the diagram.

Statement	Reason
1. $\overline{AB} \cong \overline{AC}$	1. Given ($\triangle ABC$ is equilateral)
2. $\overline{AE}$ is a median of $\triangle ABC$	2. Given
3. E is the midpoint of $\overline{BC}$	3. Medians extend from a vertex to the midpoint of the opposite side of the triangle
4. $\overline{CE} \cong \overline{BE}$	4. A midpoint divides a segment into two congruent segments
5. $\overline{AE} \cong \overline{AE}$	5. Reflexive property of congruency
6. $\triangle ACE \cong \triangle ABE$	6. SSS postulate

Note: Problems 8.41–8.42 refer to equilateral triangle ABC, which has centroid P and medians $\overline{CD}$, $\overline{BF}$, and $\overline{AE}$.

8.42 Prove $\triangle AFP \cong \triangle ADP$.

Refer to the diagram generated in Problem 8.41. You may also reference the conclusion of Problem 8.41, $\triangle ACE \cong \triangle ABE$, instead of repeating the proof here.

Statement	Reason
1. $AB = AC$	1. Given ($\triangle ABC$ is equilateral)
2. $\overline{CD}$ and $\overline{BF}$ are medians of $\triangle ABC$	2. Given
3. D is the midpoint of $\overline{AB}$; F is the midpoint of $\overline{AC}$	3. Medians extend from a vertex to the midpoint of the opposite side of the triangle
4. $2(AF) = AC$; $2(AD) = AB$	4. A midpoint bisects a segment into two smaller segments, each half as long as the original segment
5. $2(AD) = 2(AF)$	5. Substitution property of equality (Steps 1, 4)
6. $AD = AF$	6. Division property of equality
7. $\triangle ACE \cong \triangle ABE$	7. Problem 8.41
8. $\angle CAE \cong \angle BAE$	8. CPCTC
9. $\overline{AP} \cong \overline{AP}$	9. Reflexive property of congruency
10. $\triangle AFP \cong \triangle ADP$	10. SAS postulate

8.43 Given regular pentagon *VWXYZ*, prove △*VWZ* ≅ △*YXZ*.

A regular pentagon is a five-sided figure with congruent sides and congruent interior angles, as illustrated by the following diagram.

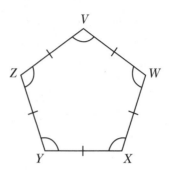

Statement	Reason
1. *VWXYZ* is a regular pentagon	1. Given
2. $\overline{VW} \cong \overline{XY} \cong \overline{YZ} \cong \overline{ZV}$	2. The sides of a regular polygon are congruent
3. ∠*V* ≅ ∠*Y*	3. The interior angles of a regular polygon are congruent
4. △*VWZ* ≅ △*YXZ*	4. SAS postulate

8.44 Prove that the angle bisector of the vertex angle of an isosceles triangle is also a median. Use the following diagram, in which $\overline{RQ}$ bisects ∠*SRT*.

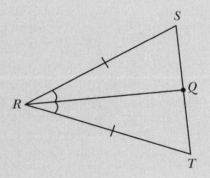

The side opposite a vertex contains the other two letters in the name of a triangle. In this case, the triangle is RST, so the side opposite R is $\overline{ST}$.

In isosceles triangle *RST*, $\overline{RS} \cong \overline{RT}$. You are given angle bisector $\overline{RQ}$ and asked to prove that the segment is also a median. A median extends from the vertex of a triangle to the midpoint of the opposite side. One endpoint of $\overline{RQ}$ is a vertex, so to prove that $\overline{RQ}$ is a median, you must prove that *Q* is the midpoint of $\overline{ST}$.

Statement	Reason
1. $\overline{RS} \cong \overline{RT}$; $\overline{RQ}$ bisects $\angle SRT$	1. Given
2. $\angle SRQ \cong \angle TRQ$	2. An angle bisector divides an angle into two congruent angles
3. $\overline{RQ} \cong \overline{RQ}$	3. Reflexive property of congruency
4. $\triangle SRQ \cong \triangle TRQ$	4. SAS postulate
5. $\overline{SQ} \cong \overline{TQ}$	5. CPCTC
6. Q is the midpoint of $\overline{ST}$	6. A midpoint is the unique point that divides a segment into two congruent segments
7. $\overline{RQ}$ is a median of $\triangle RST$	7. A median is a segment that connects the vertex of a triangle to the midpoint of the opposite side

8.45 Given $\overline{AB} \parallel \overline{DC}$ and $\triangle AED \cong \triangle CFB$ in the following diagram, prove $\triangle ABE \cong \triangle CDF$.

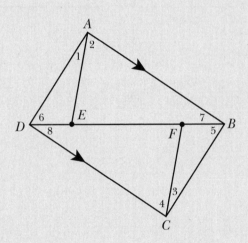

Statement	Reason
1. $\overline{AB} \parallel \overline{DC}$; $\triangle AED \cong \triangle CFB$	1. Given
2. $\angle 1 \cong \angle 3$; $\angle 5 \cong \angle 6$; $\overline{DE} \cong \overline{BF}$	2. CPCTC
3. $\angle 7 \cong \angle 8$	3. Parallel lines intersected by a transversal form congruent alternate interior angles
4. $\overline{EF} \cong \overline{FE}$	4. Reflexive property of congruency
5. $DE + EF = BF + FE$	5. Addition property of equality
6. $DE + EF = DF$; $BF + FE = BE$	6. Segment addition postulate

Step 2 states that DE = BF. Step 4 states that EF = FE, so you can add EF to the left side of the equation in Step 2 and add FE to the right side.

continues

continued

Statement	Reason
7. $DF = BE$	7. Substitution property of equality (Steps 5, 6)
8. $m\angle 6 + m\angle 1 + m\angle 2 + m\angle 7 = 180°$; $m\angle 5 + m\angle 3 + m\angle 4 + m\angle 8 = 180°$	8. The sum of the measures of the interior angles of ΔDAB is 180°; the interior angles of ΔBCD have the same sum
9. $\angle 6 + m\angle 1 + m\angle 2 + m\angle 7 = m\angle 5 + m\angle 3 + m\angle 4 + m\angle 8$	9. Transitive property of equality
10. $\angle 6 + m\angle 2 + m\angle 7 = m\angle 5 + m\angle 4 + m\angle 8$	10. Subtraction property of equality
11. $m\angle 2 + m\angle 7 = m\angle 4 + m\angle 8$	11. Subtraction property of equality
12. $m\angle 2 = m\angle 4$	12. Subtraction property of equality
13. $\Delta ABE \cong \Delta CDF$	13. AAS theorem

Step 2 states that $m\angle 1 = m\angle 3$, so subtract $m\angle 1$ from the left side and $m\angle 3$ from the right side. You'll do this two more times. In Step 11, you subtract $m\angle 5$ and $m\angle 6$, and in Step 12, you subtract $m\angle 7$ and $m\angle 8$ (equal according to Step 3).

$\angle 2 \cong \angle 4$ (Step 12); $\angle 7 \cong \angle 8$ (Step 3); and $\overline{DF} \cong \overline{BE}$ (Step 7).

Chapter 9
QUADRILATERALS

Four-sided polygons

The two preceding chapters thoroughly investigated the properties of triangles, including the means by which you prove triangles congruent. This chapter applies those concepts to the definition and construction of four-sided polygons, collectively named quadrilaterals. Of particular interest are the major classifications of quadrilaterals: parallelograms, rectangles, rhombi, squares, trapezoids, and kites.

After you can prove that triangles are congruent (work through Chapter 8 if you can't do that yet), you are ready to add a vertex and work with FOUR sides instead of three. Voilà! Quadrilaterals!

You start off with the most basic quadrilateral, the parallelogram. You then look at special kinds of parallelograms (such as rectangles), shapes that are almost but not quite parallelograms (trapezoids), and then shapes that are familiar yet weird (kites).

While you're working through this chapter, pay attention to the characteristics of each type of quadrilateral. Which ones have opposite angles that are congruent? Which ones have perpendicular diagonals? You get the idea.

Properties of Parallelograms

Pairs of parallel sides

9.1 Define parallelogram.

> Problems 9.2–9.5 prove these "other characteristics."

A quadrilateral is a parallelogram if and only if both pairs of opposite sides are parallel. A parallelogram has a number of additional characteristics regarding the relationship of opposite sides, opposite angles, and diagonals, but by definition, a parallelogram is simply a four-sided figure whose opposite sides are segments of parallel lines.

Note: Problems 9.2–9.5 refer to □ABCD, illustrated in the following diagram.

> The □ symbol stands for "parallelogram" just as the △ symbol stands for "triangle."

9.2 Identify the sides and diagonals of □*ABCD*.

The sides of a parallelogram (like the sides of any polygon) are the segments that connect consecutive vertices. Because *B* immediately follows *A* in the name of parallelogram *ABCD*, $\overline{AB}$ is a side of the parallelogram. The remaining sides are $\overline{BC}, \overline{CD},$ and $\overline{DA}$.

> *D* and *A* are technically consecutive because you "wrap around" from the last letter in the name to the first.

The diagonals of a parallelogram are the two segments connecting nonconsecutive vertices. As the following diagram illustrates, the diagonals of □*ABCD* are $\overline{AC}$ and $\overline{DB}$.

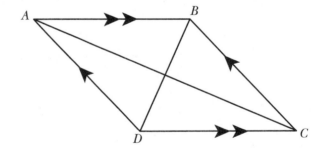

Note: Problems 9.2–9.5 refer to ▱ABCD, as illustrated in Problem 9.2.

9.3 Use ▱ABCD to prove that both pairs of opposite sides of a parallelogram are congruent.

Begin by constructing a diagram to reference in the proof. To prove that the opposite sides are congruent, you need to draw a diagonal of the parallelogram. You can use either diagonal to complete the proof; diagonal $\overline{BD}$ is used in the solution below.

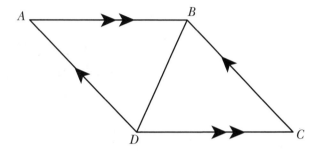

Given $\overline{AB} \parallel \overline{DC}$ and $\overline{AD} \parallel \overline{BC}$ in the above diagram, your goal is to prove that $\overline{AB} \cong \overline{CD}$ and $\overline{AD} \cong \overline{CB}$.

Statement	Reason
1. $\overline{AB} \parallel \overline{DC}$; $\overline{AD} \parallel \overline{BC}$	1. Given
2. $\angle ADB \cong \angle CBD$; $\angle ABD \cong \angle CDB$	2. Parallel lines intersected by a transversal form congruent alternate interior angles
3. $\overline{DB} \cong \overline{BD}$ ←	3. Reflexive property of congruency
4. $\triangle ABD \cong \triangle CDB$	4. ASA postulate
5. $\overline{AB} \cong \overline{CD}$; $\overline{AD} \cong \overline{CB}$	5. CPCTC

$\overline{DB}$ and $\overline{BD}$ refer to the same segment. They're named differently because you're about to prove that $\triangle ABD \cong \triangle CDB$, and the order of the vertices are different in the triangles. B comes before D in the first triangle, but D comes first in the second.

Note: Problems 9.2–9.5 refer to □ ABCD, as illustrated in Problem 9.2.

9.4   Use □ ABCD to prove that both pairs of opposite angles of a parallelogram are congruent.

You completed some of the work for this problem in Problem 9.3. Because $\triangle ABD \cong \triangle CDB$, you know that $\angle A \cong \angle C$ because corresponding angles of congruent triangles are congruent. To prove that $\angle B \cong \angle D$, you use a proof very similar to the proof in Problem 9.3, this time drawing diagonal $\overline{AC}$ and using it to prove that $\triangle ABC \cong \triangle CDA$.

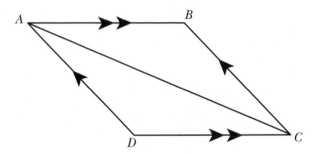

Statement	Reason
1. $\overline{AB} \parallel \overline{DC}$; $\overline{AD} \parallel \overline{BC}$	1. Given

2. $\angle BAC \cong \angle DCA$;
   $\angle DAC \cong \angle BCA$
	2. Parallel lines intersected
	   by a transversal form congruent
	   alternate interior angles

3. $\overline{AC} \cong \overline{CA}$	3. Reflexive property of congruency
4. $\triangle ABC \cong \triangle CDA$	4. ASA postulate
5. $\angle B \cong \angle D$	5. CPCTC

Note: Problems 9.2–9.5 refer to □ ABCD, as illustrated in Problem 9.2.

9.5   Use □ ABCD to prove that the diagonals of a parallelogram bisect each other. Let *M* be the point at which the diagonals intersect, as illustrated in the diagram below.

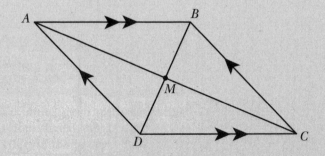

The diagonals of □ ABCD are $\overline{AC}$ and $\overline{BD}$, so your goal is to prove that each diagonal intersects the other diagonal at its midpoint.

Statement	Reason
1. $\overline{AB} \parallel \overline{DC}$; $\overline{AD} \parallel \overline{BC}$	1. Given
2. $\angle ABD \cong \angle CDB$; $\angle BAC \cong \angle DCA$	2. Parallel lines intersected by a transversal form congruent alternate interior angles
3. $\overline{AB} \cong \overline{CD}$	3. Opposite sides of a parallelogram are congruent ←
4. $\triangle AMB \cong \triangle CMD$	4. ASA postulate
5. $\overline{AM} \cong \overline{CM}$; $\overline{BM} \cong \overline{DM}$	5. CPCTC
6. M is the midpoint of $\overline{AC}$; M is the midpoint of $\overline{BD}$	6. A midpoint divides a segment into two congruent segments
7. $\overline{AC}$ bisects $\overline{BD}$; $\overline{BD}$ bisects $\overline{AC}$	7. A bisector intersects a segment at its midpoint

See Problem 9.3.

9.6 Given $\square$ LMNO in the diagram below, calculate x and y.

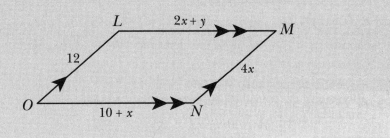

As Problem 9.3 states, the opposite sides of a parallelogram are congruent. Congruent sides have equal lengths, so create two equations by setting the lengths of the opposite sides equal to each other.

$$12 = 4x \quad \text{and} \quad 2x + y = 10 + x$$

Solve the left equation for x.

$$\frac{12}{4} = \frac{4x}{4}$$
$$3 = x$$

Substitute $x = 3$ into the remaining equation and solve for y.

$$2x + y = 10 + x$$
$$2(3) + y = 10 + 3$$
$$6 + y = 13$$
$$y = 13 - 6$$
$$y = 7$$

Therefore, $x = 3$ and $y = 7$.

9.7 Given □ *QRST* in the diagram below, calculate *x* and *y*.

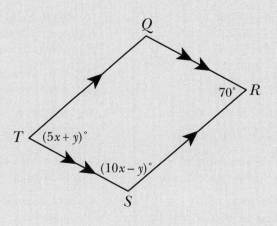

This is why ADJACENT angles of parallelograms (angles at vertices that are next to each other) are supplementary.

Opposite sides of a parallelogram are parallel. When parallel lines $\overleftrightarrow{QT}$ and $\overrightarrow{RS}$ are intersected by transversal $\overleftrightarrow{TS}$, supplementary same-side interior angles $\angle QTS$ and $\angle RST$ are formed.

$$m\angle T + m\angle S = 180°$$

Substitute the measures of the angles into the equation and solve for *x*.

$$(5x + y) + (10x - y) = 180$$
$$5x + 10x + y - y = 180$$
$$15x = 180$$
$$x = \frac{180}{15}$$
$$x = 12$$

Opposite angles of a parallelogram are congruent, so $m\angle T = m\angle R$.

$$m\angle T = m\angle R$$
$$5x + y = 70$$

Substitute *x* = 12 into the equation and solve for *y*.

$$5(12) + y = 70$$
$$60 + y = 70$$
$$y = 70 - 60$$
$$y = 10$$

Therefore, *x* = 12 and *y* = 10.

9.8   Given □ *WXYZ* below, calculate *x* and *y*.

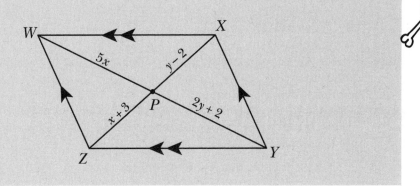

The diagonals of a parallelogram bisect each other, so *P* is the midpoint of $\overline{WY}$ and $\overline{XZ}$. Therefore, *WP* = *YP* and *ZP* = *XP*.

$$WP = YP \qquad \text{and} \qquad ZP = XP$$
$$5x = 2y + 2 \qquad\qquad x + 3 = y - 2$$

Apply the substitution technique to solve the system of equations; solve *x* + 3 = *y* − 2 for *y* by adding 2 to both sides of the equation.

$$x + 3 + 2 = y$$
$$x + 5 = y$$

> You can solve EITHER equation for x OR y and plug the result into the other equation, but this is the quickest way.

Substitute *y* = *x* + 5 into the remaining equation.

$$5x = 2y + 2$$
$$5x = 2(x + 5) + 2$$
$$5x = 2x + 10 + 2$$
$$5x = 2x + 12$$

Solve for *x*.

$$5x - 2x = 12$$
$$3x = 12$$
$$x = \frac{12}{3}$$
$$x = 4$$

Recall that *x* + 5 = *y*.

$$x + 5 = y$$
$$4 + 5 = y$$
$$9 = y$$

Therefore, *x* = 4 and *y* = 9.

Proving Parallelograms

Converses of parallelogram properties

9.9 Identify the five most common methods used to prove that a quadrilateral is a parallelogram.

Problems 9.1–9.5 identify important characteristics of the parts of a parallelogram. For instance, if a quadrilateral is a parallelogram, then both pairs of opposite angles are congruent. The converses of these statements represent four of the five methods by which to prove a quadrilateral is a parallelogram.

- ◆ Prove that both pairs of opposite sides are parallel.
- ◆ Prove that both pairs of opposite sides are congruent.
- ◆ Prove that both pairs of opposite angles are congruent.
- ◆ Prove that the diagonals bisect each other.
- ◆ Prove that one pair of opposite sides is both congruent and parallel.

Note the final method listed above, which is a combination of the first two methods.

You must use the same pair of opposite sides. You can't prove anything if one pair of opposite sides is congruent and a different pair is parallel.

Note: Problems 9.10–9.12 refer to quadrilateral FGHI below, in which $\overline{FG} \cong \overline{IH}$.

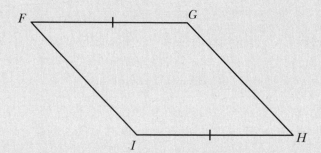

9.10 If $\overline{FI} \cong \overline{GH}$, can you conclude that *FGHI* is a parallelogram? Why or why not?

You are given $\overline{FG} \cong \overline{IH}$. If $\overline{FI} \cong \overline{GH}$, then both pairs of opposite sides are congruent. This is one of the five conditions listed in Problem 9.9, so the quadrilateral is a parallelogram.

Note: Problems 9.10–9.12 refer to the diagram in Problem 9.10, in which $\overline{FG} \cong \overline{IH}$.

9.11 If $\angle G \cong \angle I$, can you conclude that *FGHI* is a parallelogram? Why or why not?

You are given one pair of congruent opposite sides and one pair of congruent opposite angles, as illustrated in the following diagram.

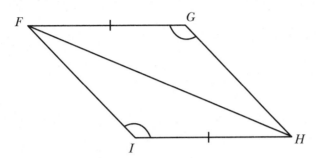

To prove that *FGHI* is a parallelogram, you need to prove that the triangles formed by the diagonal are congruent: $\triangle FGH \cong \triangle HIF$. The triangles share a common side $\left(\overline{FH}\right)$, but the congruent angles are not included between the congruent sides, so you cannot apply the SAS postulate. Because the given information alone is not sufficient to conclude that the triangles are congruent, you cannot conclude that *FGHI* must be a parallelogram.

> You have two congruent sides and a nonincluded congruent angle. No side-side-angle (SSA) postulate exists.

Note: Problems 9.10–9.12 refer to the diagram in Problem 9.10, in which $\overline{FG} \cong \overline{IH}$.

9.12 If $\overline{FG} \parallel \overline{IH}$, can you conclude that *FGHI* is a parallelogram? Why or why not?

You are given $\overline{FG} \cong \overline{IH}$ and $\overline{FG} \parallel \overline{IH}$. According to Problem 9.9, if one pair of opposite sides is both congruent and parallel, then the quadrilateral is a parallelogram. Therefore, *FGHI* must be a parallelogram.

Note: Problems 9.13–9.14 refer to the diagram that follows, in which $\overline{FI} \parallel \overline{HG}$.

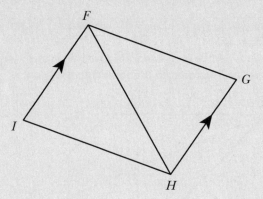

9.13 Use *FGHI* to prove that a quadrilateral with one pair of sides that is both congruent and parallel must be a parallelogram.

Note that your proof cannot cite the theorem you are proving, so a one- or two-step proof that states that a quadrilateral with one pair of congruent and parallel sides is a parallelogram is insufficient proof.

Statement	Reason
1. $\overline{FI} \parallel \overline{HG}$; $\overline{FI} \cong \overline{HG}$	1. Given
2. $\angle IFH \cong \angle GHF$	2. Parallel lines intersected by a transversal form congruent alternate interior angles
3. $\overline{FH} \cong \overline{HF}$	3. Reflexive property of congruency
4. $\triangle IFH \cong \triangle GHF$	4. SAS postulate
5. $\overline{IH} \cong \overline{GF}$	5. CPCTC
6. *FGHI* is a parallelogram	6. A quadrilateral with two pairs of congruent opposite sides is a parallelogram

Note: Problems 9.13–9.14 refer to the diagram in Problem 9.13, in which $\overline{FI} \parallel \overline{HG}$.

9.14 Given $\angle IFG \cong \angle GHI$, prove that *FGHI* is a parallelogram because both pairs of opposite angles are congruent.

One pair of opposite angles is congruent according to the given information: $\angle IFG \cong \angle GHI$. Therefore, your goal is to prove that $\angle I \cong \angle G$.

Statement	Reason
1. $\overline{FI} \parallel \overline{HG}$; $\angle IFG \cong \angle GHI$	1. Given
2. $m\angle IFH + m\angle GFH = m\angle IFG$; $m\angle IHF + m\angle GHF = m\angle GHI$	2. Angle addition postulate
3. $m\angle IFH + m\angle GFH = m\angle IHF + m\angle GHF$	3. Substitution property of equality (Steps 1, 2)

4. $\angle IFH \cong \angle GHF$	4. Parallel lines intersected by a transversal form congruent alternate interior angles
5. $m\angle GFH = m\angle IHF$	5. Subtraction property of equality
6. $\overline{FH} \cong \overline{HF}$	6. Reflexive property of congruency
7. $\triangle IFH \cong \triangle GHF$	7. ASA postulate
8. $\angle I \cong \angle G$	8. CPCTC
9. $FGHI$ is a parallelogram	9. A quadrilateral with two pairs of congruent opposite angles is a parallelogram

Subtract $m\angle IFH$ and $m\angle GHF$ from the equation in Step 3.

9.15 Given the diagram below, prove that $JKLM$ is a parallelogram.

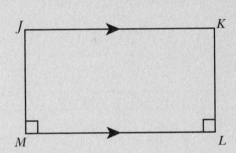

Statement	Reason
1. $\overline{JK} \parallel \overline{ML}$; $\angle M \cong \angle L$	1. Given
2. $m\angle M = m\angle L = 90°$	2. Perpendicular lines intersect at right angles
3. $\angle J$ and $\angle M$ are supplementary; $\angle K$ and $\angle L$ are supplementary	3. Parallel lines intersected by a transversal form complementary same-side interior angles
4. $\angle J \cong \angle K$	4. Supplements of congruent angles are congruent
5. $m\angle J + m\angle M = 180°$; $m\angle K + m\angle L = 180°$	5. Supplementary angles have a sum of 180°
6. $m\angle J + 90° = 180°$; $m\angle K + 90° = 180°$	6. Substitution property of equality (Steps 2, 5)
7. $m\angle J = 90°$; $m\angle K = 90°$	7. Subtraction property of equality
8. $JKLM$ is a parallelogram	8. A quadrilateral with two pairs of congruent opposite angles is a parallelogram

See Problem 5.38. You're not done yet; you haven't proven that opposite angles are congruent yet.

You're given that $\angle M$ and $\angle L$ are right angles. Step 7 concludes that $\angle J$ and $\angle K$ are also right angles. Opposite angles are congruent because ALL the angles are congruent.

9.16 Given the diagram below, in which $\overline{AP}$ is a median of ΔZAX and $\overline{BP}$ is a median of ΔWBY, prove that $WXYZ$ is a parallelogram.

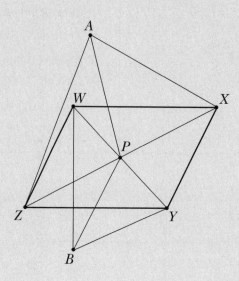

Statement	Reason
1. $\overline{AP}$ is a median of ΔZAX; $\overline{BP}$ is a median of ΔWBY	1. Given
2. P is the midpoint of $\overline{ZX}$; P is the midpoint of $\overline{WY}$	2. Medians extend from a vertex to the midpoint of the opposite side of a triangle
3. $\overline{WY}$ bisects $\overline{ZX}$; $\overline{ZX}$ bisects $\overline{WY}$	3. A segment bisector intersects the segment at its midpoint
4. $WXYZ$ is a parallelogram	4. Quadrilaterals with diagonals that bisect each other are parallelograms

Special Parallelograms
Rectangles, rhombi, and squares

Note: Problems 9.17–9.18 refer to ▱ RECT in the following diagram.

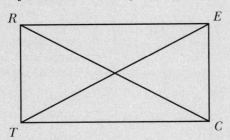

9.17 Complete the following statement and prove it using ▱*RECT*.

A rectangle is a parallelogram with four _____. Unlike the diagonals of a generic parallelogram, the diagonals of a rectangle _____ each other.

> All parallelograms (including rectangles) have diagonals that bisect each other, so that's not what the book is looking for.

A rectangle is a parallelogram with four <u>right angles</u>. Unlike the diagonals of a generic parallelogram, the diagonals of a rectangle <u>are congruent to</u> each other. To prove that the diagonals of a rectangle are congruent, you must assume that *RECT* is a parallelogram and that all four interior angles are right angles.

$$m\angle R = m\angle E = m\angle C = m\angle T = 90°$$

Your goal is to prove $\overline{RC} \cong \overline{ET}$.

Statement	Reason
1. *RECT* is a parallelogram; $m\angle C = m\angle T = 90°$	1. Given
2. $\overline{RT} \cong \overline{EC}$	2. Opposite sides of a parallelogram are congruent
3. $\overline{TC} \cong \overline{CT}$	3. Reflexive property of congruency
4. $\triangle RTC \cong \triangle ECT$	4. SAS postulate
5. $\overline{RC} \cong \overline{ET}$	5. CPCTC

Note: Problems 9.17–9.18 refer to the diagram in Problem 9.17.

9.18 Given only that $m\angle TRE = 90°$ in ▱ *RECT*, prove that *RECT* is a rectangle.

A rectangle is defined as a parallelogram with four right angles, so your goal is to prove that the three remaining interior angles are right angles as well.

Statement	Reason
1. *RECT* is a parallelogram; $m\angle TRE = 90°$	1. Given
2. $m\angle ECT = 90°$	2. Opposite angles of a parallelogram are congruent

continues

continued

Statement	Reason
3. $\overline{RT} \parallel \overline{EC}$	3. Opposite sides of a parallelogram are parallel
4. $m\angle TRE + m\angle REC = 180°$	4. Parallel lines intersected by a transversal form supplementary same-side interior angles
5. $90° + m\angle REC = 180°$	5. Substitution property of equality (Steps 1, 4)
6. $m\angle REC = 90°$	6. Subtraction property of equality
7. $m\angle RTC = 90°$	7. Opposite angles of a parallelogram are congruent
8. $\square RECT$ is a rectangle	8. A rectangle is a parallelogram that contains four right angles

Note: Problems 9.19–9.21 refer to $\square RHOM$ in the diagram below.

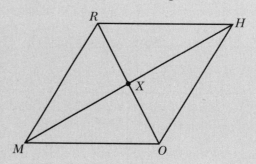

9.19 Complete the following statement and prove it using $\square RHOM$.

A rhombus is a parallelogram with four _____. Unlike the diagonals of a generic parallelogram, the diagonals of a rhombus _____ each other.

The plural form of "rhombus" is "rhombi," just as the plural form of "octopus" is "octopi."

A rhombus is a parallelogram with four <u>congruent sides</u>. Unlike the diagonals of a generic parallelogram, the diagonals of a rhombus <u>are perpendicular to</u> each other. To prove that the diagonals of $\square RHOM$ are perpendicular $\left(\overline{RO} \perp \overline{MH}\right)$, you must assume that the sides of $\square RHOM$ are congruent: $\overline{RH} \cong \overline{HO} \cong \overline{OM} \cong \overline{MR}$.

Statement	Reason
1. $\square RHOM$ is a rhombus	1. Given
2. $\overline{RH} \cong \overline{HO} \cong \overline{OM} \cong \overline{MR}$	2. The sides of a rhombus are congruent
3. $\overline{RX} \cong \overline{OX}$; $\overline{MX} \cong \overline{HX}$	3. The diagonals of a parallelogram bisect each other

4. ΔRXH ≅ ΔOXH ≅ ΔOXM ≅ ΔRXM	4. SSS postulate
5. m∠RXH + m∠OXH = 180°	5. Angle addition postulate
6. m∠RXH = m∠OXH	6. CPCTC
7. m∠RXH = m∠OXH = 90°	7. If two supplementary angles are congruent, they are right angles
8. $\overline{RO} \perp \overline{MH}$	8. Perpendicular lines intersect at right angles

The only two equal values that add up to 180 are 90 and 90.

All four of the triangles inside the rhombus are congruent. A rhombus has four congruent sides (each side has three tick marks). The diagonals (which are not congruent to the sides or to each other) are split in half at point X. Each half of the shorter diagonal has one tick mark, and each half of the longer diagonal has two.

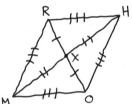

The triangles are congruent by SSS because each one is made up of the same sides—each triangle has a one-, a two-, and a three-tick-mark side.

Note: Problems 9.19–9.21 refer to the diagram in Problem 9.19.

9.20 Complete the following statement and prove its assertion using ▱ *RHOM*.

All parallelograms have diagonals that bisect each other. However, the diagonals of a rhombus also bisect _____.

All parallelograms have diagonals that bisect each other. However, the diagonals of a rhombus also bisect <u>the interior angles of the rhombus</u>. Therefore, given that ▱*RHOM* is a rhombus, you must prove that each diagonal bisects two interior angles: $\overline{RO}$ bisects ∠MRH and ∠HOM, and $\overline{MH}$ bisects ∠RHO and ∠OMR.

It's okay to reference Problem 9.19 to avoid repeating steps, because that problem started with exactly the same given information as this problem: ▱ RHOM is a rhombus.

Statement	Reason
1. ▱ *RHOM* is a rhombus	1. Given
2. ΔRXH ≅ ΔOXH ≅ ΔOXM ≅ ΔRXM	2. Problem 9.19

continues

continued

Statement	Reason
3. $\angle MRO \cong \angle HRO$; $\angle RHM \cong \angle OHM$; $\angle HOR \cong \angle MOR$; $\angle OMH \cong \angle RMH$	3. CPCTC
4. $\overline{RO}$ bisects $\angle MRH$ and $\angle HOM$; $\overline{MH}$ bisects $\angle RHO$ and $\angle OMR$	4. An angle bisector divides an angle into two congruent angles

> In this case, it's not okay to reference Problems 9.19 or 9.20 in your proof because in this problem you're not given that $\square RHOM$ is a rhombus.

Note: Problems 9.19–9.21 refer to the diagram in Problem 9.19.

9.21 Prove that $\square RHOM$ is a rhombus, given only that $\overline{RH} \cong \overline{OH}$.

You are given that $RHOM$ is a parallelogram with one pair of congruent adjacent sides $\left(\overline{RH} \cong \overline{OH}\right)$. Your goal is to prove that $\square RHOM$ is a rhombus, which means you must prove that all four sides are congruent.

Statement	Reason
1. $RHOM$ is a parallelogram; $\overline{RH} \cong \overline{OH}$	1. Given
2. $\overline{RH} \cong \overline{MO}$; $\overline{OH} \cong \overline{MR}$	2. Opposite sides of a parallelogram are congruent
3. $\overline{OH} \cong \overline{MO}$; $\overline{RH} \cong \overline{MR}$	3. Substitution property of equality (Steps 1, 2)
4. $\overline{RH} \cong \overline{OH} \cong \overline{MR} \cong \overline{MO}$	4. Transitive property of equality
5. $\square RHOM$ is a rhombus	5. A rhombus is a parallelogram with four congruent sides

9.22 Given that $\square WXYZ$ in the diagram below is a rhombus, calculate the measures of the three numbered angles.

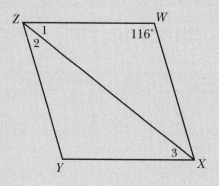

If $\square WXYZ$ is a rhombus, then all four of its sides are congruent. Therefore, $\triangle ZWX$ and $\triangle ZYX$ are isosceles triangles that share base $\overline{ZX}$. Opposite angles of a parallelogram are congruent, so $\angle W \cong \angle Y$.

Consider isosceles triangle ZYX, which has vertex angle Y. The sum of the interior angles of the triangle is $180°$.

$$m\angle 2 + m\angle 3 + m\angle Y = 180°$$

Because $\angle W \cong \angle Y$, $m\angle W = m\angle Y = 116°$.

$$m\angle 2 + m\angle 3 + 116° = 180°$$

According to the isosceles triangle theorem, the base angles of the triangle are congruent, so $m\angle 2 = m\angle 3$. Substitute $m\angle 2$ for $m\angle 3$ in the above equation and solve for $m\angle 2$.

$$m\angle 2 + m\angle 2 + 116° = 180°$$
$$2(m\angle 2) + 116° = 180°$$
$$2(m\angle 2) = 180° - 116°$$
$$2(m\angle 2) = 64°$$
$$m\angle 2 = \frac{64°}{2}$$
$$m\angle 2 = 32°$$

> You can start this problem in a lot of ways. One way is to say that ∠YZW and ∠W are supplementary (adjacent angles of a parallelogram are supplementary according to Problem 9.7), so m∠YZW = 180° – 116° = 64°. The diagonal bisects ∠YZW, so m∠1 = m∠2 = 64° ÷ 2 = 32°.

Recall that $m\angle 2 = m\angle 3$, so $m\angle 3 = 32°$ as well. According to Problem 9.20, the diagonals of a rhombus bisect the interior angles. In this rhombus, $\overline{ZX}$ bisects $\angle WZY$, so $m\angle 1 = m\angle 2 = 32°$. Therefore, $m\angle 1 = m\angle 2 = m\angle 3 = 32°$.

9.23 Given that $\square\, SQAR$ in the diagram below is a square, calculate the values of x and y.

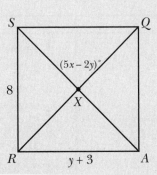

> A square has four right angles and four congruent sides, so it inherits all the properties of a parallelogram, a rectangle, AND a rhombus.

A square is a rhombus, so all four of its sides are congruent. Set the given side measurements equal and solve for y.

$$y + 3 = 8$$
$$y = 8 - 3$$
$$y = 5$$

The diagonals of a rhombus are perpendicular, so $m\angle SXQ = 90°$.

$$5x - 2y = 90$$

Substitute $y = 5$ into the equation and solve for x.

$$5x - 2(5) = 90$$
$$5x - 10 = 90$$
$$5x = 90 + 10$$
$$5x = 100$$
$$x = \frac{100}{5}$$
$$x = 20$$

Therefore, $x = 20$ and $y = 5$.

9.24 Given the following diagram, in which $ABCD$ is a rectangle and $\overline{CD}$ bisects $\angle ACE$, prove $\overline{EC} \cong \overline{DB}$.

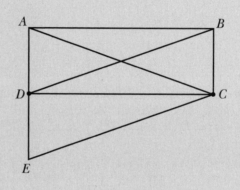

Statement	Reason
1. $ABCD$ is a rectangle; $\overline{CD}$ bisects $\angle ACE$	1. Given
2. $m\angle ADC = 90°$	2. A rectangle contains four right angles
3. $m\angle EDC = 90°$	3. The supplement of a right angle is also a right angle
4. $\angle ACD \cong \angle ECD$	4. An angle bisector splits an angle two into congruent angles
5. $\overline{DC} \cong \overline{DC}$	5. Reflexive property of congruency
6. $\triangle ADC \cong \triangle EDC$	6. ASA postulate
7. $\overline{EC} \cong \overline{AC}$	7. CPCTC
8. $\overline{AC} \cong \overline{DB}$	8. The diagonals of a rectangle are congruent
9. $\overline{EC} \cong \overline{DB}$	9. Transitive property of congruency

Consequences of Parallelograms
Affecting triangles, parallel lines, and transversals

9.25 Use the following diagram to prove that parallel lines are equidistant from each other.

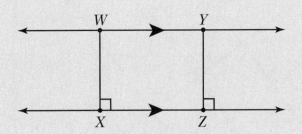

The diagram indicates that $\overleftrightarrow{WY} \parallel \overleftrightarrow{XZ}$ and includes two segments connecting the parallel lines that are both perpendicular to $\overleftrightarrow{XZ}$. To prove that the parallel lines are equidistant, you need to show that $\overline{WX} \cong \overline{YZ}$ and that those congruent segments are perpendicular to $\overleftrightarrow{WY}$ as well as $\overleftrightarrow{XZ}$.

You've got $\overleftrightarrow{XZ}$ and two points not on that line: W and Y. One line passes through each of those points that is perpendicular to $\overleftrightarrow{XZ}$.

Statement	Reason
1. $\overleftrightarrow{WY} \parallel \overleftrightarrow{XZ}$	1. Given
2. $\overline{WX} \perp \overline{XZ}$; $\overline{YZ} \perp \overline{XZ}$	2. Exactly one perpendicular line connects a line and a noncollinear point
3. $\overline{WX} \parallel \overline{YZ}$	3. Two lines perpendicular to the same line are parallel to each other
4. $WYZX$ is a parallelogram	4. A quadrilateral with two pairs of parallel sides is a parallelogram
5. $\overline{WX} \cong \overline{YZ}$	5. Opposite sides of a parallelogram are congruent
6. $m\angle XWY = 90°$; $m\angle ZYW = 90°$	6. Opposite angles of a parallelogram are congruent
7. $\square\, WYZX$ is a rectangle	7. A rectangle is a parallelogram that contains right angles
8. WX and YZ represent the distances between $\overleftrightarrow{WY}$ and $\overleftrightarrow{XZ}$	8. Distance between two lines is defined as the length of the segment that is perpendicular to both lines
9. $\overleftrightarrow{WY}$ is equidistant from $\overleftrightarrow{XZ}$	9. The distances between the lines are equal

9.26 Calculate y in the diagram below.

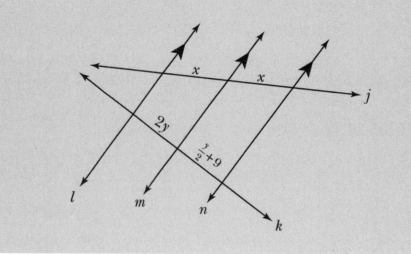

In the diagram, two transversals (j and k) intersect three parallel lines ($l \parallel m \parallel n$). Notice that the parallel lines divide j into two congruent segments, each with length x.

If a group of parallel lines divides one transversal into congruent segments, then those parallel lines divide any transversal into congruent segments. Therefore, k is divided into congruent segments as well.

$$2y = \frac{y}{2} + 9$$

Multiply the entire equation by 2 to eliminate the fraction.

$$2(2y) = 2\left(\frac{y}{2} + 9\right)$$
$$4y = \frac{2y}{2} + 2(9)$$
$$4y = y + 18$$

Solve for y.

$$4y - y = 18$$
$$3y = 18$$
$$y = \frac{18}{3}$$
$$y = 6$$

9.27 Complete the following statement and use the diagram to prove it.

A line that bisects one side of a triangle and is parallel to another side of the triangle _____ the remaining side of the triangle.

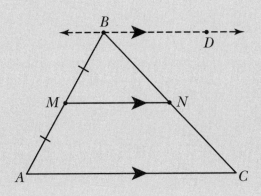

A line that bisects one side of a triangle and is parallel to another side of the triangle <u>bisects</u> the remaining side of the triangle. In the diagram, $\overline{MN}$ bisects $\overline{AB}$ because $AM = MB$. Notice that $\overline{MN}$ is parallel to $\overline{AC}$, another side of the triangle.

An additional line is included on the diagram, $\overline{BD}$, which is parallel to $\overline{AC}$ and $\overline{MN}$. Your goal is to prove that $\overline{MN}$ bisects the remaining side, $\overline{BC}$.

Statement	Reason
1. $AM = MB$; $\overline{MN} \parallel \overline{AC}$	1. Given
2. $\overline{MN}$ bisects $\overline{AB}$	2. A segment bisector divides a segment into two congruent segments
3. $BN = CN$	3. If a group of parallel lines divides one transversal into congruent segments, then it divides all transversals into congruent segments
4. $\overline{MN}$ bisects $\overline{BC}$	4. A segment bisector divides a segment into two congruent segments

See Problem 9.26.

9.28 According to the triangle midpoint theorem, what conclusions can be drawn based solely on the diagram below?

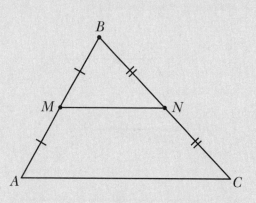

This is slightly different than Problem 9.27, where you knew that $\overline{MN}$ and $\overline{AC}$ were parallel and had to prove $\overline{BC}$ was bisected. Here, you know that the sides are bisected, and the theorem tells you that $\overline{MN} \parallel \overline{AC}$.

The triangle midpoint theorem states that the midline of a triangle (the segment connecting the midpoints of two sides—in this case, $\overline{MN}$) is parallel to the remaining side of the triangle and is exactly half as long as the remaining side. Therefore, $\overline{MN} \parallel \overline{AC}$ and $MN = \frac{1}{2}(AC)$.

9.29 Given the diagram that follows, in which $\overline{AM} \parallel \overline{LN}$, prove $MN = \frac{1}{2}AC$.

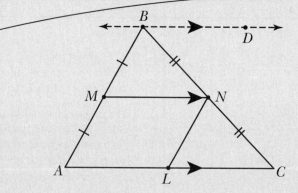

Given the first conclusion of the triangle midpoint theorem (the segment connecting two of the sides' midpoints is parallel to the third side), you're proving the second conclusion: the midline is half as long as the side it's parallel to.

Statement	Reason
1. $AM = MB$; $BN = NC$; $\overline{BD} \parallel \overline{MN} \parallel \overline{AC}$; $\overline{AM} \parallel \overline{LN}$	1. Given
2. $AMNL$ is a parallelogram	2. A quadrilateral with two pairs of parallel opposite sides is a parallelogram
3. $\angle BMN \cong \angle MAL$; $\angle MAL \cong \angle NLC$; $\angle BNM \cong \angle NCL$	3. Parallel lines intersected by a transversal form congruent corresponding angles
4. $\angle BMN \cong \angle NLC$	4. Transitive property of congruency

Look at the first two lines of Statement 3: $\angle BMN \cong \angle MAL$ and $\angle MAL \cong \angle NLC$. This means that $\angle BMN \cong \angle NLC$.

5. $\triangle MBN \cong \triangle LNC$

5. AAS postulate

6. $\overline{AL} \cong \overline{MN}$

6. Opposite sides of a parallelogram are congruent

7. $\overline{MN} \cong \overline{LC}$

7. CPCTC

8. $AL + LC = AC$

8. Segment addition postulate

9. $MN + MN = AC$

9. Substitution property of equality (Steps 6, 7, 8)

10. $2(MN) = AC$

10. Combination of like terms

11. $MN = \dfrac{1}{2}(AC)$

11. Multiplication property of equality

∠BMN ≅ ∠NLC, ∠BNM ≅ ∠NCL, and BN ≅ NC

Step 6 states that AL = MN, and Step 7 states that LC = MN. This means that you can plug in MN for both AL and LC in Step 8.

Multiply both sides by one-half.

9.30 Given $SM = MR$ and $RN = NT$ in the diagram below, calculate x and y.

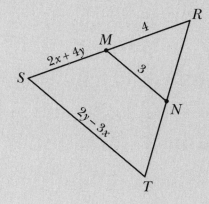

The endpoints of $\overline{MN}$ are midpoints of two sides of the triangle, so $\overline{MN}$ is a midline of the triangle. You are given $SM = MR$, so set the lengths of the segments equal: $2x + 4y = 4$.

According to the triangle midpoint theorem, the midline of a triangle is half as long as the side of the triangle to which it is parallel. Therefore, $MN = \dfrac{1}{2}(ST)$. According to the diagram, $MN = 3$ and $ST = 2y - 3x$.

$$3 = \frac{1}{2}(2y - 3x)$$

Multiply both sides of the equation by 2 to eliminate the fraction.

$$2(3) = \frac{2}{1}\left(\frac{1}{2}\right)(2y - 3x)$$
$$6 = 2y - 3x$$

Write the equation in standard form: $3x - 2y = -6$. Combine this equation with the equation you generated at the beginning of the problem, $2x + 4y = 4$. You now have a system of equations in two variables.

$$\begin{cases} 2x + 4y = 4 \\ 3x - 2y = -6 \end{cases}$$

Solve the system of equations. To apply the elimination technique, multiply the second equation by 2, add the resulting equations, and solve for x.

$$\begin{array}{rrrrr} 2x & + & 4y & = & 4 \\ 6x & - & 4y & = & -12 \\ \hline 8x & & & = & -8 \end{array}$$

$$8x = -8$$

$$x = \frac{-8}{8}$$

$$x = -1$$

To calculate y, substitute $x = -1$ into either equation of the system.

$$2x + 4y = 4$$
$$2(-1) + 4y = 4$$
$$-2 + 4y = 4$$
$$4y = 4 + 2$$
$$4y = 6$$
$$y = \frac{6}{4}$$
$$y = \frac{3}{2}$$

Therefore, $x = -1$ and $y = \frac{3}{2}$.

9.31 Prove that the midpoint of the hypotenuse of a right triangle is equidistant from the vertices of the triangle.

Construct a diagram of a right triangle and add congruent sides opposite the legs (the dotted lines in the diagram below) to form a rectangle.

> This is the trickiest part of the proof, adding lines that aren't part of the original figure. It helps to make the right triangle into a rectangle because the diagonals of a rectangle are congruent.

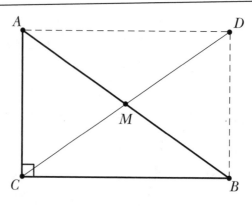

Transforming the right triangle ABC into rectangle $ADBC$ makes the proof significantly easier to complete. Your goal is to demonstrate that $AM = BM = CM$.

Statement	Reason
1. $ADBC$ is a rectangle; M is the midpoint of $\overline{AB}$	1. Given
2. $AM + MB = AB$; $CM + MD = CD$	2. Segment addition postulate
3. $AM = MB$; $CM = MD$	3. The diagonals of a parallelogram bisect each other
4. $AM + AM = AB$; $CM + CM = CD$	4. Substitution property of equality (Steps 2, 3)
5. $2(AM) = AB$; $2(CM) = CD$	5. Combination of like terms
6. $AB = CD$	6. The diagonals of a rectangle are congruent
7. $2(AM) = 2(CM)$	7. Substitution property of equality (Steps 5, 6)
8. $AM = CM$	8. Division property of equality
9. $AM = BM = CM$	9. Transitive property of equality ←

> You already knew that AM = BM from Step 3. Now you know that AM also equals CM, so BM and CM must be congruent as well by the transitive property.

9.32 Given that M is the midpoint of $\overline{BC}$ in the diagram below, calculate x and y.

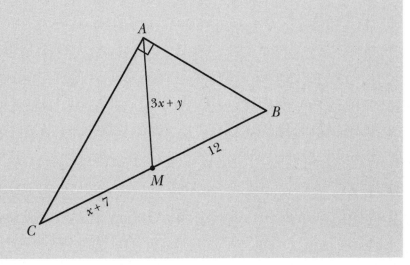

As Problem 9.31 states, the midpoint of the hypotenuse of a right triangle is equidistant from the vertices of the triangle. In this diagram, M is the midpoint of hypotenuse $\overline{BC}$, so $AM = BM = CM$.

Calculate x by setting $BM = CM$.

$$12 = x + 7$$
$$12 - 7 = x$$
$$5 = x$$

Set $AM = BM$ and substitute $x = 5$ into the equation to solve for y.

$$3x + y = 12$$
$$3(5) + y = 12$$
$$15 + y = 12$$
$$y = 12 - 15$$
$$y = -3$$

Therefore, $x = 5$ and $y = -3$.

Properties of Trapezoids

Halfallelograms

9.33 What characteristic differentiates a trapezoid from a general quadrilateral?

A trapezoid is a quadrilateral with *exactly one* pair of parallel sides. Note that a trapezoid is not a parallelogram because a parallelogram has two pairs of parallel sides.

9.34 Identify the bases, the legs, the median, and the altitude of trapezoid $TRAP$ below, given that M is the midpoint of $\overline{TP}$ and N is the midpoint of $\overline{RA}$.

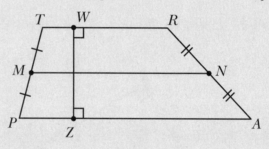

> The median of a trapezoid is a lot like the midline of a triangle, as described in Problems 9.27– 9.30.

The bases of a trapezoid are the parallel sides. In $TRAP$, $\overline{TR} \parallel \overline{PA}$, so $\overline{TR}$ and $\overline{PA}$ are the bases of the trapezoid. The nonparallel sides are the legs of the trapezoid; the legs of $TRAP$ are $\overline{TP}$ and $\overline{RA}$.

The median of a trapezoid is the segment that connects the midpoints of the legs. Therefore, $\overline{MN}$ is the median of trapezoid $TRAP$. The altitude of a trapezoid is a segment that both connects and is perpendicular to the bases; in this diagram, $\overline{WZ}$ is an altitude.

> A trapezoid has only one median, but it can have an infinite number of altitudes. Any line segment drawn between the bases that is perpendicular to both bases is an altitude.

Note: Problems 9.35–9.36 refer to the diagram below, in which trapezoid ABCD has median $\overline{MN}$.

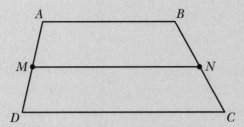

9.35 Given $AB = 5$ and $DC = 11$, calculate MN.

If a trapezoid has bases b_1 and b_2, the length of the median is $\dfrac{b_1 + b_2}{2}$. In other

words, the median is half as long as the sum of the lengths of the bases.

It doesn't matter which base you plug in for b_1 and which you plug in for b_2.

$$MN = \frac{b_1 + b_2}{2}$$
$$= \frac{5 + 11}{2}$$
$$= \frac{16}{2}$$
$$= 8$$

Therefore, $MN = 8$.

Note: Problems 9.35–9.36 refer to the diagram in Problem 9.35, in which trapezoid ABCD has median $\overline{MN}$.

9.36 If $MN = 15$ and the length of $\overline{DC}$ is 3 inches less than twice the length of $\overline{AB}$, what are the lengths of the bases?

If the length of $\overline{DC}$ is 3 inches less than twice the length of $\overline{AB}$, then $DC = 2(AB) - 3$. Substitute $b_1 = AB$, $b_2 = 2(AB) - 3$, and $MN = 15$ into the median formula.

$$MN = \frac{b_1 + b_2}{2}$$
$$15 = \frac{AB + [2(AB) - 3]}{2}$$
$$15 = \frac{3(AB) - 3}{2}$$

Multiply both sides of the equation by 2 to eliminate the fraction and solve for AB.

$$2(15) = \frac{2}{1}\left[\frac{3(AB) - 3}{2}\right]$$

$$30 = 3(AB) - 3$$

$$30 + 3 = 3(AB)$$

$$33 = 3(AB)$$

$$\frac{33}{3} = AB$$

$$11 = AB$$

Base $\overline{AB}$ is 11 inches long. To determine the length of base $\overline{DC}$, recall that $DC = 2(AB) - 3$.

$$DC = 2(11) - 3$$

$$DC = 22 - 3$$

$$DC = 19$$

The lengths of the bases of the trapezoid are $AB = 11$ and $DC = 19$.

9.37 Use the diagram below to prove that both pairs of base angles of an isosceles trapezoid are congruent.

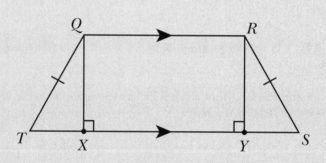

Isosceles trapezoids have two pairs of base angles. Each pair is formed by the intersection of one base and the legs of the trapezoid. In this diagram, $\angle TQR$ and $\angle SRQ$ represent one pair of base angles, and $\angle QTS$ and $\angle RST$ represent the other pair.

Given altitudes $\overline{QX}$ and $\overline{RY}$ on isosceles trapezoid $QRST$, your goal is to prove (1) $\angle TQR \cong \angle SRQ$ and (2) $\angle QTS \cong \angle RST$.

Statement	Reason
1. $\overline{QR} \parallel \overline{TS}$; $\overline{QT} \cong \overline{RS}$; $\overline{QX} \perp \overline{TS}$; $\overline{RY} \perp \overline{TS}$	1. Given
2. $\overline{QX} \parallel \overline{RY}$	2. Two lines perpendicular to the same line are parallel to each other

3. *QRYX* is a parallelogram	3. A parallelogram is a quadrilateral with parallel opposite sides
4. $\overline{QX} \cong \overline{RY}$	4. Opposite sides of a parallelogram are congruent
5. $\triangle QTX \cong \triangle RSY$	5. HL theorem
6. $\angle QTS \cong \angle RST$	6. CPCTC ←
7. $\Box\ QRYX$ is a rectangle	7. If one interior angle of a parallelogram is a right angle, the parallelogram is a rectangle
8. $m\angle XQR = m\angle YRQ = 90°$	8. The interior angles of a rectangle are right angles
9. $m\angle TQX = m\angle SRY$	9. CPCTC
10. $m\angle TQX + m\angle XQR = $ $m\angle SRY + m\angle YRQ$	10. Addition property of equality
11. $m\angle TQR = m\angle TQX + m\angle XQR$; $m\angle SRQ = m\angle SRY + m\angle YRQ$	11. Angle addition postulate
12. $m\angle TQR = m\angle SRQ$	12. Substitution property of equality (Steps 10, 11) ←

This proves that the bottom pair of base angles are congruent.

∠TQX and ∠SRY are congruent because they're corresponding parts of congruent triangles. If you add each of the right angles at the top of the rectangle to the little congruent angles, the sums stay equal. Those sums just happen to be the upper base angles.

9.38 Use the diagram below to prove that opposite angles of an isosceles trapezoid are supplementary.

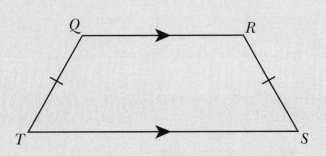

Given the isosceles trapezoid *QRST*, your goal is to prove (1) $m\angle S + m\angle Q = 180°$ and (2) $m\angle T + m\angle R = 180°$.

Statement	Reason
1. $\overline{QR} \parallel \overline{TS}$; $\overline{QT} \cong \overline{RS}$	1. Given
2. $m\angle T + m\angle Q = 180°$; $m\angle S + m\angle R = 180°$	2. Two parallel lines intersected by a transversal form supplementary same-side interior angles

continues

continued

Statement	Reason
3. $m\angle T = m\angle S$	3. Base angles of an isosceles trapezoid are congruent
4. $m\angle S + m\angle Q = 180°$; $m\angle T + m\angle R = 180°$	4. Substitution property of equality (Steps 2, 3)

Step 3 says that $\angle S \cong \angle T$, so you can swap S and T in the Step 2 equations.

9.39 Use the following diagram to prove that the diagonals of an isosceles trapezoid are congruent.

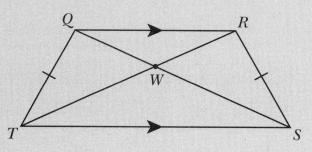

Given the isosceles trapezoid $QRST$, your goal is to prove $\overline{QS} \cong \overline{RT}$.

Statement	Reason
1. $\overline{QR} \parallel \overline{TS}$; $\overline{QT} \cong \overline{RS}$	1. Given
2. $\angle QTS \cong \angle RST$	2. Base angles of an isosceles triangle are congruent
3. $\overline{TS} \cong \overline{ST}$	3. Reflexive property of congruency
4. $\triangle QTS \cong \triangle RST$	4. SAS postulate
5. $\overline{QS} \cong \overline{RT}$	5. CPCTC

9.40 Given $\angle 1 \cong \angle 2$ and $\angle 3 \cong \angle 4$ in the diagram below, prove that $LMNO$ is an isosceles trapezoid.

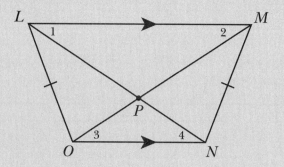

Proving that $LMNO$ is a trapezoid is a trivial matter, because it contains exactly one pair of parallel sides. To prove that $LMNO$ is isosceles, you can use one of three techniques: prove that the legs are congruent, prove both pairs of base angles are congruent, or prove that the diagonals are congruent.

Statement	Reason
1. $\overline{LM} \parallel \overline{ON}$; $\overline{LO} \cong \overline{MN}$; $\angle 1 \cong \angle 2$; $\angle 3 \cong \angle 4$	1. Given
2. $LMNO$ is a trapezoid	2. A trapezoid is a quadrilateral with exactly one pair of parallel sides
3. $\overline{LP} \cong \overline{MP}$; $\overline{PO} \cong \overline{PN}$	3. Converse of the isosceles triangle theorem
4. $LP + PN = MP + PO$	4. Addition property of equality
5. $LP + PN = LN$; $MP + PO = MO$	5. Segment addition postulate
6. $LN = MO$	6. Substitution property of equality (Steps 4, 5)
7. $LMNO$ is isosceles	7. If the diagonals of a trapezoid are congruent, then the trapezoid is isosceles

If two angles of a triangle are congruent, then the sides opposite those angles are congruent.

You know that LP = MP. You can add PN to the left side and PO to the right side because PO = PN.

9.41 Given the isosceles trapezoid and its median in the diagram below, calculate x, y, and z.

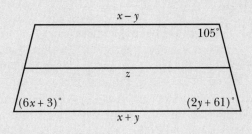

According to Problem 9.38, opposite angles of an isosceles trapezoid are supplementary. Add the lower-left angle measure to the upper-right angle measure and set the sum equal to 180.

$$(6x + 3) + 105 = 180$$
$$6x + 108 = 180$$
$$6x = 180 - 108$$
$$6x = 72$$
$$x = \frac{72}{6}$$
$$x = 12$$

Problem 9.37 proves that base angles of an isosceles trapezoid are congruent.

$$6x + 3 = 2y + 61$$

Substitute $x = 12$ into the equation and solve for y.

$$6(12) + 3 = 2y + 61$$
$$72 + 3 = 2y + 61$$
$$75 = 2y + 61$$
$$75 - 61 = 2y$$
$$14 = 2y$$
$$7 = y$$

Calculate the lengths of the bases.

$$x - y = 12 - 7 = 5 \qquad x + y = 12 + 7 = 19$$

The length of the median (z in this diagram) is equal to half the sum of the bases.

$$z = \frac{b_1 + b_2}{2} = \frac{5 + 19}{2} = \frac{24}{2} = 12$$

Therefore, $x = 12$, $y = 7$, and $z = 12$.

Properties of Kites
(Besides that they're colorful and can fly)

9.42 Define a geometric kite. Under what circumstances is a kite a parallelogram?

A kite is a quadrilateral with exactly two pairs of adjacent congruent sides. Note that a parallelogram has opposite congruent sides, whereas the congruent sides of kites are adjacent. Therefore, a kite is also a parallelogram only when both pairs of adjacent congruent sides of the kite are congruent to each other, making the kite a rhombus.

> The congruent sides of a kite touch, unlike the congruent sides of a parallelogram.

Note: Use the diagram below to prove the theorems presented in Problems 9.43–9.45. The diagram illustrates convex kite ABCD, in which $\overline{AB} \cong \overline{BC}$ and $\overline{AD} \cong \overline{CD}$.

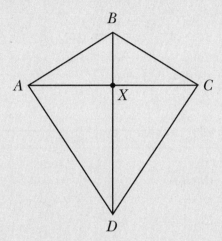

9.43 Prove that the diagonals of *a kite* are perpendicular.

You are given the congruent, consecutive sides of the kite: $\overline{AB} \cong \overline{BC}$ and $\overline{AD} \cong \overline{CD}$. Your goal is to prove $\overline{AC} \perp \overline{BD}$.

Statement	Reason
1. *ABCD* is a kite; $\overline{AB} \cong \overline{BC}$; $\overline{AD} \cong \overline{CD}$	1. Given
2. $\overline{BD} \cong \overline{BD}$	2. Reflexive property of congruency
3. $\triangle BAD \cong \triangle BCD$	3. SSS postulate
4. $\angle ADX \cong \angle CDX$	4. CPCTC
5. $\overline{DX} \cong \overline{DX}$	5. Reflexive property of congruency
6. $\triangle ADX \cong \triangle CDX$	6. SAS postulate
7. $\angle AXD \cong \angle CXD$	7. CPCTC
8. $m\angle AXD + m\angle CXD = 180°$	8. Angle addition postulate
9. $m\angle AXD = m\angle CXD = 90°$	9. If two angles are both congruent and supplementary, then they are right angles
10. $\overline{AC} \perp \overline{BD}$	10. Perpendicular lines intersect at right angles

If the kite is also a rhombus, then both diagonals bisect each other. However, ABCD is definitely not a rhombus.

Note: Problems 9.43–9.45 refer to the diagram in Problem 9.43, in which $\overline{AB} \cong \overline{BC}$ and $\overline{AD} \cong \overline{CD}$.

9.44   Prove that exactly one of the diagonals bisects the other.

Unlike a parallelogram, in which both diagonals bisect each other, only one diagonal of a kite is bisected by the other diagonal. In this diagram, it appears that $\overline{AC}$ is bisected by $\overline{BD}$, because X appears to be the midpoint of $\overline{AC}$, not the midpoint of $\overline{BD}$.

Statement	Reason
1. *ABCD* is a kite; $\overline{AB} \cong \overline{BC}$; $\overline{AD} \cong \overline{CD}$	1. Given
2. $\overline{BD} \cong \overline{BD}$	2. Reflexive property of congruency
3. $\triangle BAD \cong \triangle BCD$	3. SSS postulate
4. $\angle ADX \cong \angle CDX$	4. CPCTC
5. $\overline{DX} \cong \overline{DX}$	5. Reflexive property of congruency
6. $\triangle ADX \cong \triangle CDX$	6. SAS postulate
7. $\overline{AX} \cong \overline{CX}$	7. CPCTC
8. $\overline{BD}$ bisects $\overline{AC}$	8. A segment bisector intersects a segment at its midpoint, dividing the segment into two congruent segments

Note: Problems 9.43–9.45 refer to the diagram in Problem 9.43, in which $\overline{AB} \cong \overline{BC}$ and $\overline{AD} \cong \overline{CD}$.

9.45   Prove that the angles not formed by the adjacent congruent sides of a kite are congruent.

The angle formed by congruent sides $\overline{AB}$ and $\overline{BC}$ is $\angle ABC$; the angle formed by congruent sides $\overline{AD}$ and $\overline{CD}$ is $\angle ADC$. Your goal is to prove that the angles *not* formed by the adjacent congruent sides—the two remaining interior angles of the kite—are congruent: $\angle BAD \cong \angle BCD$.

Statement	Reason
1. *ABCD* is a kite; $\overline{AB} \cong \overline{BC}$; $\overline{AD} \cong \overline{CD}$	1. Given
2. $\overline{BD} \cong \overline{BD}$	2. Reflexive property of congruency
3. $\triangle BAD \cong \triangle BCD$	3. SSS postulate
4. $\angle BAD \cong \angle BCD$	4. CPCTC

9.46 Draw a nonconvex kite.

The proofs in Problems 9.43–9.45 refer to a convex kite. To determine whether a polygon is convex, extend its sides infinitely in each direction. If the resulting lines do not intersect the interior of the polygon, then that polygon is convex; otherwise, the polygon is described as nonconvex.

In other words, stretch the line segments that are the sides of the polygon into actual lines. See Problem 7.42 for more information.

Consider kite *LMNO* in the diagram below. *LMNO* is a kite because it consists of two pairs of adjacent congruent sides: $\overline{OL} \cong \overline{LM}$ and $\overline{ON} \cong \overline{NM}$. If you extended sides $\overline{ON}$ and $\overline{MN}$, the resulting lines would intersect the interior of the kite. Therefore, *LMNO* is a nonconvex kite.

Or "concave."

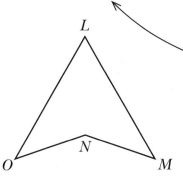

9.47 Given kite *QRST* below, calculate *x* and *y*.

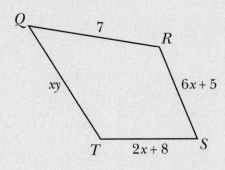

Because *QRST* is a kite, it consists of two pairs of adjacent congruent sides. In this kite, $\overline{RS} \cong \overline{ST}$ and $\overline{QR} \cong \overline{QT}$. Set $RS = ST$ and solve the resulting equation for *x*.

$$6x + 5 = 2x + 8$$
$$6x - 2x = 8 - 5$$
$$4x = 3$$
$$x = \frac{3}{4}$$

Set $QT = QR$ and substitute $x = \dfrac{3}{4}$ into the equation to calculate y.

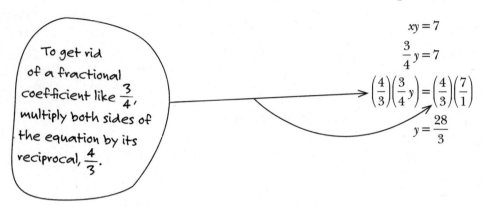

To get rid of a fractional coefficient like $\dfrac{3}{4}$, multiply both sides of the equation by its reciprocal, $\dfrac{4}{3}$.

$$xy = 7$$

$$\frac{3}{4}y = 7$$

$$\left(\frac{4}{3}\right)\left(\frac{3}{4}y\right) = \left(\frac{4}{3}\right)\left(\frac{7}{1}\right)$$

$$y = \frac{28}{3}$$

Chapter 10
INEQUALITIES AND INDIRECT REASONING

Things that aren't equal or aren't true

Whereas the majority of the proofs so far have relied primarily on congruence, you have options when faced with inequality. A strict set of properties govern the manipulation of unequal quantities, much like the familiar properties of equality (such as the transitive, addition, and substitution properties of equality, to name a few) govern the manipulation of equal quantities. After you investigate these axioms, you can apply them to triangles; these proofs are occasionally approached indirectly by the method of contradiction.

In Chapters 8 and 9, you spend a lot of time proving triangles congruent. In this chapter, you won't always know that sides or angles are congruent, so SAS, SSS, ASA, AAS, and HL won't come into play as often. Instead, you'll be given one (or two) triangles and you'll be asked to identify the largest side or the largest angle.

Before you get into that, you need to nail down the properties of inequality. They're based on the equality properties, so many of them feel familiar. The bad news is that many properties don't have well-known names like the equality properties do, which makes citing them in proofs kind of a pain.

One other note: this chapter also includes indirect proof, which, for our purposes, is proof by contradiction. To prove "if p, then q," you assume the opposite of q and show that's impossible. It's a little weird to end a proof on a logical impossibility; getting the wrong answer isn't usually your goal in math, but in this case it is.

The points are described as "distinct," which means they are different. If X, Y, and Z can't overlap, then all the segments will have some sort of positive length. If $XY \neq 0$ and $YZ \neq 0$, then both are shorter than XZ.

Properties of Inequalities
Distant cousins of equality properties

10.1 Complete the following statement and justify your answer.

Given distinct collinear points X, Y, and Z, assume XY + YZ = XZ. Therefore, XY _____ XZ.

The segment addition postulate enables you to express a segment, such as $\overline{XZ}$, in terms of its parts. In this case, combining $\overline{XY}$ and $\overline{YZ}$, two collinear segments that share endpoint Y, produces the longer segment $\overline{XZ}$. Logically, the whole segment $\overline{XZ}$ is longer than either of its component parts, so *XY* <u>is less than</u> (<) *XZ*.

10.2 Complete the following statement and justify your answer.

If a > b and b > c, then _____.

According to the transitive property of inequality, if $a > b$ and $b > c$, then <u>$a > c$</u>. To better visualize this property, assign temporary values to a, b, and c. For instance, let $a = 10$, $b = 5$, and $c = 2$. It is true that $a > b$ ($10 > 5$) and $b > c$ ($5 > 2$). Therefore, it is also true that $a > c$ ($10 > 2$).

10.3 Complete the following statement and justify your answer.

If j < k and j = g, then _____.

The substitution property of inequality allows you to substitute one equivalent quantity for another in an inequality statement. In this case, $j = g$, so you can replace j with g in the statement: if $j < k$ and $j = g$, then <u>$g < k$</u>.

Think of a playground seesaw where the kid on the left weighs less than the kid on the right. Two identical twins (weighing exactly the same as each other) jump onto the seesaw, one on each side. The total weight of the left side is still less than the total weight of the right side.

10.4 Complete the following statement and justify your answer.

If m∠1 < m∠2 and m∠3 = m∠4, then m∠1 + m∠3 _____ m∠2 + m∠4.

If equal quantities ($m\angle 3 = m\angle 4$) are added to unequal quantities ($m\angle 1 < m\angle 2$), then the resulting sums are unequal in the same order. Therefore, if $m\angle 1 < m\angle 2$ and $m\angle 3 = m\angle 4$, then $m\angle 1 + m\angle 3$ <u>*is less than*</u> (<) $m\angle 2 + m\angle 4$.

10.5 Complete the following statement and justify your answer.

If AB < CD and EF < GH, then _____ + _____ < _____ + _____.

If unequal quantities are added to unequal quantities in the same order (that is, the lesser quantities are combined and the greater quantities are combined), then the sums are unequal in the same order. In this case, you are given $AB < CD$ and $EF < GH$. The sum of the lesser values in each inequality statement ($AB + EF$) is less than the sum of the greater values in each inequality statement ($CD + GH$). Therefore, if $AB < CD$ and $EF < GH$, then <u>$AB + EF < CD + GH$</u>.

10.6 Complete the following statement and justify your answer.

If a = b and c < d, then c − a _____ d − b.

If equal quantities are subtracted from unequal quantities, then the differences are unequal in the same order. Visualize a seesaw with sandbags on either seat, such that the total weight of the left side is less than the weight on the right (*c < d*). If an equal number of sandbags are taken from each side (*a* are taken from the left side, *b* are taken from the right, and *a = b*), then the left side remains the lighter side. Therefore, if *a = b* and *c < d*, then *c − a* <u>is less than</u> (<) *d − b*.

10.7 Complete the following statement and justify your answer.

If x = y and m < n, then x − m _____ y − n.

To better understand the mechanics of this statement, assign integer values to the variables. For instance, *x* = 10 and *y* = 10 satisfy the first inequality statement (*x = y*); and *m* = 3 and *n* = 5 satisfy the second (*m < n*). Calculate the differences as directed.

$$x - m \underline{\quad} y - n$$
$$10 - 3 \underline{\quad} 10 - 5$$
$$7 > 5$$

You're subtracting *m* and *n* from *x* and *y*, so make *x* and *y* larger than *m* and *n*, unless you feel like dealing with negative numbers.

When you subtract unequal values from equal values, the differences are unequal in the opposite order. In this care, you were given *m < n*, so the left value was less than the right value. When you calculate the differences, the opposite is true; the left difference is greater than the right difference.

Therefore, if *x = y* and *m < n*, then *x − m* <u>is greater than</u> (>) *y − n*.

This is the opposite of Problem 10.6, in which you subtract equal values from unequal values.

10.8 Complete the following statement and justify your answer.

If C > D and X is a positive number, then CX ___ DX.

As suggested by previous problems, assign real number values to the variables. For instance, *C* = 4 and *D* = 1 satisfy the inequality because 4 > 1. Let *X* be a positive number, such as *X* = 9. Calculate the products as directed.

$$CX \underline{\quad} DX$$
$$4(9) \underline{\quad} 1(9)$$
$$36 > 9$$

Think of it this way: the bigger the number you subtract, the smaller the result.

The original inequality (4 > 1) and the product statement (36 > 9) have the same sign (>), and the statements are unequal in the same order. Therefore, if unequal values are multiplied by the same positive number, the result is unequal in the same order. Symbolically, if *C > D* and *X* is a positive number, then *CX* <u>is greater than</u> (>) *DX*.

Think of a number line. Whichever number is farther left is the lesser number. Because –20 is farther left than –10, –20 < –10 and –10 > –20.

10.9 Complete the following statement and justify your answer.

If A < B and Y is a negative number, then AY ___ BY.

Choose real number values to represent the variables. The values $A = 5$ and $B = 10$ satisfy the first inequality ($5 < 10$). Let Y be any negative number, such as $Y = -2$. Calculate the products as directed.

$$AY \underline{\quad} BY$$
$$5(-2) \underline{\quad} 10(-2)$$
$$-10 > -20$$

The original inequality statement ($5 < 10$) and the product ($-10 > -20$) have opposite inequality symbols. Therefore, when you multiply equal values by a negative number, the products are unequal in the opposite order. Symbolically, if $A < B$ and Y is a negative number, then AY <u>is greater than</u> (>) BY.

10.10 Complete the following statement and justify your answer.

If a > 0, b < 0, and x ≥ y, then $\dfrac{x}{a} \underline{\quad} \dfrac{y}{a}$ and $\dfrac{x}{b} \underline{\quad} \dfrac{y}{b}$.

Dividing both sides of an equation by a or b is equivalent to multiplying both sides of the equation by $\dfrac{1}{a}$ or $\dfrac{1}{b}$. Therefore, the same inequality properties that applied to multiplication in Problems 10.8 and 10.9 apply to division.

Dividing unequal values by a positive number produces unequal quotients in the same order, whereas dividing unequal values by a negative number produces unequal values in the opposite order. In this case, dividing by the positive number a results in a statement with the same inequality symbol as the original statement ($\geq$); dividing by the negative number b produces a statement with the opposite inequality symbol ($\leq$).

Therefore, if $a > 0$, $b < 0$, and $x \geq y$, then $\dfrac{x}{a} \geq \dfrac{y}{a}$ and $\dfrac{x}{b} \leq \dfrac{y}{b}$.

10.11 Explain why dividing both sides of an inequality by a negative number reverses the inequality symbol. Use an inequality containing real numbers to illustrate your answer.

Consider the inequality statement $2 < 8$. If you divide both sides of the inequality by 2, the result is $1 < 4$. Dividing by a positive number does not affect the inequality symbol; the left side remains the lesser value when both sides are divided by a positive number.

However, when you divide both sides of the original inequality by –2, the result is $-1 > -4$. Disregarding the sign, the "larger" a negative number is, the more negative its value. Because 4 is greater than 1, the opposite is true when you take the opposite of each value: –4 is less than –1.

10.12 Given $a > b$, $b > c$, and $c = 5$, complete the following proof to demonstrate that $a > 5$.

Statement	Reason
1. $a > b$; $b > c$; $c = 5$	1. Given
2. $b > 5$	2.
3. $a > 5$	3.

Statement	Reason
1. $a > b$; $b > c$; $c = 5$	1. Given
2. $b > 5$	2. Substitution property of inequality (Step 1)
3. $a > 5$	3. Transitive property of inequality

Substitute $c = 5$ into $b > c$.

If $a > b$ and $b > 5$, then $a > 5$.

10.13 Complete the following proof to solve the inequality $2x + 3 \leq 4$.

Statement	Reason
1. $2x + 3 \leq 4$	1. Given
2. $2x \leq 1$	2.
3. $x \leq \frac{1}{2}$	3.

Statement	Reason
1. $2x + 3 \leq 4$	1. Given
2. $2x \leq 1$	2. If equal quantities are subtracted from unequal quantities, the differences are unequal in the same order
3. $x \leq \frac{1}{2}$	3. If unequal values are divided by a positive number, the quotients are unequal in the same order

Subtracting 3 from both sides of $2x + 3 \leq 4$ doesn't change the inequality symbol (see Problem 10.6).

Dividing both sides of an inequality by a positive number doesn't change the inequality symbol (see Problem 10.10).

10.14 Construct a proof to solve the inequality: $-3(x-2) \geq x-5$

Statement	Reason
1. $-3(x-2) \geq x-5$	1. Given
2. $-3x+6 \geq x-5$	2. Distributive property of equality
3. $-3x-x \geq -5-6$	3. If equal quantities are subtracted from unequal quantities, the differences are unequal in the same order
4. $-4x \geq -11$	4. Combination of like terms
5. $x \leq \dfrac{11}{4}$	5. If unequal quantities are divided by a negative number, the quotients are unequal in the opposite order

> Even though this is an inequality problem, Step 2 states that $-3(x-2)$ EQUALS $-3x+6$; distribution is a property of equality.

10.15 Given N is the midpoint of $\overline{LP}$ and $LM < NO$ in the diagram below, prove $MN > OP$.

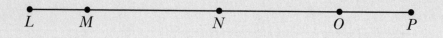

Statement	Reason
1. N is the midpoint of $\overline{LP}$; $LM < NO$	1. Given
2. $LN = NP$	2. A midpoint divides a segment into two congruent segments
3. $LM + MN = LN$; $NO + OP = NP$	3. Segment addition postulate
4. $LM + MN = NO + OP$	4. Substitution property of equality (Steps 2, 3)
5. $MN > OP$	5. If unequal values are subtracted from equal values, the differences are unequal in the opposite order

> This is the property stated in Problem 10.7. Step 4 contains two equal sums. Then you subtract LM from the left side and NO from the right side. You subtracted a smaller value on the left (LM), which makes the result bigger: MN > OP.

Indirect Proof

Proof by contradiction

Note: Problems 10.16–10.17 refer to rectangle ABCD below.

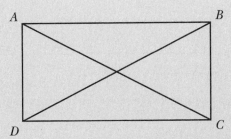

10.16 When proving that the diagonals of *ABCD* are congruent, what additional assumption is used in an indirect proof, if compared to a direct proof of the same theorem?

The majority of the proofs in the preceding chapters of this book are direct proofs, which begin by stating the given information. In this case, you are given only that *ABCD* is a rectangle. You then progress through a series of statements and reasons to determine that $\overline{AC} \cong \overline{BD}$ (the diagonals are congruent).

An indirect proof takes a slightly different approach. Instead of proving that the diagonals are congruent, an indirect proof demonstrates that it is impossible for the diagonals *not* to be congruent. In an indirect proof, you include one additional assumption—the opposite of what you are attempting to prove.

In this case, besides assuming that *ABCD* is a rectangle, you also assume that $\overline{AC} \not\cong \overline{BD}$.

> The direct proof that the diagonals of a rectangle are congruent is in Problem 9.17.

Note: Problems 10.16–10.17 refer to the diagram in Problem 10.16.

10.17 Use an indirect proof to demonstrate that the diagonals of *ABCD* must be congruent.

As Problem 10.16 states, to prove indirectly that the diagonals of rectangle *ABCD* are congruent, you temporarily assume that the diagonals are not congruent. You then proceed with the proof in an attempt to reach a logical contradiction, which demonstrates that the opposite of the temporary assumption must be true.

Statement	Reason
1. *ABCD* is a rectangle; $\overline{AC} \not\cong \overline{BD}$	1. Given
2. $\overline{AD} \cong \overline{BC}$	2. The opposite sides of parallelograms (including rectangles) are congruent
3. $\angle ADC \cong \angle BCD$	3. The interior angles of a rectangle measure 90° and are therefore congruent

> To prove that I am NOT the tallest person in the world, I would temporarily assume that I WAS the tallest person in the world. However, as soon as I found someone taller than me, it would violate the temporary assumption and prove that I am NOT the tallest person alive.

continues

continued

Statement	Reason
4. $\overline{DC} \cong \overline{CD}$	4. Reflexive property of congruency
5. $\triangle ADC \cong \triangle BCD$	5. SAS postulate
6. $\overline{AC} \cong \overline{BD}$	6. CPCTC

> An indirect proof always BEGINS with the opposite of what you're trying to prove, but it doesn't always END with it (as it does in this problem). It's usually a bit more subtle, as you'll see in the next few problems.

The proof states that $\overline{AC} \cong \overline{BD}$, which directly contradicts the assumption that $\overline{AC} \not\cong \overline{BD}$. Therefore, you must conclude that the opposite of the temporary assumption is true: $\overline{AC} \cong \overline{BD}$.

10.18 Prove that a triangle cannot contain two obtuse angles.

The problem provides no diagram, nor is one needed. However, providing the name of a triangle is useful for reference. For instance, the following proof refers to a generic triangle XYZ. You are asked to demonstrate that $\triangle XYZ$ cannot contain two obtuse angles. To prove this statement indirectly, temporarily assume that $\triangle XYZ$ contains two obtuse angles. For instance, assume that $\angle X$ and $\angle Y$ are obtuse.

Statement	Reason
1. $\angle X$ and $\angle Y$ in $\triangle XYZ$ are obtuse	1. Given
2. $m\angle X + m\angle Y + m\angle Z = 180°$	2. The sum of the interior angle of a triangle is 180°
3. $m\angle X > 90°$; $m\angle Y > 90°$	3. Obtuse angles have measures between 90° and 180°
4. $m\angle X + m\angle Y > 180°$	4. If unequal values are added to unequal values in the same order, the sums are unequal in the same order
5. $m\angle X + m\angle Y > m\angle X + m\angle Y + m\angle Z$	5. Substitution property of inequality (Steps 2, 4)
6. $0 > m\angle Z$	6. If equal values are subtracted from unequal values, the differences are unequal in the same order

> Step 2 states that the sum of all three angles equals 180°. Substitute that sum for 180° in Step 4.

> Subtract $m\angle X$ and $m\angle Y$ from both sides of the inequality.

Step 6 of the proof states that $0 > m\angle Z$; in other words, $m\angle Z < 0$, so $m\angle Z$ is a negative number. Angle measurements are positive values, so this is a logical contradiction and the proof is complete. No triangle may have two obtuse angles, or the measure of the third angle in the triangle would have to be negative for the sum of the angles to be 180°.

10.19 Use the diagram below to prove that the exterior angle of a triangle cannot be less than either of the remote interior angles.

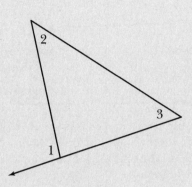

The diagram contains an exterior angle (∠1) and the corresponding remote interior angles (∠2 and ∠3). An indirect proof assumes the opposite of that which you are trying to prove, so temporarily assume that the exterior angle *can* be less than one of the remote interior angles: $m\angle 1 < m\angle 2$.

Statement	Reason
1. ∠1 is the exterior angle of a triangle; ∠2 and ∠3 are the remote interior angles; $m\angle 1 < m\angle 2$	1. Given
2. $m\angle 1 = m\angle 2 + m\angle 3$	2. The measure of an exterior angle is equal to the sum of the remote interior angles
3. $0 > m\angle 3$	3. If you subtract unequal values from equal values, the differences are unequal in the opposite order ←

See Problem 7.29.

$m\angle 1 < m\angle 2$, so if you subtract $m\angle 1$ from the left side of the equation in Step 2 and subtract $m\angle 2$ from the right side, you must reverse the inequality symbol.

As in Problem 10.18, you have reached the impossible conclusion that one of the angles of the triangle has a negative measurement. To prove that ∠1 cannot be smaller than ∠3, simply replace ∠2 with ∠3 (and vice versa) in the proof.

10.20 Prove that no convex quadrilateral *ABCD* exists such that $m\angle A = 100°$, $m\angle B = 105°$, $m\angle C = 80°$, and $m\angle D = 70°$.

To prove this statement indirectly, temporarily assume that some convex quadrilateral *ABCD* exists with those exterior angles.

Statement	Reason
1. *ABCD* is a convex quadrilateral; $m\angle A = 100°$; $m\angle B = 105°$; $m\angle C = 80°$; $m\angle D = 70°$	1. Given

continues

continued

Statement	Reason
2. $m\angle A + m\angle B + m\angle C + m\angle D = 360°$	2. The sum of the interior angles of a convex polygon with n sides is $180°(n-2)$
3. $100° + 105° + 80° + 70° = 360°$	3. Substitution property of inequality (Steps 1, 2)
4. $355° = 360°$	4. Combination of like terms

> See Problem 7.47.

The sum of the interior angles of a convex quadrilateral is 360°, but the sum of this particular set of interior angles is 355°. The proof reaches the contradictory statement 355° = 360°, so the opposite of the temporary assumption is true: no such quadrilateral *ABCD* exists.

10.21 Use the diagram below to prove the following statement: if a trapezoid has supplementary opposite angles, then it is isosceles.

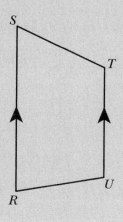

Temporarily assume the opposite of the conclusion: given a trapezoid with supplementary opposite angles, the legs are not necessarily congruent. In the case of trapezoid *RSTU*, assume that $\angle R$ and $\angle T$ are supplementary, $\angle S$ and $\angle U$ are supplementary, and $\overline{ST} \not\cong \overline{RU}$.

> $\angle R$ and $\angle S$ are both supplements of $\angle T$; $\angle T$ and $\angle U$ are both supplements of $\angle R$.

Statement	Reason
1. $\overline{SR} \parallel \overline{TU}$; $\overline{ST} \not\cong \overline{RU}$; $\angle R$ and $\angle T$ are supplementary; $\angle S$ and $\angle U$ are supplementary	1. Given
2. $\angle S$ and $\angle T$ are supplementary; $\angle R$ and $\angle U$ are supplementary	2. Two parallel lines intersected by a transversal form supplementary same-side interior angles
3. $\angle R \cong \angle S$; $\angle T \cong \angle U$	3. Two angles supplementary to the same angle are congruent

4. *RSTU* is an isosceles trapezoid	4. If both pairs of base angles of a trapezoid are congruent, the trapezoid is isosceles ←

See Problem 9.40. By the way, you can't do a one-step proof and say "if opposite angles are supplementary, it's isosceles," because that's exactly what the book is asking you to prove. That would be like proving that a banana is yellow by saying "bananas are yellow." You can't prove something by repeating it.

Step 4 contradicts the temporary assumption that the legs of the trapezoid are not congruent; an isosceles trapezoid has congruent legs, so $\overline{ST} \cong \overline{RU}$. Therefore, the proof is complete.

10.22 Use the diagram below to prove that a scalene triangle cannot contain two congruent angles.

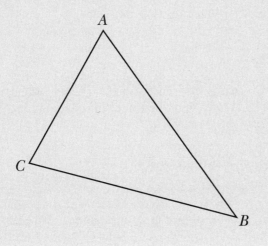

Temporarily assume that $\angle A \cong \angle B$.

Statement	Reason
1. $\triangle ABC$ is scalene; $\angle A \cong \angle B$	1. Given
2. $\overline{AB} \not\cong \overline{CB}$; $\overline{AB} \not\cong \overline{AC}$; $\overline{CB} \not\cong \overline{AC}$	2. The legs of a scalene triangle have different lengths
3. $\overline{CB} \cong \overline{AC}$	3. Converse of the isosceles triangle theorem

The proof contains the contradictory statements $\overline{CB} \cong \overline{AC}$ and $\overline{CB} \not\cong \overline{AC}$, so the proof is complete. A scalene triangle cannot contain a pair of congruent angles, as the sides opposite those angles would be congruent. By definition, a scalene triangle has no congruent sides.

Single Triangle Inequality Theorems
Bigger angles mean bigger sides

10.23 State the triangle inequality theorem.

The sum of the lengths of two sides of a triangle must be greater than the length of the third side.

10.24 Use the diagram below to construct three inequality statements based on the triangle inequality theorem.

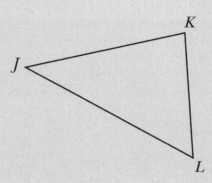

The triangle inequality theorem states that the sum of any two sides of a triangle must be greater than the length of the remaining side: $JK + KL > JL$, $KL + LJ > JK$, and $LJ + JK > KL$.

10.25 Complete the following statement.

If two sides of a triangle have different lengths, then the angles opposite those sides _____. Specifically, the measure of the angle opposite the larger side is _____ than the measure of the angle opposite the smaller side.

If two sides of a triangle have different lengths, then the angles opposite those sides <u>are not congruent</u>. Specifically, the measure of the angle opposite the larger side is <u>greater</u> than the measure of the angle opposite the smaller side.

> The larger the side of a triangle is, the larger the opposite angle. That means the largest angle of a triangle is opposite the largest side of the triangle (and vice versa).

10.26 Is it possible to construct a triangle with sides of the following lengths: 4 cm, 10 cm, and 12 cm?

According to the triangle inequality theorem, adding the lengths of any two sides of a triangle must produce a sum that is greater than the length of the remaining side. There are three ways to choose two out of a triangle's three sides, so three sums must be calculated. Verify that each sum is greater than the length of the side not included in the sum.

> If a triangle has side lengths A, B, and C, then you must make sure that $A + B > C$, $B + C > A$, and $A + C > B$.

$$4 + 10 > 12 \qquad 10 + 12 > 4 \qquad 4 + 12 > 10$$
$$14 > 12 \qquad\qquad 22 > 4 \qquad\qquad 16 > 10$$
$$\text{True} \qquad\qquad \text{True} \qquad\qquad \text{True}$$

All three sums satisfy the triangle inequality theorem, so it is possible to construct a triangle with sides that measure 4 cm, 10 cm, and 12 cm.

10.27 Is it possible to construct a triangle with sides of the following lengths: 13 cm, 29 cm, and 42 cm?

Follow the procedure outlined in Problem 10.26, adding the lengths of two sides of the triangle to verify that the sum is greater than the remaining side.

$$13 + 29 > 42 \qquad 29 + 42 > 13 \qquad 13 + 42 > 29$$
$$42 > 42 \qquad\quad 71 > 13 \qquad\quad 55 > 29$$
$$\textbf{False} \qquad\qquad \text{True} \qquad\qquad \text{True}$$

Because $13 + 29 \not> 42$, the given side lengths violate the triangle inequality theorem and no such triangle can be constructed.

10.28 Given a triangle with sides 6 and 11 units long, identify the range of possible lengths for the third side of the triangle.

Let x represent the length of the remaining side of the triangle. According to the triangle inequality theorem, the sum of any two sides of the triangle must be greater than the third side.

$$x + 6 > 11 \qquad x + 11 > 6 \qquad 6 + 11 > x$$

Solve each of the inequalities for x.

$$x > 11 - 6 \qquad x > 6 - 11 \qquad 17 > x$$
$$x > 5 \qquad\quad x > -5 \qquad\quad x < 17$$

> When you reverse the sides of an inequality, you also must reverse the inequality symbol.

Remove the unnecessary repetition from the final inequality statements. If x must be greater than −5 *and* greater than 5, it is sufficient to state that $x > 5$ because any number greater than 5 is also greater than −5.

Therefore, $x > 5$ and $x < 17$. (You may write this as the compound inequality statement $5 < x < 17$.) The third side of the triangle must be longer than 5 units and shorter than 17 units.

10.29 Given a triangle with sides 4 and 7 units long, identify the range of possible lengths for the third side of the triangle.

Apply the technique described in Problem 10.28: use the triangle inequality theorem to construct three inequality statements (in which x represents the length of the third side) and solve those statements.

$$x + 4 > 7 \qquad x + 7 > 4 \qquad 4 + 7 > x$$
$$x > 7 - 4 \qquad x > 4 - 7 \qquad 11 > x$$
$$x > 3 \qquad\quad x > -3 \qquad\quad x < 11$$

> You can also say that $x > 3$ and $x < 11$, but be sure to include the word "and" because x must satisfy both of those inequality statements.

Notice that any number greater than 3 is also greater than −3, so the statement $x > -3$ may be omitted from the solution. The range of possible values for the third side of the triangle is $3 < x < 11$.

So don't base your answer on how big the angles look. Otherwise, you wouldn't have to know anything about geometry to get this problem correct.

10.30 List the measures of the angles below from least to greatest. Note that the diagram may not be drawn to scale.

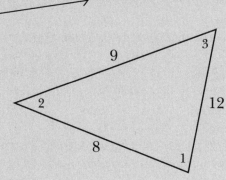

According to Problem 10.25, the longer a triangle's side, the larger the opposite angle. This triangle has sides of length 8, 9, and 12. The shortest side has length 8, so the smallest angle is opposite that side: $\angle 3$. The side with length 9 is slightly larger, so its opposite angle ($\angle 1$) is as well. The longest side of the triangle measures 12 units, so its opposite angle is the largest angle of the triangle: $\angle 2$. Hence, $m\angle 3 < m\angle 1 < m\angle 2$.

10.31 List the lengths of the sides of the triangle below in order, from least to greatest. Note that the diagram may not be drawn to scale.

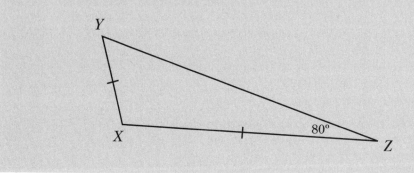

According to the diagram, $\overline{XY} \cong \overline{XZ}$, so ΔXYZ is isosceles. According to the isosceles triangle theorem, base angles are congruent, so $m\angle Y = m\angle Z = 80°$. Recall that the measures of the angles of a triangle have a sum of 180°.

$$m\angle X + m\angle Y + m\angle Z = 180°$$

Substitute the measures of $\angle Y$ and $\angle Z$ into the equation and solve for $m\angle X$.

$$m\angle X + 80° + 80° = 180°$$
$$x\angle X + 160° = 180°$$
$$m\angle X = 180° - 160°$$
$$m\angle X = 20°$$

Because $\angle X$ is the smallest angle of the triangle, $\overline{YZ}$ (the side opposite $\angle X$) is the shortest side of the triangle. The legs of the isosceles triangle are congruent $\overline{XY} \cong \overline{XZ}$. Therefore, $YZ < XY$ and $XY = XZ$. ⟵

10.32 List the lengths of the sides of the triangle below in order, from least to greatest. Note that the diagram may not be drawn to scale.

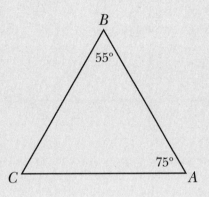

You can't technically list all three from least to greatest, because two of the sides are the same length. Basically, the answer is that ZY is the shortest length because XY and XZ are equal and bigger than YZ.

The diagram provides two of the angle measurements. Calculate the measure of the third angle, recalling that all three measurements must have a sum of 180°.

$$m\angle A + m\angle B + m\angle C = 180°$$
$$75° + 55° + m\angle C = 180°$$
$$130° + m\angle C = 180°$$
$$m\angle C = 180° - 130°$$
$$m\angle C = 50°$$

The measures of the angles, in order from least to greatest, are $m\angle C = 50°$, $m\angle B = 55°$, and $m\angle A = 75°$. The lengths of the sides opposite these angles increase in the same order. Therefore, $AB < AC < BC$.

10.33 Use the diagram below to prove that the shortest distance between a line and a point not on that line is the perpendicular segment connecting them.

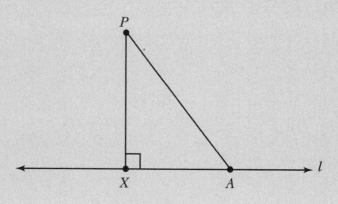

As long as A and X are separate points. If they've the same, then both distances are the same.

Your goal is to prove that the shortest distance between line l and point P is XP, the length of the perpendicular segment connecting them. A second distance between the point and the line is provided (AP), but A may be located anywhere on the line and the proof will remain true. Therefore, any nonperpendicular segment $\overline{AP}$ may be used without affecting the truth of the assertion that the perpendicular distance is the shortest.

See Problem 10.18. If you have two right angles or two obtuse angles, the sum of the angles of a triangle can't be 180°.

Statement	Reason
1. $\overline{XP} \perp l$	1. Given
2. $\angle PXA$ is a right angle	2. Perpendicular lines intersect at right angles
3. $m\angle PAX < m\angle PXA$	3. A triangle may include no more than one right angle
4. $PX < PA$	4. If two angles of one triangle are not congruent, the side opposite the smaller angle is shorter than the side opposite the larger angle

In the diagram, PX and PA represent two different ways to measure distance between the point and the line. The shortest distance is PX, the distance measured along the line perpendicular to l that passes through P.

Note: Problems 10.34–10.37 refer to the diagram below.

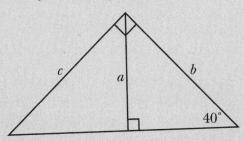

10.34 Determine which value is greater: *a* or *b*. Note that the diagram is not necessarily drawn to scale.

Consider the right triangle in the diagram with hypotenuse *b* and leg *a*, as illustrated below.

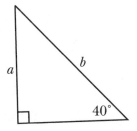

Side *a* is opposite a 40° angle, whereas *b* is opposite a 90° angle. Therefore, *b* > *a*.

Note: Problems 10.34–10.37 refer to the diagram in Problem 10.34.

10.35 Determine which value is greater: *b* or *c*. Note that the diagram is not necessarily drawn to scale.

Consider the right triangle in the diagram with legs *b* and *c*, as illustrated below.

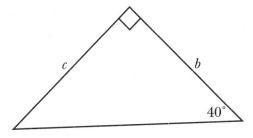

Calculate the missing angle measurement opposite side *b* by adding the interior angles of the right triangle and setting the sum equal to 180°. Let *x* represent the measure of the unknown angle.

$$x + 40° + 90° = 180°$$
$$x + 130° = 180°$$
$$x = 180° - 130°$$
$$x = 50°$$

Side c is opposite a 40° angle, and side b is opposite a 50° angle. Therefore, $b > c$.

Note: Problems 10.34–10.37 refer to the diagram in Problem 10.34.

10.36 Determine which value is greater: a or c. Note that the diagram is not necessarily drawn to scale.

Consider the right triangle with hypotenuse c and leg a, as illustrated below. Notice that the 50° angle, as calculated in Problem 10.35, is included.

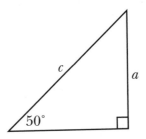

Side c is opposite a 90° angle, so it must be longer than side a, which is opposite a 50° angle: $c > a$.

Note: Problems 10.34–10.37 refer to the diagram in Problem 10.34.

10.37 List a, b, and c in order, from least to greatest. Note that the diagram is not necessarily drawn to scale.

According to Problem 10.34, $a < b$; Problem 10.36 states that $a < c$, so a is the smallest value. Problem 10.35 states that $c < b$, so b is the largest value. Therefore, $a < c < b$.

By doing Problems 10.34–10.36, you find out that a is smaller than everything (b > a and c > a). You also find out that b is bigger than everything (b > a and b > c). That means c falls in the middle: a is less than c, which is less than b.

10.38 Given $m\angle 1 < m\angle 3$ and $m\angle 2 > m\angle 4$ in the following diagram, list OM, MN, and LM in ascending order. Note that the diagram is not necessarily drawn to scale.

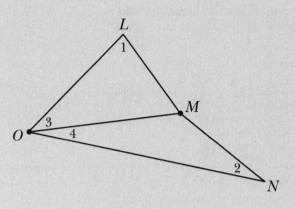

If $m\angle 1 < m\angle 3$ in $\triangle LMO$, then the side opposite $\angle 1$ is shorter than the side opposite $\angle 3$: $OM < LM$. If $m\angle 2 > m\angle 4$ in $\triangle OMN$, then the side opposite $\angle 2$ is longer than the side opposite $\angle 4$: $OM > MN$. Combine the inequality statements into a single compound inequality: $MN < OM < LM$.

10.39 Given $\square ABCD$ in the diagram below and $AD > CY$, prove that $m\angle 1 < m\angle 2$. Note that the diagram is not necessarily drawn to scale.

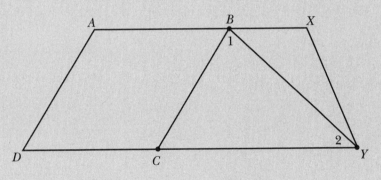

Statement	Reason
1. $ABCD$ is a parallelogram; $AD > CY$	1. Given
2. $AD = BC$	2. Opposite sides of a parallelogram are congruent
3. $BC > CY$	3. Substitution property of inequality (Steps 1, 2)
4. $m\angle 1 < m\angle 2$	4. If two sides of one triangle are not congruent, the angle opposite the smaller side is smaller than the angle opposite the larger side

Dual Triangle Inequality Theorems

SAS, SSS inequality theorems

10.40 Given the diagram below, determine which value is greater: *x* or *y*. Note that the diagram is not necessarily drawn to scale.

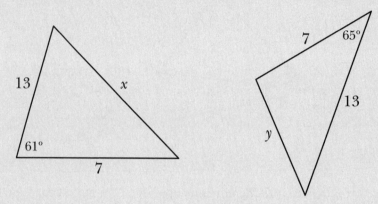

According to the SAS inequality theorem, if two sides of one triangle are congruent to two sides of another triangle, then the side opposite the larger included angle is longer than the side opposite the smaller included angle. In this diagram, two triangles share two congruent sides; both have sides with length 7 and 13. Consider the angles formed by those sides in each triangle: 61° and 65°. Because 65 > 61, the side opposite the 65° angle is larger than the side opposite the 61° angle: $y > x$.

10.41 Given the diagram below, determine which value is greater: *x* or *y*. Note that the diagram is not necessarily drawn to scale.

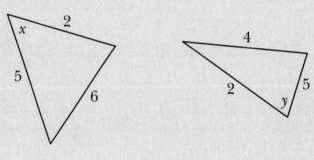

According to the SSS inequality theorem, if two sides of one triangle are congruent to two sides of another triangle, then the angle opposite the longer remaining side is larger than the angle opposite the shorter remaining side. In this diagram, both triangles have sides that measure 2 and 5 units in length. However, the remaining side of the left triangle has length 6, whereas the remaining side of the right triangle has length 4. Because 6 > 4, the angle opposite the longer side is larger than the angle opposite the shorter side: $x > y$.

Note: Problems 10.42–10.43 refer to the following diagram, in which a = d.

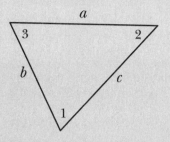 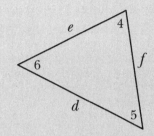

To apply the SAS inequality theorem, the congruent sides must form the angles you're referencing. In this case, a and b form ∠3, and d and e form ∠6.

10.42 What must be true of the sides and angles of the two triangles to conclude that $c < f$ according to the SAS inequality theorem?

To conclude that $c < f$, the angle opposite c must be smaller than the angle opposite f: $m\angle3 < m\angle6$. Furthermore, the sides that form $\angle3$ must be congruent to the sides that form $\angle6$. You are given $a = d$, so $b = e$ as well. ←

Note: Problems 10.42–10.43 refer to the diagram in Problem 10.42, in which a = d.

10.43 What must be true of the sides and angles of the two triangles to conclude that $m\angle2 > m\angle5$ according to the SSS inequality theorem?

If $m\angle2 > m\angle5$, then the side opposite $\angle2$ must be larger than the side opposite $\angle5$: $b > e$. The sides that form $\angle2$ must be congruent to the sides that form $\angle5$. You are given $a = d$, so $c = f$ as well.

10.44 If the perimeter of $\triangle ABC$ is greater than the perimeter of $\triangle XYZ$ in the diagram below, what must be true according to the SSS inequality theorem?

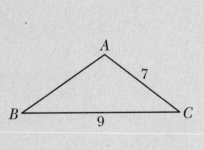 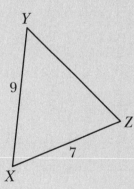

The perimeter of a geometric figure is the sum of the lengths of its sides.

perimeter of $\triangle ABC$: $AB + 7 + 9 = AB + 16$

perimeter of $\triangle XYZ$: $YZ + 7 + 9 = YZ + 16$

The perimeter of $\triangle ABC$ is greater than the perimeter of $\triangle XYZ$.

$AB + 16 > YZ + 16$

If an equal value (16) is subtracted from unequal values, the differences are unequal in the same order. Thus, the inequality symbol remains unchanged.

$$AB > YZ$$

AC = XZ = 7 and BC = XY = 9

Two sides of $\triangle ABC$ are congruent to two sides of $\triangle XYZ$. According to the SSS inequality theorem, the angle opposite the longer remaining side ($\angle C$ is opposite $\overline{AB}$) is larger than the angle opposite the shorter remaining side ($\angle X$ is opposite $\overline{YZ}$). Therefore, $m\angle C > m\angle X$.

10.45 Given $JK = JL = MN = MO$ in the following diagram, prove $KL > NO$.

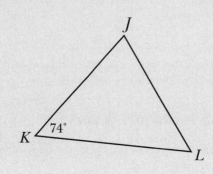
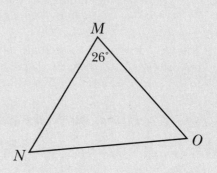

Statement	Reason
1. $JK = JL = MN = MO$; $m\angle K = 74°$; $m\angle M = 26°$	1. Given
2. $\angle K \cong \angle L$	2. Isosceles triangle theorem
3. $m\angle J + m\angle K + m\angle L = 180°$	3. The sum of the measures of the angles of a triangle is 180°
4. $m\angle J + 74° + 74° = 180°$	4. Substitution property of equality (Steps 2, 3)
5. $m\angle J + 148° = 180°$	5. Combination of like terms
6. $m\angle J = 32°$	6. Subtraction property of equality
7. $KL > NO$	7. SAS inequality theorem

You CAN'T conclude that $\angle K$, $\angle L$, $\angle N$, and $\angle O$ are congruent. Otherwise the triangles would be congruent, and if KL > NO, that can't be true.

Two sides of $\triangle JKL$ are congruent to two sides of $\triangle MNO$ (actually, all four of those sides are congruent). Included angle J is larger than included angle M (32 > 26), so the side opposite J is longer than the side opposite M.

Chapter 11
SIMILARITY

Figures that are the same shape but different sizes

Congruent figures are alike in every geometric sense—corresponding angles and sides are congruent, so the resulting figures are congruent. Similar figures share one of these defining characteristics: corresponding angles of similar figures are congruent. The corresponding sides of similar figures, however, are in the same proportion.

This chapter begins with a discussion of ratio and proportion and then graduates to the study of similar polygons. Of specific interest are similar triangles and the consequences of similar triangles on other geometric concepts.

Congruent figures are exact copies of each other. Similar figures are also copies, assuming that you played with the zoom setting on the copier beforehand. Similar figures have the same shape, but one is usually a bigger version of the other.

Mathematically speaking, two figures are similar if they have congruent angles in the same corresponding locations and segments that are in the same proportion. In other words, the sides of polygon A might be twice as long as the corresponding sides of polygon B. In that case, the ratio of the lengths is 2:1.

The chapter begins with ratios, so you'll learn what things like "2:1" mean. It then moves to proportions (two fractions set equal) and similar triangles (three ways you can prove that two triangles are similar), and ends with a grab bag of other similarity theorems.

Ratio and Proportion

x to y, $x{:}y$, or $\dfrac{x}{y}$

> *Problems 11.1–11.4 refer to a farm that breeds livestock. The entire animal population consists of 12 horses, 16 cows, and 36 pigs.*

In other words, write the proportion as a fraction.

11.1 Express the ratio of horses to cows as a rational number in lowest terms.

The farm has 12 horses and 16 cows. Express the ratio as the quotient of the horse and cow population.

$$\frac{\text{horses}}{\text{cows}} = \frac{12}{16}$$

Reduce the fraction to lowest terms.

$$\frac{12}{16} = \frac{3}{4}$$

The ratio of horses to cows is $\dfrac{3}{4}$.

> *Problems 11.1–11.4 refer to a farm that breeds livestock. The entire animal population consists of 12 horses, 16 cows, and 36 pigs.*

a:b is read "a to b."

11.2 Express the ratio of pigs to horses in the form $a{:}b$. Write the ratio in simplest form.

The livestock population includes 36 pigs and 12 horses, so the ratio of pigs to horses is 36:12. To express the ratio in simplest form, divide both numbers by 12, the greatest common factor.

$$\frac{36}{12} : \frac{12}{12} = 3 : 1$$

For every horse at the farm, there are 3 pigs.

The ratio of pigs to horses is 3:1.

> *Problems 11.1–11.4 refer to a farm that breeds livestock. The entire animal population consists of 12 horses, 16 cows, and 36 pigs.*

11.3 Express the ratio of horses to the total livestock population as a rational number in lowest terms.

The farm has 12 horses. To determine the total livestock population, add the number of horses, cows, and pigs.

$$\text{Total livestock population} = 12 + 16 + 36 = 64$$

The ratio of horses to the total population is equal to the quotient of those populations.

$$\frac{\text{horses}}{\text{total livestock}} = \frac{12}{64}$$

Reduce the fraction to lowest terms.

$$\frac{12 \div 4}{64 \div 4} = \frac{3}{16}$$

Problems 11.1–11.4 refer to a farm that breeds livestock. The entire animal population consists of 12 horses, 16 cows, and 36 pigs.

11.4 Express the ratio of pigs to the other livestock population in the form *a:b*. Write the ratio in simplest form.

The farm has 36 pigs. To determine the population of the other livestock, add the number of horses and cows: $12 + 16 = 28$.

The ratio of pigs to other animals is 36:28. Divide both numbers by 4, the greatest common factor, to reduce the ratio.

$$\frac{36}{4} : \frac{28}{4} = 9 : 7$$

11.5 If *AB* = 15 in the following diagram and the ratio of *AX* to *XB* is 3:2, calculate *XB*.

If the ratio of *AX* to *XB* is 3:2, then $AX = 3n$ and $XB = 2n$, where *n* is a constant. To calculate either length, you must first determine the value of *n*. Apply the segment addition postulate, because *AB* is equal to the sum of its parts.

$$AX + XB = AB$$

Substitute $AX = 3n$, $XB = 2n$, and $AB = 15$ into the equation and solve for *n*.

$$3n + 2n = 15$$
$$5n = 15$$
$$n = \frac{15}{5}$$
$$n = 3$$

The problem asks you to calculate *XB*; recall that $XB = 2n$.

$$XB = 2n = 2(3) = 6$$

11.6 Calculate the lengths of the sides of a triangle with perimeter 34 if they are in the ratio 3:6:8.

If the lengths of the sides are in the ratio 3:6:8, then those lengths are equal to $3n$, $6n$, and $8n$, where n is a constant. The perimeter of the triangle is 34, so construct an equation stating that the sum of the lengths of the sides of the triangle is 34.

$$3n + 6n + 8n = 34$$

Solve the equation for n.

$$17n = 34$$
$$n = \frac{34}{17}$$
$$n = 2$$

Substitute $n = 2$ into the expressions representing the lengths of the sides: $3n = 3(2) = 6$; $6n = 6(2) = 12$; and $8n = 8(2) = 16$. Therefore, the lengths of the sides are 6, 12, and 16.

11.7 Calculate the measures of the supplementary angles that are in the ratio 7:29.

Two angles in the ratio 7:29 are equal to $7n$ and $29n$, where n is a constant. The sum of the measures of supplementary angles is 180°.

$$7n + 29n = 180$$

Solve the equation for n.

$$36n = 180$$
$$n = \frac{180}{36}$$
$$n = 5$$

The measures of the supplementary angles are $7(5) = 35°$ and $29(5) = 145°$.

11.8 Identify the measures of the base angles of an isosceles triangle if the ratio of the vertex angle to a base angle is $\frac{2}{5}$.

An isosceles triangle contains two congruent angles called the base angles; the remaining angle is called the vertex angle. If the ratio of the vertex angle to a base angle is $\frac{2}{5}$, then the vertex angle has measure $2n$ and each of the base angles has measure $5n$, where n is a constant.

The measures of the angles of a triangle have a sum of 180°.

$$\text{vertex angle} + \text{base angle } \#1 + \text{base angle } \#2 = 180$$
$$2n + 5n + 5n = 180$$
$$12n = 180$$
$$n = \frac{180}{12}$$
$$n = 15$$

If $n = 15$, then the vertex angle of the isosceles triangle measures $2n = 2(15) = 30°$. Each of the base angles measures $5n = 5(15) = 75°$.

Properties of Proportions
Rearrange fractions so they stay equal

11.9 Identify the means and extremes of the proportions $a{:}b = c{:}d$ and $\dfrac{w}{x} = \dfrac{y}{z}$.

Consider the first proportion, $a{:}b = c{:}d$. The values a and d are the "extremes" of the proportion because they are located at the extreme left and right sides of the equation. The remaining two values, b and c, are the "means" of the proportion.

Rewrite $\dfrac{w}{x} = \dfrac{y}{z}$ in the same form as the previous proportion: $w{:}x = y{:}z$. Once again, the extremes are the values farthest from the equals sign, w and z; the means are closest to the equals sign, x and y.

11.10 Apply the means-extremes property to rewrite the proportion $x{:}5 = 9{:}y$ as an equation that does not contain ratios. ←

According to the means-extremes property, the product of the means is equal to the product of the extremes.

$$5 \cdot 9 = x \cdot y$$
$$45 = xy$$

So the equation won't contain "a:b," "a to b," or "$\frac{a}{b}$."

11.11 Write the equation $\dfrac{x}{12} = 5$ as a proportion and apply the means-extremes property to solve it.

> "Means-extremes" is usually called "cross-multiplication" when you're dealing with fractions. Multiply the top of the left fraction by the bottom of the right fraction, and vice versa. Then set the products equal.

To convert the given equation into a proportion, express both sides of the equation as a fraction. In this case, 5 should be written as a rational number. Dividing a real number by 1 does not affect its value: $5 = \dfrac{5}{1}$.

$$\frac{x}{12} = \frac{5}{1}$$

Apply the means-extremes property to rewrite the proportion as an equation that does not contain ratios.

$$x \cdot 1 = 12 \cdot 5$$
$$x = 60$$

11.12 Apply the means-extremes property to solve the following equation.

$$\frac{2}{x+1} = \frac{5}{4x-7}$$

Set the product of the means equal to the product of the extremes and apply the distributive property.

$$2(4x-7) = 5(x+1)$$
$$8x - 14 = 5x + 5$$

Solve the equation for x.

$$8x - 5x = 5 + 14$$
$$3x = 19$$
$$x = \frac{19}{3}$$

Note: Problems 11.13–11.14 refer to the proportion $\dfrac{5}{x} = \dfrac{9}{4x-1}$.

11.13 Use the means-extremes property to solve the proportion.

Set the product of the means equal to the product of the extremes and solve for x.

$$5(4x-1) = x \cdot 9$$
$$20x - 5 = 9x$$
$$20x - 9x = 5$$
$$11x = 5$$
$$x = \frac{5}{11}$$

Note: Problems 11.13–11.14 refer to the proportion $\dfrac{5}{x} = \dfrac{9}{4x-1}$.

11.14 Apply the means reversal property and verify that the solution is unaffected.

According to the means reversal property, you can exchange the positions of the means in the proportion without affecting the equality of the statement. The means in this proportion are x and 9; reverse the means and solve for x.

$$\frac{5}{9} = \frac{x}{4x-1}$$
$$5(4x-1) = 9x$$
$$20x - 5 = 9x$$
$$20x - 9x = 5$$
$$11x = 5$$
$$x = \frac{5}{11}$$

> In case you forgot which is which, E stands for extreme and M stands for mean: $\dfrac{E}{M} = \dfrac{M}{E}$.

Problems 11.13 and 11.14 have the same solution, verifying that the means reversal property changes neither the equality of the statement nor the solution.

11.15 Assuming that a, b, and d are nonzero real numbers, prove that $\dfrac{a}{b} = \dfrac{c}{d}$ is equivalent to $\dfrac{d}{b} = \dfrac{c}{a}$.

> This is called the extremes reversal property. If you swap the extremes (in this case, a and d), the equation is still true.

Statement	Reason
1. $\dfrac{a}{b} = \dfrac{c}{d}$	1. Given
2. $ad = bc$	2. Means-extremes property
3. $\dfrac{ad}{b} = c$	3. Division property of equality
4. $\dfrac{d}{b} = \dfrac{c}{a}$	4. Division property of equality

> In Step 3, you divide both sides by b. In Step 4, you divide both sides by a.

Because you are able to generate the proportion $\dfrac{d}{b} = \dfrac{c}{a}$ from the proportion $\dfrac{a}{b} = \dfrac{c}{d}$, the proportions are equivalent.

11.16 Apply the extremes reversal property to solve the following proportion. Assume $x \neq 0$.

$$\frac{x}{18} = \frac{4x^2}{9}$$

Reverse the extremes of the proportion (x and 9).

$$\frac{9}{18} = \frac{4x^2}{x}$$

Reduce both fractions to lowest terms.

$$\frac{\cancel{9}}{\cancel{9}\cdot 2}=\frac{4\cdot\cancel{x}\cdot x}{\cancel{x}}$$

$$\frac{1}{2}=\frac{4x}{1}$$

Apply the means-extremes property and solve for x.

$$1(1)=2(4x)$$

$$1=8x$$

$$\frac{1}{8}=x$$

11.17 Demonstrate the reciprocal property of proportions by verifying that $\dfrac{4}{3x}=\dfrac{1}{x+2}$ and $\dfrac{3x}{4}=\dfrac{x+2}{1}$ have the same solution.

You can flip both fractions of a proportion upside down, and the upside-down fractions will still be equal.

Apply the means-extremes property to each proportion and solve the resulting equations for x.

$$\frac{4}{3x}=\frac{1}{x+2} \qquad\qquad \frac{3x}{4}=\frac{x+2}{1}$$

$$4(x+2)=3x(1) \qquad\qquad 3x(1)=4(x+2)$$

$$4x+8=3x \qquad\qquad 3x=4x+8$$

$$4x-3x=-8 \qquad\qquad -8=4x-3x$$

$$x=-8 \qquad\qquad -8=x$$

Taking the reciprocals of the fractions of a proportion merely reverses the sides of the equation. Thus, the equality of the statements is unaffected.

11.18 Complete and verify the following statement.

According to the denominator addition property of proportions, if $\dfrac{2}{8}=\dfrac{5}{20}$,

then $\dfrac{\Box}{8}=\dfrac{\Box}{20}$.

Replace the top of each fraction with the sum of the top and bottom of each fraction. Leave the denominators alone.

The denominator addition property of proportions states that adding the denominator of each fraction to the corresponding numerator will not affect the equality of the proportion.

In this example, adding 8 to the numerator of the left fraction and 20 to the numerator of the right fraction produces a new, valid proportion.

$$\frac{2+8}{8}=\frac{5+20}{20}$$

$$\frac{\boxed{10}}{8}=\frac{\boxed{25}}{20}$$

If you are not convinced that the rational numbers are equal, reduce them to lowest terms.

$$\frac{10 \div 2}{8 \div 2} = \frac{25 \div 5}{20 \div 5}$$

$$\frac{5}{4} = \frac{5}{4}$$

11.19 Use the proportion $\frac{a}{b} = \frac{c}{d}$ to verify the denominator subtraction property of proportions. Assume that b and d are nonzero real numbers.

Similar to the denominator addition property presented in Problem 11.18, the denominator subtraction property states that subtracting the denominator of each fraction from the numerator does not affect the equality of the proportion. In this example, subtracting b from a in the left fraction and d from c in the right fraction produces a valid proportion.

$$\frac{a-b}{b} = \frac{c-d}{d}$$

Write each fraction as a sum. Recall that $\frac{x}{z} - \frac{y}{z} = \frac{x-y}{z}$ because rational numbers with common denominators may be combined. Therefore, it is equally true that $\frac{x-y}{z} = \frac{x}{z} - \frac{y}{z}$.

$$\frac{a}{b} - \frac{b}{b} = \frac{c}{d} - \frac{d}{d}$$

Any nonzero number divided by itself is equal to 1. Therefore, $\frac{b}{b} = \frac{d}{d} = 1$.

$$\frac{a}{b} - 1 = \frac{c}{d} - 1$$

Apply the addition property of equality, adding 1 to both sides of the equation.

$$\frac{a}{b} = \frac{c}{d}$$

This is the original proportion, so it is equivalent to the proportion generated by the denominator subtraction property.

> The denominator subtraction property works because the net effect is that you subtract 1 from both sides of an equation.

11.20 Given $\dfrac{AB}{BC} = \dfrac{3}{8}$ in the following figure, calculate $\dfrac{AC}{BC}$.

Apply the denominator addition property (as described in Problem 11.18) to the given proportion.

$$\frac{AB + BC}{BC} = \frac{3 + 8}{8}$$

According to the segment addition property, $AB + BC = AC$.

$$\frac{AC}{BC} = \frac{3 + 8}{8}$$
$$\frac{AC}{BC} = \frac{11}{8}$$

> The symbol ~ means "is similar to."

Similar Polygons
Congruent shapes at different magnifications

Note: Problems 11.21–11.26 refer to the following diagram, in which ABCDE ~ VWXYZ.

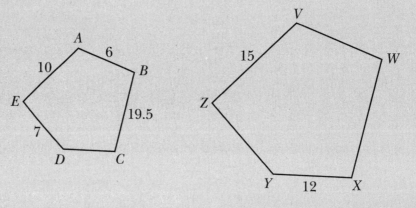

11.21 If $\angle D = 110°$, what can be concluded about *VWXYZ*?

The pentagons are similar figures, so all of the corresponding angles are congruent. Angles *D* and *Y* are in the same corresponding position in the similar pentagons; therefore, $m\angle D = m\angle Y = 110°$.

Note: Problems 11.21–11.26 refer to the diagram in Problem 11.21, in which ABCDE ~ VWXYZ.

11.22 What is the scale factor from *VWXYZ* to *ABCDE*?

The corresponding sides of similar figures are in proportion. The ratio that describes pairs of corresponding sides is called the scale factor. To calculate the scale factor, identify one pair of corresponding sides for which the lengths are given. In this diagram, *VZ* = 15 and *AE* = 10. Calculate the quotient.

$$\text{scale factor} = \frac{VZ}{AE} = \frac{15}{10} = \frac{3}{2}$$

Note: Problems 11.21–11.26 refer to the diagram in Problem 11.21, in which ABCDE ~ VWXYZ.

11.23 Use a property of proportions to determine the scale factor from *ABCDE* to *VWXYZ*.

See Problem 11.17.

Problem 11.22 states that the ratio of side lengths of *VWXYZ* to the corresponding side lengths of *ABCDE* is $\frac{VZ}{AE} = \frac{3}{2}$. The reciprocal property of proportions states that a proportion maintains its equality if both ratios are replaced by their reciprocals.

$$\frac{AE}{VZ} = \frac{2}{3}$$

The ratio of side lengths of *ABCDE* to side lengths of *VWXYZ* is $\frac{2}{3}$. Therefore, the scale factor from *ABCDE* to *VWXYZ* is $\frac{2}{3}$.

Note: Problems 11.21–11.26 refer to the diagram in Problem 11.21, in which ABCDE ~ VWXYZ.

11.24 Calculate *DC*.

According to Problem 11.22, the scale factor from *VWXYZ* to *ABCDE* is $\frac{3}{2}$. Therefore, the length of any side of *VWXYZ* divided by the length of the corresponding side of *ABCDE* is equal to $\frac{3}{2}$. You are asked to determine the length of *DC*, which corresponds to side *YX*.

$$\frac{YX}{DC} = \frac{3}{2}$$

The diagram states that *YX* = 12.

$$\frac{12}{DC} = \frac{3}{2}$$

Apply the means-extremes property to solve for *DC*.

$$12(2) = 3(DC)$$
$$24 = 3(DC)$$
$$\frac{24}{3} = DC$$
$$8 = DC$$

Make sure you get the order right. This scale factor is from *VWXYZ* to *ABCDE*. Whenever you set up a proportion, the numerator must be a length in the first figure (from *VWXYZ*) and the denominator is the same side of the second figure (to *ABCDE*).

Note: Problems 11.21–11.26 refer to the diagram in Problem 11.21, in which *ABCDE ~ VWXYZ*.

11.25 Calculate *VW*.

Side $\overline{VW}$ corresponds with side $\overline{AB}$. The pentagons are similar, with a scale factor of $\frac{3}{2}$ from *VWXYZ* to *ABCDE*.

$$\frac{VW}{AB} = \frac{3}{2}$$

The diagram states that *AB* = 6. Substitute this value into the proportion and solve for *VW*.

$$\frac{VW}{6} = \frac{3}{2}$$
$$2(VW) = 6(3)$$
$$2(VW) = 18$$
$$VW = \frac{18}{2}$$
$$VW = 9$$

Note: Problems 11.21–11.26 refer to the diagram in Problem 11.21, in which *ABCDE ~ VWXYZ*.

11.26 Calculate *WX*.

Notice that $\overline{WX}$ and $\overline{BC}$ are corresponding sides of similar figures. Because the scale factor from *VWXYZ* to *ABCDE* is $\frac{3}{2}$, the ratio of any side length of *VWXYZ* to the corresponding length in *ABCDE* is equal to the scale factor.

$$\frac{WX}{BC} = \frac{3}{2}$$

Substitute the known value, *BC* = 19.5, into the equation and solve for *WX*.

$$\frac{WX}{19.5} = \frac{3}{2}$$

$$2(WX) = 3(19.5)$$

$$2(WX) = 58.5$$

$$WX = \frac{58.5}{2}$$

$$WX = 29.25$$

Note: Problems 11.27–11.29 refer to a pair of similar convex quadrilaterals: JKLM ~ NOPQ. The scale factor JKLM to NOPQ is 5:3.

11.27 Given $m\angle J = 121°$, $m\angle L = 65°$, and $m\angle M = 85°$, calculate $m\angle O$.

The sum of the interior angles of a convex polygon is $180°(n-2)$, where n represents the number of sides of the polygon. Substitute $n = 4$ to calculate the sum of the interior angles of a convex quadrilateral.

$$180°(n-2) = 180°(4-2)$$

$$= 180°(2)$$

$$= 360°$$

Add the interior angles of *JKLM* and set the sum equal to 360°.

$$m\angle J + m\angle K + m\angle L + m\angle M = 360°$$

Substitute the known angle measurements into the equation.

$$121° + m\angle K + 65° + 85° = 360°$$

$$271° + m\angle K = 360°$$

Solve for $m\angle K$.

$$m\angle K = 360° - 271°$$

$$= 89°$$

Note that $\angle K$ and $\angle O$ are corresponding angles of similar polygons. Similar polygons have congruent corresponding angles, so $m\angle O = m\angle K = 89°$.

K and O are the second letters in the names of the polygons, in the same corresponding position. This means that ∠K and ∠O correspond as well.

Note: Problems 11.27–11.29 refer to a pair of similar convex quadrilaterals: JKLM ~ NOPQ. The scale factor JKLM to NOPQ is 5:3.

11.28 Given $MJ = 30$, calculate QN.

The scale factor from $JKLM$ to $NOPQ$ is $\dfrac{5}{3}$, so the ratio of any side length of $JKLM$ to the corresponding side length of $NOPQ$ is equal to $\dfrac{5}{3}$. Note that $\overline{MJ}$ and $\overline{QN}$ are corresponding sides because their endpoints correspond in the names of the polygons.

$$\frac{MJ}{QN} = \frac{5}{3}$$

Substitute $MJ = 30$ into the proportion and solve for QN.

$$\frac{30}{QN} = \frac{5}{3}$$
$$30(3) = 5(QN)$$
$$90 = 5(QN)$$
$$\frac{90}{5} = QN$$
$$18 = QN$$

Note: Problems 11.27–11.29 refer to a pair of similar convex quadrilaterals: JKLM ~ NOPQ. The scale factor JKLM to NOPQ is 5:3.

11.29 Calculate the perimeter of $NOPQ$, given the perimeter of $JKLM$ is 105.

The perimeters of similar figures are in the same proportion as the pairs of corresponding sides. Thus, the ratio of the perimeter of $JKLM$ to the perimeter of $NOPQ$ is 5:3.

$$\frac{\text{perimeter of } JKLM}{\text{perimeter of } NOPQ} = \frac{5}{3}$$

The perimeter of $JKLM$ is 105. Let x represent the perimeter of $NOPQ$.

$$\frac{105}{x} = \frac{5}{3}$$
$$105(3) = 5(x)$$
$$315 = 5x$$
$$\frac{315}{5} = x$$
$$63 = x$$

The perimeter of $NOPQ$ is 63.

11.30 Is the following statement true or false? Justify your answer, including a diagram that supports your response.

All rectangles are similar.

Similar figures must have congruent corresponding angles and corresponding sides that are in proportion. All rectangles contain four right angles, so the interior angles of all rectangles are congruent. Thus, the first condition of similarity is met.

However, rectangles need not have proportional sides. Consider square *ABCD* and rectangle *XYCD* in the following diagram.

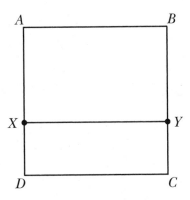

By definition, all squares are rectangles, so *ABCD* is a rectangle as well. However, all four sides of a square are congruent, so for the corresponding sides of a rectangle to be in the same proportion, the sides of the rectangle must be congruent. Clearly, the sides of *XYCD* are not congruent, so these rectangles are not similar.

A single counterexample is sufficient to conclude that the statement "All rectangles are similar" is false.

Triangle Similarity Theorems
AA, SSS, and SAS

11.31 Identify the two theorems and one postulate most often used to prove triangles similar, and give a brief description of each.

The *angle-angle (AA) similarity postulate* states that if two angles of one triangle are congruent to two angles of another triangle, then the triangles are similar.

The *side-side-side (SSS) similarity theorem* states that two triangles whose corresponding sides are in proportion are similar.

Finally, according to the *side-angle-side (SAS) similarity theorem*, if two sides of one triangle are in proportion with two sides of another triangle and the included angles are congruent, then the triangles are similar.

11.32 True or false: All equilateral triangles are similar. Justify your answer.

> "Equiangular" describes a polygon with interior angles that are all congruent.

An equilateral triangle is constructed of congruent sides. According to the isosceles triangle theorem, angles opposite congruent sides of a triangle are congruent. Therefore, equilateral triangles are equiangular as well. The sum of the interior angles of a triangle is 180°. Let x represent the measure of one interior angle of an equilateral triangle.

$$x + x + x = 180°$$
$$3x = 180°$$
$$x = \frac{180°}{3}$$
$$x = 60°$$

> The AA similarity postulate requires only that two angles in one triangle are congruent to two angles in another triangle. In this case, all three angles of both triangles are congruent.

The interior angles of all equilateral triangles measure 60°. Therefore, all equilateral triangles are similar by the AA similarity postulate.

Note: Problems 11.33–11.34 refer to the following diagram.

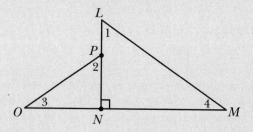

11.33 Given $\dfrac{OP}{ML} = \dfrac{ON}{MN}$, can you conclude that the triangles are similar? Justify your answer and correctly identify the triangles if they are similar.

The diagram specifies that $m\angle LNM = 90°$, so $m\angle ONP$ (the supplement of a right angle) measures 90° as well. Because $\dfrac{OP}{ML} = \dfrac{ON}{MN}$, you know that two sides of $\triangle PON$ are in proportion to two sides of $\triangle LMN$.

> The included angles are ∠3 and ∠4.

Even though two sides of one triangle are in proportion with two sides of another triangle, and one angle of each triangle is congruent to one angle of the other, you cannot conclude that the triangles are similar. To apply the SAS similarity theorem, the congruent angle must be included between the proportional sides, and the right angles are not.

Note: Problems 11.33–11.34 refer to the diagram in Problem 11.33.

11.34 Given $m\angle 2 = 60°$ and $m\angle 3 = \frac{1}{2} m\angle 1$, can you conclude that the triangles are similar? Justify your answer and correctly identify the triangles if they are similar.

The measures of two angles of $\triangle PON$ are known: $m\angle 2 = 60°$ and $m\angle ONP = 90°$. The sum of the measures of the angles of $\triangle PON$ is $180°$.

$$m\angle 2 + m\angle 3 + m\angle ONP = 180°$$

Calculate $m\angle 3$.

$$60° + m\angle 3 + 90° = 180°$$
$$150° + m\angle 3 = 180°$$
$$m\angle 3 = 180° - 150°$$
$$m\angle 3 = 30°$$

The problem states that $m\angle 3 = \frac{1}{2} m\angle 1$. Substitute $m\angle 3 = 30°$ into the equation.

$$30° = \frac{1}{2} m\angle 1$$
$$2(30°) = 2\left(\frac{1}{2} m\angle 1\right)$$
$$60° = m\angle 1$$

Two angles of $\triangle PON$ are congruent to two angles of $\triangle LMN$: $m\angle 1 = m\angle 2 = 60°$ and $m\angle PNO = m\angle LNM = 90°$. Therefore, $\triangle PON \sim \triangle LMN$, according to the AA similarity postulate.

> Make sure you name the vertices in the correct order so that the angles correspond. Another correct answer is $\triangle OPN \sim \triangle MLN$.

11.35 Identify the pair of similar triangles in the following figure and calculate the scale factor from the smaller to the larger triangle.

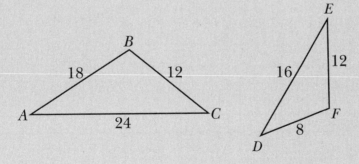

Triangles *DFE* and *CBA* are similar according to the SSS similarity theorem because the corresponding sides are in proportion.

$$\frac{CB}{DF} = \frac{12}{8} = \frac{3}{2} \qquad \frac{BA}{FE} = \frac{18}{12} = \frac{3}{2} \qquad \frac{AC}{ED} = \frac{24}{16} = \frac{3}{2}$$

> Pair up the small, medium, and large legs of each triangle.

The scale factor from $\triangle GHI$ to $\triangle DFE$ is 3:2.

11.36 Given $\overline{RT} \parallel \overline{AB}$ in the following diagram, prove that $\triangle ASB \sim \triangle RST$.

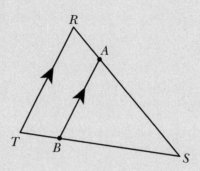

You are asked to prove that overlapping triangles are similar. Notice that $\triangle ASB$ and $\triangle RST$ share $\angle S$.

Statement	Reason
1. $\overline{RT} \parallel \overline{AB}$	1. Given
2. $\angle RTS \cong \angle ABS$	2. Parallel lines intersected by a transversal form congruent corresponding angles
3. $\angle S \cong \angle S$	3. Reflexive property of congruency
4. $\triangle ASB \sim \triangle RST$	4. AA similarity postulate

11.37 Assuming that d is a nonzero real number in the following diagram, prove that the triangles are similar.

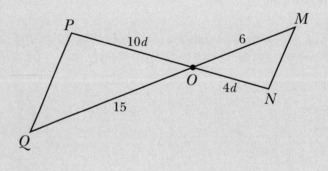

Statement	Reason
1. $d \neq 0$; $\dfrac{PO}{NO} = \dfrac{10d}{4d}$; $\dfrac{QO}{MO} = \dfrac{15}{6}$	1. Given
2. $\dfrac{PO}{NO} = \dfrac{10\cancel{d}}{4\cancel{d}} = \dfrac{10}{4} = \dfrac{5}{2}$; $\dfrac{QO}{MO} = \dfrac{15}{6} = \dfrac{5}{2}$	2. Reduce fractions to lowest terms

3. $\dfrac{PO}{NO} = \dfrac{QO}{MO}$	3. Transitive property of equality ←
4. $\angle POQ \cong \angle NOM$	4. Vertical angles are congruent
5. $\triangle QOP \sim \triangle MON$	5. SAS similarity theorem

> They both equal $\dfrac{5}{2}$, so they equal each other. By the way, this also proves that two sides of $\triangle QOP$ are in proportion with two sides of $\triangle MON$.

11.38 Given the following figure, in which $\triangle ABC \sim \triangle XYZ$, prove that medians $\overline{BM}$ and $\overline{YN}$ are in the same proportion as the sides of the similar triangles.

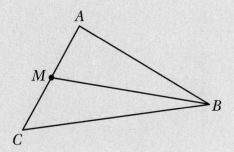 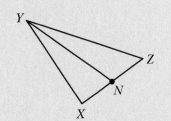

The problem states that the triangles are similar, so corresponding sides are in the same proportion: $\dfrac{AB}{XY} = \dfrac{CB}{ZY} = \dfrac{AC}{XZ}$. You also know that $\overline{BM}$ and $\overline{YN}$ are medians, segments that extend from a vertex of a triangle to the midpoint of the opposite side.

Statement	Reason
1. $\triangle ABC \sim \triangle XYZ$; $\overline{BM}$ and $\overline{YN}$ are medians	1. Given
2. $AM = \dfrac{1}{2} AC; \ XN = \dfrac{1}{2} XZ$	2. Medians bisect one side of a triangle, dividing it into congruent segments that are each half the length of the original side
3. $2(AM) = AC; \ 2(XN) = XZ$	3. Multiplication property of equality
4. $\dfrac{AC}{XZ} = \dfrac{AB}{XY}$	4. CSSTP ←
5. $\dfrac{2(AM)}{2(XN)} = \dfrac{AB}{XY}$	5. Substitution property of equality (Steps 3, 4)
6. $\dfrac{AM}{XN} = \dfrac{AB}{XY}$	6. Reduce the fraction to lowest terms ←
7. $\angle A \cong \angle X$	7. Corresponding angles of similar triangles are congruent
8. $\triangle AMB \sim \triangle XNY$	8. SAS similarity postulate
9. $\dfrac{BM}{YN} = \dfrac{AB}{XY}$	9. CSSTP

> Multiply both sides of each equation by 2.

> Corresponding sides of similar triangles are in proportion.

> $\overline{AM}$ and $\overline{XN}$ are corresponding sides of triangles AMB and XNY, and they're in the same proportion as the corresponding sides of $\triangle ABC$ and $\triangle XYZ$.

Because $\dfrac{BM}{YN} = \dfrac{AB}{XY}$, the lengths of the medians are in the same proportions as the lengths of the sides of $\triangle ABC$ and $\triangle XYZ$. Note that all triangle lengths follow this pattern—all corresponding medians, altitudes, perpendicular bisectors, angle bisectors, and so on of similar triangles have the same scale factor as the similar triangles upon which they're based.

11.39 Given $\overline{GH} \parallel \overline{JK}$ and $\overline{GI} \parallel \overline{JL}$ in the following diagram, prove $GM \cdot KL = JM \cdot HI$.

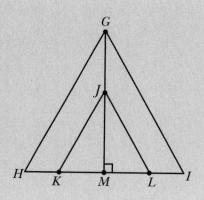

Statement	Reason
1. $\overline{GH} \parallel \overline{JK}$; $\overline{GI} \parallel \overline{JL}$	1. Given
2. $\angle GHM \cong \angle JKM$; $\angle GIM \cong \angle JLM$	2. Parallel lines intersected by a transversal form congruent corresponding angles
3. $\triangle HGI \sim \triangle KJL$	3. AA similarity postulate
4. $\overline{GM}$ and $\overline{JM}$ are altitudes of $\triangle HGI$ and and $\triangle KJL$, respectively	4. Altitudes that extend from a vertex of a triangle are perpendicular to the opposite side
5. $\dfrac{GM}{JM} = \dfrac{HI}{KL}$	5. Corresponding altitudes of similar triangles are in the same proportion as corresponding sides of similar triangles
6. $GM \cdot KL = JM \cdot HI$	6. Means-extremes property of proportions

See Problem 11.38.

Segment Proportionality Theorems

Including the triangle proportionality theorem

Note: Problems 11.40–11.41 refer to the following diagram, in which point A divides segments $\overline{WX}$ and $\overline{YZ}$ proportionally.

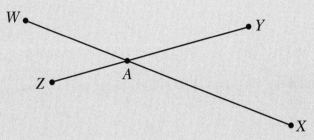

11.40 Construct a proportionality statement relating the lengths of the segments of $\overline{WX}$ to the lengths of the segments of $\overline{YZ}$.

Point A divides the segments proportionally, so the ratio of the segments created on $\overline{WX}$ is equal to the corresponding ratio of the segments created on $\overline{YZ}$.

$$\frac{WA}{AX} = \frac{ZA}{AY}$$

> The ratio of the short piece of $\overline{WX}$ to the long piece is equal to the ratio of the short piece of $\overline{YZ}$ to the long piece.

Note: Problems 11.40–11.41 refer to the diagram in Problem 11.40, in which point A divides segments $\overline{WX}$ and $\overline{YZ}$ proportionally.

11.41 Given $WA = 6$, $ZA = 4$, and AY is 3 more than half of AX, calculate AX and AY.

If AY is 3 more than half of AX, then $AY = \frac{1}{2}(AX) + 3$. According to Problem 11.40, $\frac{WA}{AX} = \frac{ZA}{AY}$. Substitute all known values into the proportion.

$$\frac{6}{AX} = \frac{4}{(1/2)(AX) + 3}$$

Apply the means-extremes property.

$$6\left[\frac{1}{2}(AX) + 3\right] = 4(AX)$$

$$\frac{6}{2}(AX) + 6(3) = 4(AX)$$

$$3(AX) + 18 = 4(AX)$$

Solve for AX.

$$18 = 4(AX) - 3(AX)$$

$$18 = AX$$

Recall that $AY = \frac{1}{2}(AX) + 3$.

$$AY = \frac{1}{2}(18) + 3$$
$$= 9 + 3$$
$$= 12$$

Therefore, $AX = 18$ and $AY = 12$.

11.42 Use the following diagram to prove the triangle proportionality theorem.

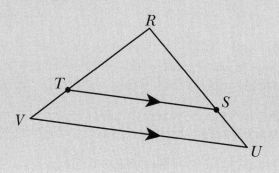

According to the triangle proportionality theorem, if a line is parallel to one side of a triangle and intersects the other sides, then it splits those sides proportionally. In this diagram, $\overline{ST}$ is parallel to side $\overline{UV}$ and intersects sides $\overline{RU}$ and $\overline{RV}$.

To prove the triangle proportionality theorem, your goal is to demonstrate that $\overline{ST}$ splits those sides proportionally: $\dfrac{RS}{SU} = \dfrac{RT}{TV}$.

Statement	Reason
1. $\overline{ST} \parallel \overline{UV}$	1. Given
2. $\angle RST \cong \angle RUV$; $\angle RTS \cong \angle RVU$	2. Parallel lines intersected by a transversal produce congruent corresponding angles
3. $\angle R \cong \angle R$	3. Reflexive property of congruency
4. $\triangle TRS \sim \triangle VRU$	4. AA similarity postulate
5. $RS + SU = RU$; $RT + TV = RV$	5. Segment addition postulate
6. $\dfrac{RS}{RU} = \dfrac{RT}{RV}$	6. CSSTP
7. $\dfrac{RU}{RS} = \dfrac{RV}{RT}$	7. Reciprocal property of proportions

8. $\dfrac{RS+SU}{RS}=\dfrac{RT+TV}{RT}$

9. $\dfrac{SU}{RS}=\dfrac{TV}{RT}$

10. $\dfrac{RS}{SU}=\dfrac{RT}{TV}$

8. Substitution property of equality (Steps 5, 7)

9. Denominator subtraction property of proportions ←

10. Reciprocal property of proportions

> You can subtract the denominator from the numerator of each fraction, so subtract RS from the left numerator and subtract RT from the right numerator.
> RS + SU – RS = SU
> RT + TV – RT = TV

Note: Problems 11.43–11.44 refer to the following diagram, in which $\overline{ZV} \parallel \overline{YW}$.

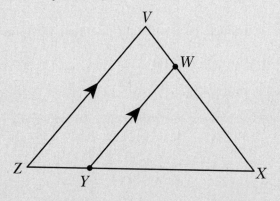

11.43 Given $WX = a - 2$, $VW = 2a - 9$, $ZY = 9$, and $YX = 12$, calculate a.

According to the triangle proportionality theorem, a line parallel to one side of a triangle that intersects the other sides splits those sides proportionally.

$$\frac{VW}{WX}=\frac{ZY}{YX}$$

Substitute the given values into the proportion.

$$\frac{2a-9}{a-2}=\frac{9}{12}$$

Apply the means-extremes property and solve for a.

$$12(2a-9)=9(a-2)$$
$$12(2a)+12(-9)=9(a)+9(-2)$$
$$24a-108=9a-18$$
$$24a-9a=-18+108$$
$$15a=90$$
$$a=\frac{90}{15}$$
$$a=6$$

Note: Problems 11.43–11.44 refer to the diagram in Problem 11.43, in which $\overline{ZV} \parallel \overline{YW}$.

11.44 Given $VW = c$, $VX = d$, $YX = d + 1$, and $ZY = 2c - 3$, express d in terms of c.

According to the segment addition postulate, $VX = VW + WX$. Substitute known values into the equation and solve for WX.

$$VX = VW + WX$$
$$d = c + WX$$
$$d - c = WX$$

> *Your goal is to end with "d = something containing the variable c."*

According to the triangle proportionality theorem, $\overline{WY}$ splits $\overline{VX}$ and $\overline{ZX}$ proportionally.

$$\frac{VW}{WX} = \frac{ZY}{YX}$$

Substitute known values into the proportion, including $WX = d - c$, as calculated earlier.

$$\frac{c}{d - c} = \frac{2c - 3}{d + 1}$$

Apply the means-extremes property.

$$c(d + 1) = (d - c)(2c - 3)$$
$$cd + c = 2cd - 3d - 2c^2 + 3c$$

Combine like terms.

$$0 = (2cd - cd) - 3d - 2c^2 + (3c - c)$$
$$0 = cd - 3d - 2c^2 + 2c$$

> *Add 3d to, and subtract cd from, both sides of the equation.*

Move all terms containing d to the left side of the equation.

$$3d - cd = -2c^2 + 2c$$

Factor d out of the expression left of the equals sign.

$$d(3 - c) = -2c^2 + 2c$$

Solve for d.

$$d = \frac{-2c^2 + 2c}{3 - c}$$
$$d = \frac{2c^2 - 2c}{c - 3}$$

> *The book multiplies the numerator and denominator by –1, but you don't have to do that.*

Note: Problems 11.45–11.46 refer to the following diagram, in which $\overline{AB} \parallel \overline{CD} \parallel \overline{EF}$.

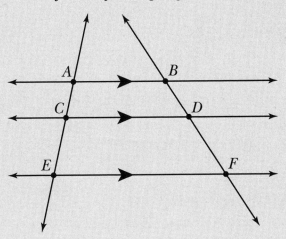

11.45 Complete the following statement.

Three or more parallel lines divide two transversals _____. *In this diagram,* $\dfrac{AC}{\Box} = \dfrac{BD}{\Box}$.

Three or more parallel lines divide two transversals <u>proportionally</u>. In this diagram, $\dfrac{AC}{\boxed{CE}} = \dfrac{BD}{\boxed{DF}}$. This is reminiscent of the theorem presented in Problem 9.26, which states that a group of parallel lines that divides one transversal into congruent segments divides any transversal into congruent segments.

Note: Problems 11.45–11.46 refer to the diagram in Problem 11.45, in which $\overline{AB} \parallel \overline{CD} \parallel \overline{EF}$.

11.46 Given $AC = x + 1$, $DF = 27$, $CE = 6x$, and $BD = 6$, calculate x.

Substitute the given values into the proportion constructed in Problem 11.45.

$$\frac{AC}{CE} = \frac{BD}{DF}$$

$$\frac{x+1}{6x} = \frac{6}{27}$$

Apply the means-extremes property and solve for x.

$$27(x+1) = 6(6x)$$
$$27x + 27 = 36x$$
$$27 = 36x - 27x$$
$$27 = 9x$$
$$\frac{27}{9} = x$$
$$3 = x$$

Note: Problems 11.47–11.48 refer to the following diagram, in which $\overline{RV}$ bisects $\angle SRT$.

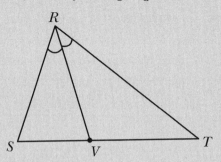

11.47 Complete the following statement.

> *According to the triangle angle bisector theorem, an angle bisector of a triangle divides _____ into segments that are _____ to the _____.*
> *In this diagram, $\dfrac{SV}{VT} = \dfrac{\square}{\square}$.*

According to the triangle angle bisector theorem, an angle bisector of a triangle divides <u>the opposite side</u> into segments that are <u>proportional</u> to the <u>adjacent</u> sides. In this diagram, $\dfrac{SV}{VT} = \dfrac{\boxed{SR}}{\boxed{TR}}$.

> Each piece of $\overline{ST}$ is proportional to the side of the triangle it touches: $\overline{SV}$ touches $\overline{SR}$, and $\overline{VT}$ touches $\overline{TR}$.

Note: Problems 11.47–11.48 refer to the diagram in Problem 11.47, in which $\overline{RV}$ bisects $\angle SRT$.

11.48 Given $ST = 2x - 1$, $TR = 2x + 6$, $SR = 2x - 2$, and $VT = x + 1$, calculate x.

According to the segment addition postulate, $SV + VT = ST$. Substitute $ST = 2x - 1$ and $VT = x + 1$ into the equation to calculate SV.

$$SV + (x + 1) = 2x - 1$$
$$SV = (2x - x) + (-1 - 1)$$
$$SV = x - 2$$

Apply the triangle angle bisector theorem, substituting the known lengths into the proportion constructed in Problem 11.47.

$$\frac{SV}{VT} = \frac{SR}{TR}$$
$$\frac{x - 2}{x + 1} = \frac{2x - 2}{2x + 6}$$

Apply the means-extremes property and solve for x.

$$(x-2)(2x+6)=(x+1)(2x-2)$$
$$2x^2+6x-4x-12=2x^2-2x+2x-2$$
$$2x^2+2x-12=2x^2-2$$
$$2x^2-2x^2+2x-12=-2$$
$$2x=-2+12$$
$$2x=10$$
$$x=\frac{10}{2}$$
$$x=5$$

Chapter 12
RIGHT TRIANGLES

Including the geometric mean and basic trigonometry

Although a right triangle is nothing more than a triangle that contains a right angle, it is surprisingly versatile geometrically. This chapter begins by defining the geometric mean, a value derived from a proportion and visually represented by the altitude drawn to the hypotenuse of a right triangle.

The proceeding sections of this chapter investigate the relationships between the lengths of the sides of a right triangle (including the Pythagorean Theorem), the angles of a right triangle (including special right triangles), and how the sides and angles affect one another (including trigonometric ratios).

You've probably heard of the Pythagorean Theorem. (Even the scarecrow from the movie The Wizard of Oz tries to quote it after he gets his "brain" at the end.) It applies to the lengths of right triangles, but it's just the tip of the right triangle theorem iceberg.

In this chapter, you'll learn more than you ever cared to know about right triangles. You'll even expand on the Pythagorean Theorem, modifying it a little so you can tell whether a triangle is right, acute, or obtuse, given just the lengths of its sides.

The chapter ends with a little taste of trigonometry, a word that basically means "the study of right triangles." In other words, there's much more to learn about right triangles than can fit into a single geometry course.

Geometric Mean
A value based on similar right triangles

> A proportion contains means and extremes (see Problem 11.9). To calculate a geometric mean between a and b, make a and b the extremes, make both of the means x, and solve for x.

12.1 Calculate the geometric mean between 4 and 49.

If a and b are positive real numbers, then $\dfrac{a}{x} = \dfrac{x}{b}$, where x is defined as the geometric mean between a and b. In this problem, set $a = 4$ and $b = 49$. (Note that it is equally correct to set $a = 49$ and $b = 4$ according to the extremes reversal property.)

$$\frac{4}{x} = \frac{x}{49}$$

Apply the means-extremes property and solve for x.

$$x^2 = 4 \cdot 49$$
$$\sqrt{x^2} = \sqrt{4 \cdot 49}$$
$$x = \sqrt{4} \cdot \sqrt{49}$$
$$x = 2 \cdot 7$$
$$x = 14$$

The geometric mean between 4 and 49 is 14.

12.2 Calculate the geometric mean between 12 and 50.

Let x represent the geometric mean between 12 and 50, and construct a proportion as directed by Problem 12.1.

$$\frac{12}{x} = \frac{x}{50}$$
$$x^2 = 12 \cdot 50$$
$$\sqrt{x^2} = \sqrt{12 \cdot 50}$$
$$x = \sqrt{12} \cdot \sqrt{50}$$
$$x = 2\sqrt{3} \cdot 5\sqrt{2}$$
$$x = 10\sqrt{6}$$

12.3 Construct a formula for the geometric mean between positive real numbers a and b.

Let x represent the geometric mean between a and b. Construct a proportion as directed by Problem 12.1.

$$\frac{a}{x} = \frac{x}{b}$$

Apply the means-extremes property and solve for *x*.

$$x^2 = a \cdot b$$

$$\sqrt{x^2} = \sqrt{ab}$$

$$x = \sqrt{ab}$$

The geometric mean between positive real numbers *a* and *b* is $\sqrt{ab}$.

Note: Use the following diagram to complete the proofs in Problems 12.4–12.6.

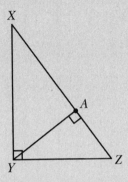

12.4 Prove that a right triangle is similar to the two right triangles formed by the altitude drawn to its hypotenuse.

An altitude is a segment perpendicular to one side of a triangle that extends to the opposite vertex. In this diagram, $\overline{YA}$ is the altitude extending from the right angle $\angle XYZ$ to the opposite side $\overline{XZ}$. The altitude divides the right triangle into two smaller right triangles, $\triangle XAY$ and $\triangle YAZ$. Your goal is to prove that each of these triangles is similar to $\triangle XYZ$.

Statement	Reason
1. XYZ is a right triangle; $\overline{YA} \perp \overline{XZ}$	1. Given
2. $m\angle XAY = 90°$; $m\angle YAZ = 90°$; $m\angle XYZ = 90°$	2. Perpendicular lines form 90° angles
3. $\angle X \cong \angle X$; $\angle Z \cong \angle Z$	3. Reflexive property of congruency
4. $\triangle XAY \sim \triangle XYZ$; $\triangle YAZ \sim \triangle XYZ$	4. AA similarity postulate

An altitude usually doesn't split the opposite side in half—the median does that. When the altitude and the median are the same, the triangle is isosceles.

Each of the smaller triangles shares one angle with the big triangle XYZ. Each of the smaller triangles also has a right angle.

Note: Use the diagram in Problem 12.4 to complete the proofs in Problems 12.4–12.6.

12.5 Prove that the altitude drawn to the hypotenuse of a right triangle forms similar right triangles.

In this diagram, $\overline{YA}$ is the altitude extending from the right angle $\angle XYZ$ to the opposite side $\overline{XZ}$. The altitude divides the right triangle into two smaller right triangles; your goal is to prove that these triangles are similar: $\triangle XAY \sim \triangle YAZ$.

This is the theorem proven by Problem 12.4.

Statement	Reason
1. XYZ is a right triangle; $\overline{YA} \perp \overline{XZ}$	1. Given
2. $\triangle XAY \sim \triangle XYZ$; $\triangle YAZ \sim \triangle XYZ$	2. A right triangle is similar to the two right triangles formed by the altitude drawn to its hypotenuse
3. $\angle AXY \cong \angle AYZ$	3. Corresponding angles of similar triangles are congruent
4. $\triangle XAY \sim \triangle YAZ$	4. AA similarity postulate

Note: Use the diagram in Problem 12.4 to complete the proofs in Problems 12.4–12.6.

12.6 Prove that the length of the altitude drawn to the hypotenuse of a right triangle is the geometric mean between the lengths of the segments of the hypotenuse.

In this diagram, $\overline{YA}$ is the altitude extending from the right angle $\angle XYZ$ to the opposite side $\overline{XZ}$. Your goal is to prove that YA is the geometric mean of XA and AZ.

Statement	Reason
1. XYZ is a right triangle; $\overline{YA} \perp \overline{XZ}$	1. Given
2. $\triangle XAY \sim \triangle YAZ$	2. The altitude drawn to the hypotenuse of a right triangle forms similar right triangles
3. $\dfrac{XA}{YA} = \dfrac{YA}{AZ}$	3. CSSTP
4. YA is the geometric mean between XA and AZ	4. The geometric mean x between positive real numbers a and b is defined by the proportion $\dfrac{a}{x} = \dfrac{x}{b}$

Note: Problems 12.7–12.9 refer to the following diagram.

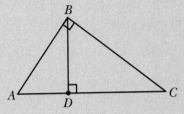

12.7 Given $AD = 6$ and $DC = 16$, calculate BD.

The length of the altitude drawn to the hypotenuse of a right triangle is the geometric mean of the lengths of the segments of the hypotenuse. In this diagram, $\overline{AD}$ and $\overline{DC}$ are segments created by the intersection of the hypotenuse $\overline{AC}$ and the altitude $\overline{BD}$. Therefore, BD is the geometric mean between AD and DC.

$$\frac{AD}{BD} = \frac{BD}{DC}$$

Substitute $AD = 6$ and $DC = 16$ into the proportion and solve for BD.

$$\frac{6}{BD} = \frac{BD}{16}$$
$$6(16) = (BD)^2$$
$$\sqrt{6 \cdot 16} = \sqrt{(BD)^2}$$
$$4\sqrt{6} = BD$$

Note: Problems 12.7–12.9 refer to the diagram in Problem 12.7.

12.8 Given $AD = 7$ and $BD = 2\sqrt{21}$, calculate DC.

As Problem 12.7 states, $\overline{AD}$ and $\overline{DC}$ are segments created by the intersection of the hypotenuse and an altitude. Therefore, BD is the geometric mean of AD and DC.

$$\frac{AD}{BD} = \frac{BD}{DC}$$

Substitute $AD = 7$ and $BD = 2\sqrt{21}$ into the proportion and solve for DC.

$$\frac{7}{2\sqrt{21}} = \frac{2\sqrt{21}}{DC}$$
$$7(DC) = 2^2 \left(\sqrt{21}\right)^2$$
$$7(DC) = 4(21)$$
$$7(DC) = 84$$
$$DC = \frac{84}{7}$$
$$DC = 12$$

Note: Problems 12.7–12.9 refer to the diagram in Problem 12.7.

12.9 Given $AC = 20$ and $BD = 8$, calculate AD and DC.

According to the segment addition postulate, $AD + DC = AC$. Let x represent the length of one segment of the hypotenuse: $x = AD$. Substitute that value and $AC = 20$ into the equation and solve for DC.

$$AD + DC = AC$$
$$x + DC = 20$$
$$DC = 20 - x$$

The length of the altitude drawn to the hypotenuse is the geometric mean between the segments of the hypotenuse it creates.

$$\frac{AD}{BD} = \frac{BD}{DC}$$

Substitute $AD = x$, $DC = 20 - x$, and $BD = 8$ into the proportion and solve for x.

$$\frac{x}{8} = \frac{8}{20 - x}$$
$$x(20 - x) = 8 \cdot 8$$
$$20x - x^2 = 64$$
$$0 = x^2 - 20x + 64$$

> The book solves this quadratic equation by factoring, so one side of the equation must equal 0. See Problems 3.41–3.42 for more information.

Solve the quadratic equation.

$$0 = (x - 16)(x - 4)$$
$$x - 16 = 0 \quad \text{or} \quad x - 4 = 0$$
$$x = 16 \qquad\qquad x = 4$$

> Adding –16 and –4 results in the x-coefficient of $x^2 - 20x + 64$ (in other words, –20); multiplying –16 and –4 gives you the constant (+64).

If the figure is drawn to scale, then $AD < DC$, $x = AD = 4$, and $DC = 20 - 4 = 16$. If the figure is not necessarily drawn to scale, then the lengths of the segments may be reversed. In either case, the segments of the hypotenuse have lengths 4 and 16.

Note: Problems 12.10–12.12 refer to the following diagram.

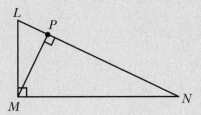

> The segments of the hypotenuse are $\overline{LP}$ and $\overline{NP}$. Each one is adjacent to the leg it intersects: $\overline{LP}$ is adjacent to $\overline{ML}$, and $\overline{NP}$ is adjacent to $\overline{MN}$.

12.10 Use the diagram to prove that each leg of a right triangle is the geometric mean between the hypotenuse and the segment of the hypotenuse adjacent to that leg.

Your goal is to prove that (1) *ML* is the geometric mean between *LP* and *LN*, and (2) *MN* is the geometric mean between *NP* and *LN*.

Statement	Reason
1. $m\angle LMN = 90°$; $\overline{MP} \perp \overline{LN}$	1. Given
2. $\triangle LPM \sim \triangle LMN$; $\triangle MPN \sim \triangle LMN$	2. A right triangle is similar to the two right triangles formed by the altitude drawn to its hypotenuse
3. $\dfrac{LP}{ML} = \dfrac{ML}{LN}$; $\dfrac{NP}{MN} = \dfrac{MN}{LN}$	3. CSSTP ←
4. *ML* is the geometric mean between *LP* and *LN*; *MN* is the geometric mean between *NP* and *LN*	4. The geometric mean x between positive real numbers a and b is defined by the proportion $\dfrac{a}{x} = \dfrac{x}{b}$

> The shorter legs and the hypotenuses of $\triangle LPM$ and $\triangle LMN$ are in proportion. The longer legs and the hypotenuses of $\triangle MPN$ and $\triangle LMN$ are also in proportion.

Note: Problems 12.10–12.12 refer to the diagram in Problem 12.10.

12.11 Given *LP* = 5 and *PN* = 7, calculate the lengths of the legs of $\triangle LMN$.

Apply the segment addition postulate to calculate the length of the hypotenuse.

$$LP + PN = LN$$
$$5 + 7 = LN$$
$$12 = LN$$

The lengths of the legs of right triangle *LMN* are the geometric means between the hypotenuse and the adjacent segment of the hypotenuse.

$$\frac{LP}{ML} = \frac{ML}{LN} \qquad \frac{PN}{MN} = \frac{MN}{LN}$$

Substitute $LP = 5$, $PN = 7$, and $LN = 12$ into the proportions. Solve the left proportion for ML and solve the right proportion for MN.

$$\frac{5}{ML} = \frac{ML}{12} \qquad\qquad \frac{7}{MN} = \frac{MN}{12}$$

$$(ML)^2 = 5 \cdot 12 \qquad\qquad (MN)^2 = 7 \cdot 12$$

$$\sqrt{(ML)^2} = \sqrt{5 \cdot 12} \qquad\qquad \sqrt{(MN)^2} = \sqrt{7 \cdot 12}$$

$$ML = \sqrt{5 \cdot 4 \cdot 3} \qquad\qquad MN = \sqrt{7 \cdot 4 \cdot 3}$$

$$ML = 2\sqrt{15} \qquad\qquad MN = 2\sqrt{21}$$

The lengths of the legs of $\triangle LMN$ are $ML = 2\sqrt{15}$ and $MN = 2\sqrt{21}$.

Note: Problems 12.10–12.12 refer to the diagram in Problem 12.10.

12.12 Given $ML = 3$, $MN = 4$, and $LN = 5$, calculate MP.

Apply the technique demonstrated in Problem 12.11 to calculate the lengths of the segments of the hypotenuse.

> Start by calculating LP and PN. The length of the altitude (MP) is the geometric mean between those numbers.

$$\frac{LP}{3} = \frac{3}{5} \qquad\qquad \frac{PN}{4} = \frac{4}{5}$$

$$5(LP) = 3^2 \qquad\qquad 5(PN) = 4^2$$

$$5(LP) = 9 \qquad\qquad 5(PN) = 16$$

$$LP = \frac{9}{5} \qquad\qquad PN = \frac{16}{5}$$

The length of the altitude drawn to the hypotenuse of a right triangle is the geometric mean between the lengths of the segments of the hypotenuse.

$$\frac{LP}{MP} = \frac{MP}{PN}$$

$$\frac{9/5}{MP} = \frac{MP}{16/5}$$

Solve the proportion for MP.

$$(MP)^2 = \left(\frac{9}{5}\right)\left(\frac{16}{5}\right)$$

$$(MP)^2 = \frac{144}{25}$$

$$\sqrt{(MP)^2} = \sqrt{\frac{144}{25}}$$

$$MP = \frac{12}{5}$$

Pythagorean Theorem

$a^2 + b^2 = c^2$

12.13 Calculate the length of the hypotenuse of a right triangle with legs 7 cm and 24 cm long.

The Pythagorean Theorem states that the square of the length of the hypotenuse of a right triangle is equal to the sum of the squares of the lengths of the legs. More commonly, if a and b represent the lengths of the legs and c represents the length of the hypotenuse, then $a^2 + b^2 = c^2$.

In this problem, substitute $a = 7$ and $b = 24$ into the equation and solve for c. ←

> It doesn't matter which leg you plug into a and which you plug into b, but make sure c is the hypotenuse.

$$7^2 + 24^2 = c^2$$
$$49 + 576 = c^2$$
$$625 = c^2$$
$$\sqrt{625} = c^2$$
$$25 = c$$

The length of the hypotenuse is 25 cm.

12.14 If the hypotenuse of a right triangle is 6 inches long and one of the legs measures 2 inches, how long is the other leg?

Let a and b represent the lengths of the legs. One of those lengths is known: $a = 2$. Let c represent the length of the hypotenuse: $c = 6$. Apply the Pythagorean Theorem and solve for b.

$$a^2 + b^2 = c^2$$
$$2^2 + b^2 = 6^2$$
$$4 + b^2 = 36$$
$$b^2 = 36 - 4$$
$$b^2 = 32$$
$$\sqrt{b^2} = \sqrt{32}$$
$$b = \sqrt{16 \cdot 2}$$
$$b = 4\sqrt{2}$$

The remaining leg of the right triangle measures $4\sqrt{2}$ inches.

> That's about 5.657 inches long.

12.15 Calculate the length of the legs of an isosceles right triangle if the length of the hypotenuse is 12 units.

The legs of an isosceles triangle are congruent. Apply the Pythagorean Theorem.

> The legs are usually written as a and b, but if they're the same length, then a = b and you can substitute a for b in the Pythagorean Theorem. That gives you $a^2 + a^2 = c^2$.

$$a^2 + a^2 = c^2$$
$$2a^2 = 12^2$$
$$2a^2 = 144$$
$$a^2 = \frac{144}{2}$$
$$a^2 = 72$$
$$\sqrt{a^2} = \sqrt{72}$$
$$a = \sqrt{36 \cdot 2}$$
$$a = 6\sqrt{2}$$

Note: Problems 12.16–12.17 refer to the following diagram, in which BD = 5 and AD = 12.

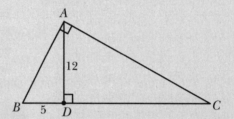

12.16 Calculate *AB*.

> Lowercase a and b (the lengths of the legs) represent different things than uppercase A and B (points on the diagram).

Apply the Pythagorean Theorem to the sides of right triangle *ABD*; let $a = 5$ and $b = 12$, and solve for c, the length *AB* of the hypotenuse.

$$a^2 + b^2 = c^2$$
$$5^2 + 12^2 = c^2$$
$$25 + 144 = c^2$$
$$169 = c^2$$
$$\sqrt{169} = c$$
$$13 = c$$

Note: Problems 12.16–12.17 refer to the diagram in Problem 12.16, in which BD = 5 and AD = 12.

12.17 Calculate *AC* and *DC*.

When a right triangle is split by the altitude drawn to its hypotenuse, the resulting right triangles are similar to each other. Therefore, $\triangle BDA \sim \triangle ADC$. Corresponding sides of similar triangles are in proportion.

$$\frac{BD}{AD} = \frac{BA}{AC}$$

$$\frac{5}{12} = \frac{13}{AC}$$

Solve the proportion for *AC*.

$$5(AC) = 12 \cdot 13$$

$$5(AC) = 156$$

$$AC = \frac{156}{5}$$

The altitude drawn to the hypotenuse of a right triangle is the geometric mean of the segments of the hypotenuse.

$$\frac{BD}{AD} = \frac{AD}{DC}$$

$$\frac{5}{12} = \frac{12}{DC}$$

$$5(DC) = 144$$

$$DC = \frac{144}{5}$$

12.18 Classify the triangle with sides of length 4, 8, and $4\sqrt{5}$ as acute, obtuse, or right.

The converse of the Pythagorean Theorem is true. Therefore, if *a*, *b*, and *c* are the lengths of sides of a triangle such that *c* represents the longest side and $a^2 + b^2 = c^2$, the sides form a right triangle.

In this problem, the longest side is $c = 4\sqrt{5} \approx 8.944$. Let $a = 4$ and $b = 8$, and apply the Pythagorean Theorem.

$$a^2 + b^2 \stackrel{?}{=} c^2$$

$$4^2 + 8^2 \stackrel{?}{=} \left(4\sqrt{5}\right)^2$$

$$16 + 64 \stackrel{?}{=} 16 \cdot 5$$

$$80 = 80$$

> You don't know whether $a^2 + b^2$ will actually equal c^2, so you use this question mark/equals sign until you can replace it with something more definitive, like =, <, or >.

Because $a^2 + b^2 = c^2$, the sides form a right triangle.

12.19 Classify the triangle with sides of length 3, 6, and $5\sqrt{3}$ as acute, obtuse, or right.

Let c represent the longest side, $5\sqrt{3} \approx 8.660$. Substitute $a = 3$, $b = 6$, and $c = 5\sqrt{3}$ into the Pythagorean Theorem.

$$a^2 + b^2 \stackrel{?}{=} c^2$$
$$3^2 + 6^2 \stackrel{?}{=} \left(5\sqrt{3}\right)^2$$
$$9 + 36 \stackrel{?}{=} 25 \cdot 3$$
$$45 < 75$$

When $a^2 + b^2 < c^2$, the sides form an obtuse triangle.

12.20 Classify the triangle with sides of length 4, 5, and 6 as acute, obtuse, or right.

Let $a = 4$, $b = 5$, and $c = 6$, and apply the Pythagorean Theorem.

$$a^2 + b^2 \stackrel{?}{=} c^2$$
$$4^2 + 5^2 \stackrel{?}{=} 6^2$$
$$16 + 25 \stackrel{?}{=} 36$$
$$41 > 36$$

When $a^2 + b^2 > c^2$, the sides form an acute triangle.

12.21 Classify the triangle with sides of length 7, 9, and 16 as acute, obtuse, or right.

Substitute $a = 7$, $b = 9$, and $c = 16$ into the Pythagorean Theorem.

$$a^2 + b^2 \stackrel{?}{=} c^2$$
$$7^2 + 9^2 \stackrel{?}{=} 16^2$$
$$49 + 81 \stackrel{?}{=} 256$$
$$130 < 256$$

See Problems 10.23 and 10.24 for more information.

Because $a^2 + b^2 < c^2$, you may be tempted to conclude that the sides form an obtuse triangle. However, the lengths of the sides violate the triangle inequality theorem because $7 + 9 \not> 16$. Therefore, no triangle with sides of length 7, 9, and 16 exists, and it is not logical to classify a nonexistent triangle as acute, obtuse, or right.

Don't you hate trick questions? What kind of triangle is this? Gotcha! It's not a triangle at all!

12.22 Complete the Pythagorean triples below and verify that each set of side lengths corresponds to a right triangle.

____, 4, ____   ____, ____, 13

8, ____, ____   ____, 24, ____

Pythagorean triples are commonly known sets, each consisting of three values that satisfy the Pythagorean Theorem.

$\underline{3}$, 4, $\underline{5}$	$\underline{5}$, $\underline{12}$, 13	8, $\underline{15}$, $\underline{17}$	$\underline{7}$, 24, $\underline{25}$
$3^2 + 4^2 = 5^2$	$5^2 + 12^2 = 13^2$	$8^2 + 15^2 = 17^2$	$7^2 + 24^2 = 25^2$
$9 + 16 = 25$	$25 + 144 = 169$	$64 + 225 = 289$	$49 + 576 = 625$
$25 = 25$	$169 = 169$	$289 = 289$	$625 = 625$

Note that multiples of the Pythagorean triples are also right triangles. For instance, doubling the Pythagorean triple 3, 4, 5 produces the equally valid Pythagorean triple 6, 8, 10.

12.23 Draw $\square QRST$ such that the diagonals intersect at point P. If $QS = 16$, $RS = 17$, and $RT = 30$, what kind of parallelogram is $QRST$?

Construct a diagram of $\square QRST$ and use P to mark the point at which the diagonals intersect. Remember that the diagonals of a parallelogram bisect each other. Thus, if $QS = 16$, then $QP = PS = 8$; if $RT = 30$, then $TP = PR = 15$.

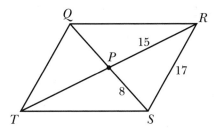

Consider $\triangle RPS$, constructed of sides that are 8, 15, and 17 units long. Problem 12.22 identifies these three numbers as a Pythagorean triple, so $\triangle RPS$ is a right triangle.

The largest angle of a triangle is opposite the longest side, so $m\angle RPS = 90°$ and the diagonals of $\square QRST$ are perpendicular: $\overline{QS} \perp \overline{RT}$. Therefore, $\square QRST$ is a rhombus.

The right angle is the largest angle of a right triangle.

See Problem 9.19.

12.24 Assume that $\overline{YM}$ is a median of the right triangle XYZ below. Given $YM = 9$ and $YZ = 12$, prove $m\angle YMZ < m\angle XMY$.

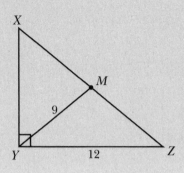

Statement	Reason
1. $m\angle XYZ = 90°$; $YM = 9$; $YZ = 12$; $\overline{YM}$ is a median	1. Given
2. M is the midpoint of $\overline{XZ}$	2. A median extends from a vertex of a triangle to the midpoint of the opposite side
3. $XM = YM = ZM = 9$	3. The midpoint of the hypotenuse of a right triangle is equidistant from the vertices of the triangle
4. $XM + MZ = XZ$	4. Segment addition postulate
5. $9 + 9 = XZ$	5. Substitution property of equality (Steps 3, 4)
6. $18 = XZ$	6. Combination of like terms
7. $(XY)^2 + (YZ)^2 = (XZ)^2$	7. Pythagorean Theorem
8. $(XY)^2 + (12)^2 = (18)^2$	8. Substitution property of equality (Steps 1, 6, and 7)
9. $(XY)^2 + 144 = 324$	9. Substitution property of equality
10. $(XY)^2 = 180$	10. Subtraction property of equality
11. $XY = \sqrt{180} = \sqrt{36 \cdot 5} = 6\sqrt{5}$	11. Properties of radical expressions and equations
12. $YZ < XY$	12. Step 1 states that $YZ = 12$; Step 11 states that $XY = 6\sqrt{5} \approx 13.416$
13. $m\angle YMZ < m\angle XMY$	13. SSS inequality theorem

See Problem 9.31.

It's not written explicitly, but you can substitute $12^2 = 144$ and $18^2 = 324$ into the equation.

Two sides of $\triangle XMY$ are congruent to two sides of $\triangle YMZ$. Compare the remaining sides of each triangle. Whichever is larger has the larger opposite angle.

Special Right Triangles
45°-45°-90° and 30°-60°-90° triangles

Note: Problems 12.25–12.28 refer to the following diagram, in which m∠C = 45°.

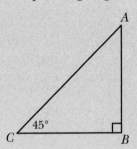

12.25 Use the diagram to prove that a right triangle containing one 45° angle must be isosceles.

The diagram indicates that $m\angle C = 45°$. Your goal is to prove that the legs of the right triangle are congruent: $\overline{AB} \cong \overline{BC}$.

Statement	Reason
1. $m\angle B = 90°$; $m\angle C = 45°$	1. Given
2. $m\angle A + m\angle B + m\angle C = 180°$	2. The sum of the interior angles of a triangle is 180°
3. $m\angle A + 90° + 45° = 180°$	3. Substitution property of equality (Steps 1, 2)
4. $m\angle A + 135° = 180°$	4. Combination of like terms
5. $m\angle A = 45°$	5. Subtraction property of equality
6. $m\angle A = 45° = m\angle C$	6. Transitive property of equality
7. $\overline{AB} \cong \overline{BC}$	7. Converse of the isosceles triangle theorem
8. $\triangle XYZ$ is isosceles	8. A triangle is isosceles if at least two of its sides are congruent

> If two angles of a triangle are congruent, the sides opposite those angles are congruent.

Note: Problems 12.25–12.28 refer to the diagram in Problem 12.25, in which m∠C = 45°.

12.26 Assume that the legs of ΔABC have length s. Generate a formula in terms of s to calculate the length of the hypotenuse h.

According to Problem 12.25, the legs of ΔABC are congruent. Let s represent the length of each leg and let h represent the length of the hypotenuse. Apply the Pythagorean Theorem to right triangle ABC.

$$s^2 + s^2 = h^2$$
$$2s^2 = h^2$$

Solve the equation for h.

$$\sqrt{2s^2} = \sqrt{h^2}$$
$$\sqrt{2} \cdot \sqrt{s^2} = h$$
$$s\sqrt{2} = h$$

> To figure out the hypotenuse's length in a 45-45-90 triangle, multiply the length of one leg by √2.

If s is the length of a leg of an isosceles right triangle (also referred to as a 45°-45°-90° triangle because those are the measures of its interior angles), then the length of the hypotenuse is $s\sqrt{2}$.

Note: Problems 12.25–12.28 refer to the diagram in Problem 12.25, in which m∠C = 45°.

12.27 Calculate AC, given $AB = BC = 3\sqrt{2}$.

According to Problem 12.26, if the length of one leg of a 45°-45°-90° triangle is s, then the length of the hypotenuse is $s\sqrt{2}$. In this problem, the lengths of the legs are given $\left(AB = BC = 3\sqrt{2}\right)$; multiply that length by $\sqrt{2}$ to calculate the length of the hypotenuse.

$$AC = (AB)\left(\sqrt{2}\right)$$
$$= \left(3\sqrt{2}\right)\left(\sqrt{2}\right)$$
$$= 3\sqrt{2 \cdot 2}$$
$$= 3\sqrt{4}$$
$$= 3 \cdot 2$$
$$= 6$$

Note: Problems 12.25–12.28 refer to the diagram in Problem 12.25, in which m∠C = 45°.

12.28 Calculate *AB*, given *AC* = 8.

The hypotenuse of a 45°-45°-90° right triangle is $\sqrt{2}$ times as long as a leg of the triangle.

$$AC = \left(\sqrt{2}\right)(AB)$$

Substitute *AC* = 8 into the equation and solve for *AB*.

$$8 = \left(\sqrt{2}\right)(AB)$$

$$\frac{8}{\sqrt{2}} = AB$$

Rationalize the denominator.

See Problem 3.20 for help in rationalizing expressions with a square root in the denominator.

$$\frac{8 \cdot \sqrt{2}}{\sqrt{2} \cdot \sqrt{2}} = AB$$

$$\frac{8\sqrt{2}}{\sqrt{4}} = AB$$

$$\frac{8\sqrt{2}}{2} = AB$$

Reduce the fraction to lowest terms.

$$\frac{8}{2}\sqrt{2} = AB$$

$$4\sqrt{2} = AB$$

Note: Problems 12.29–12.31 refer to the 30°-60°-90° triangle below, in which RT = 4.

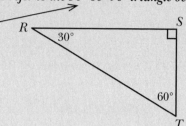

> Like the 45-45-90 triangle, the "special" 30-60-90 triangle is named after its angle measures.

12.29 Calculate *ST*.

The shortest side of a triangle is opposite the smallest angle. Thus, the shortest side of a 30°-60°-90° triangle is opposite the 30° angle; the shortest side of △*RST* is $\overline{ST}$.

A 30°-60°-90° triangle is classified as a "special triangle" (in conjunction with 45°-45°-90° triangles, discussed in Problems 12.25–12.28) because, given the length of only one side, you can quickly determine the lengths of the remaining sides.

Specifically, the length of the side opposite the 30° angle in a 30°-60°-90° right triangle is half the length of the hypotenuse. In this problem, the hypotenuse *RT* is 4 units long. Therefore, $ST = 4 \div 2 = 2$.

Note: Problems 12.29–12.31 refer to the diagram in Problem 12.29, in which RT = 4.

12.30 Calculate *RS*.

> The hypotenuse is always the longest side because it's opposite the largest angle, the right angle.

As Problem 12.29 explains, the lengths of the sides of a 30°-60°-90° triangle can be calculated given the length of only one side. In this problem, you are given the length of the hypotenuse: $RT = 4$. The length of the side opposite the 60° angle (in this case, $\overline{RS}$) is $\sqrt{3}$ times the length of the shortest side (that is, the side opposite the 30° angle—in this case, $\overline{ST}$).

$$RS = \sqrt{3}\,(ST)$$

According to Problem 12.29, $ST = 2$.

$$RS = \sqrt{3}\,(2) = 2\sqrt{3}$$

12.31 Assume that *LMN* is a right triangle with hypotenuse $\overline{LN}$. Let *x* represent the length of the hypotenuse and $\frac{1}{2}x$ represent *MN*, the length of the shortest side. Justify the assertion of Problem 12.30 by demonstrating that the length of the remaining side, *LM*, is $\sqrt{3}(MN)$.

If $\overline{MN}$ is the shortest side of the 30°-60°-90° triangle, then according to Problem 12.29, it is half as long as the hypotenuse.

$$MN = \frac{1}{2}LN = \frac{1}{2}x$$

Your goal in this problem is to prove $LM = \sqrt{3}(MN)$. Substitute the value of *MN* into the equation.

$$LM = \sqrt{3}\left(\frac{1}{2}x\right) = \frac{\sqrt{3}}{1}\left(\frac{1}{2}x\right) = \frac{\sqrt{3}}{2}x$$

To verify that *LM* has the predicted length of $\frac{\sqrt{3}}{2}x$, apply the Pythagorean Theorem.

$$(LM)^2 + (MN)^2 = (LN)^2$$

Substitute $LN = x$ and $MN = \frac{1}{2}x$ into the equation.

$$(LM)^2 + \left(\frac{1}{2}x\right)^2 = \left(x^2\right)$$

$$(LM)^2 + \frac{1^2}{2^2}x^2 = x^2$$

$$(LM)^2 + \frac{1}{4}x^2 = x^2$$

Solve for *LM*.

$$(LM)^2 = 1x^2 - \frac{1}{4}x^2$$

$$(LM)^2 = \frac{4}{4}x^2 - \frac{1}{4}x^2$$

$$(LM)^2 = \frac{3}{4}x^2$$

$$\sqrt{(LM)^2} = \frac{\sqrt{3}}{\sqrt{4}}\sqrt{x^2}$$

$$LM = \frac{\sqrt{3}}{2}x$$

The final value of *LM* matches the predicted value, validating the assertion that *LM* is $\sqrt{3}$ times as long as the shortest side of the triangle, *MN*.

Note: Problems 12.32–12.33 refer to right triangle ABC below.

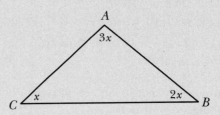

If x > 1, then
3x > 2x > x, and
you can assume
that the largest
angle of the
triangle is 3x.

12.32 Calculate the measures of the angles of $\triangle ABC$. Assume $x > 1$.

The problem indicates that $\triangle ABC$ is a right triangle, so $\angle A$ (the largest angle) must be a right angle. Use this information to calculate x.

$$m\angle A = 90°$$
$$3x = 90$$
$$x = \frac{90}{3}$$
$$x = 30$$

If $x = 30$, then $m\angle B = 2x = 2(30°) = 60°$. Therefore, $\triangle ABC$ is a 30°-60°-90° triangle because $m\angle A = 30°$, $m\angle B = 60°$, and $m\angle C = 90°$.

Note: Problems 12.32–12.33 refer to the diagram in Problem 12.32.

12.33 Given $AC = \frac{2}{3}$, calculate AB and BC.

Side $\overline{AC}$ is opposite the 60° angle, so it is $\sqrt{3}$ times as long as $\overline{AB}$, the side opposite the 30° angle.

$$AC = \sqrt{3}\,(AB)$$

Substitute $AC = \frac{2}{3}$ into the equation and solve for AB.

Dividing
the right
side by √3 and
multiplying the
left side by $\frac{1}{\sqrt{3}}$
are equivalent
operations. They're
two different
ways of doing
the same
thing.

$$\frac{2}{3} = \sqrt{3}\,(AB)$$
$$\frac{2}{3}\left(\frac{1}{\sqrt{3}}\right) = \frac{\sqrt{3}\,(AB)}{\sqrt{3}}$$
$$\frac{2}{3\sqrt{3}} = AB$$

Rationalize the denominator.

$$\frac{2}{3\sqrt{3}}\left(\frac{\sqrt{3}}{\sqrt{3}}\right) = AB$$

$$\frac{2\sqrt{3}}{3\sqrt{3^2}} = AB$$

$$\frac{2\sqrt{3}}{3 \cdot 3} = AB$$

$$\frac{2\sqrt{3}}{9} = AB$$

The longest side of a 30°-60°-90° triangle has twice the length of the shortest side. In this diagram, $BC = 2(AB)$.

$$BC = 2\left(\frac{2\sqrt{3}}{9}\right) = \frac{4\sqrt{3}}{9}$$

Note: Problems 12.34–12.35 refer to kite WXYZ below, in which $\overline{WX} \perp \overline{XY}$, WY = 10, and $m\angle WZY = 60°$.

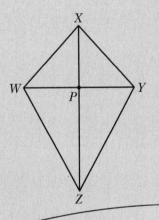

12.34 Calculate the perimeter of $\triangle WXY$.

> The perimeter is the sum of the lengths of the sides.

A kite is a quadrilateral with two pairs of adjacent congruent sides ($\overline{WX} \cong \overline{XY}$ and $\overline{WZ} \cong \overline{ZY}$). Therefore, $\triangle WXY$ and $\triangle WZY$ are isosceles. The problem indicates that $\overline{WX} \perp \overline{XY}$, so $m\angle WXY = 90°$ and $\triangle WXY$ is a right triangle.

Isosceles right triangles are 45°-45°-90° triangles, so the hypotenuse of $\triangle WXY$ is $\sqrt{2}$ times as long as a leg.

$$WY = \sqrt{2}\,(WX)$$

Substitute $WY = 10$ into the equation and solve for WX.

$$10 = \sqrt{2}\,(WX)$$

$$\frac{10}{\sqrt{2}} = WX$$

Rationalize the denominator.

$$\frac{10}{\sqrt{2}} \cdot \frac{\sqrt{2}}{\sqrt{2}} = WX$$

$$\frac{10\sqrt{2}}{\sqrt{2^2}} = WX$$

$$\frac{10\sqrt{2}}{2} = WX$$

Reduce the fraction to lowest terms.

$$\frac{10}{2}\sqrt{2} = WX$$

$$5\sqrt{2} = WX$$

The perimeter of $\triangle WXY$ is the sum of the lengths of its sides: $WX + XY + WY$. Recall that the triangle is isosceles, so its legs are the same length: $WX = XY$.

$$\text{perimeter of } \triangle WXY = WX + XY + WY$$
$$= 5\sqrt{2} + 5\sqrt{2} + 10$$
$$= 10\sqrt{2} + 10$$

Note: Problems 12.34–12.35 refer to kite WXYZ in Problem 12.34, in which $\overline{WX} \perp \overline{XY}$, WY = 10, and m∠WZY = 60°.

12.35 Calculate the perimeter of $\triangle PYZ$.

> See Problem 9.44.

The diagonal of a kite that connects the vertices of the congruent opposite angles is bisected by the other diagonal of a kite. In this kite, $\angle XWZ \cong \angle XYZ$, so $\overline{WY}$ is bisected by $\overline{XZ}$ and $WP = PY = 5$.

> $\triangle XWZ \cong \triangle XYZ$ by the SSS postulate, and $\angle WXP$ and $\angle YXP$ are corresponding angles of congruent triangles.

The noncongruent opposite angles of a kite are bisected by the diagonals that intersect them. In this diagram, $\angle WXY$ and $\angle WZY$ are bisected by $\overline{XZ}$. Therefore, $m\angle WZP = m\angle YZP = 30°$, half the measure of $\angle WZY$.

Recall that the diagonals of a kite are perpendicular, so $m\angle ZPY = 90°$ and $\triangle PYZ$ is a 30°-60°-90° triangle. The shortest side of the triangle is $\overline{PY}$ because it is opposite the 30° angle; $PY = 5$, as calculated earlier. The hypotenuse of the triangle is twice as long as the shortest side, so $YZ = 10$.

> If two angles of a triangle measure 30° and 90°, the last angle has to be 60° because all three angles must add up to 180°.

Side $\overline{PZ}$ is opposite the 60° angle, so its length is $\sqrt{3}$ times the length of the shortest side.

$$PZ = \sqrt{3}(PY) = \sqrt{3}(5) = 5\sqrt{3}$$

Calculate the perimeter of $\triangle PYZ$.

$$\text{perimeter of } \triangle PYZ = PY + YZ + PZ$$
$$= 5 + 10 + 5\sqrt{3}$$
$$= 15 + 5\sqrt{3}$$

12.36 Given $AF = 12$ in the diagram below, calculate EF.

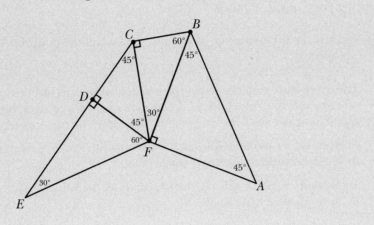

$\overline{AF}$ is a leg of the 45°-45°-90° triangle AFB. The legs of the isosceles triangle are congruent, so $AF = BF = 12$.

$\overline{BF}$ is the hypotenuse of the 30°-60°-90° triangle FCB, so it is twice as long as the shortest side: $BF = 2(CB)$. Thus, $CB = BF \div 2 = 12 \div 2 = 6$. Side $\overline{CF}$ is $\sqrt{3}$ times as long as side $\overline{CB}$, so $CF = 6\sqrt{3}$.

$\overline{CF}$ is the hypotenuse of 45°-45°-90° triangle FDC, so it is $\sqrt{2}$ times as long as leg $\overline{DF}$.

$$CF = \sqrt{2}\,(DF)$$
$$6\sqrt{3} = \sqrt{2}\,(DF)$$
$$\frac{6\sqrt{3}}{\sqrt{2}} = DF$$

Rationalize the denominator.

$$\frac{6\sqrt{3}}{\sqrt{2}} \cdot \frac{\sqrt{2}}{\sqrt{2}} = DF$$
$$\frac{6\sqrt{6}}{\sqrt{2^2}} = DF$$
$$\frac{6\sqrt{6}}{2} = DF$$

Reduce the fraction to lowest terms.

$$\frac{6}{2}\sqrt{6} = DF$$
$$3\sqrt{6} = DF$$

$\overline{DF}$ is the shortest leg of the 30°-60°-90° triangle DEF, so the hypotenuse $\overline{EF}$ is twice as long.

$$EF = 2(DF) = 2\left(3\sqrt{6}\right) = 6\sqrt{6}$$

Right Triangle Trigonometry

Tangent, sine, and cosine

12.37 Identify the three basic trigonometric ratios and the quotients by which they are defined.

> The adjacent side is one of the segments that forms the angle—the segment that's not the hypotenuse.

The three basic trigonometric ratios are tangent, sine, and cosine (abbreviated tan, sin, and cos, respectively). A trigonometric ratio relates three quantities of a right triangle: one of the angles and the lengths of two sides of the triangle.

The two sides used to define each ratio vary, but they are referenced based upon their relationship to the angle used.

Trigonometric ratios refer to the hypotenuse, the side opposite an angle, and the side adjacent to an angle.

In each of the following ratios, "hypotenuse," "opposite," and "adjacent" refer to the lengths of the sides of a right triangle. For instance, the numerator of cos D is the length of the leg of a right triangle that is adjacent to $\angle D$.

$$\text{tangent} = \frac{\text{opposite}}{\text{adjacent}} \qquad \text{sine} = \frac{\text{opposite}}{\text{hypotenuse}} \qquad \text{cosine} = \frac{\text{adjacent}}{\text{hypotenuse}}$$

Note: Problems 12.38–12.40 refer to the following diagram.

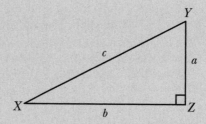

12.38 Evaluate tan X.

> You won't be calculating the tangent, sine, and cosine values of the right angle in the triangle.

The tangent ratio is defined as the quotient of the length of the side opposite an acute angle of a right triangle and the length of the side adjacent to that angle. In this diagram, the side opposite $\angle X$ has length a, and the side adjacent to $\angle X$ has length b.

$$\tan X = \frac{\text{length of the side opposite } \angle X}{\text{length of the side adjacent to } \angle X} = \frac{a}{b}$$

Note: Problems 12.38–12.40 refer to the diagram in Problem 12.38.

12.39 Evaluate cos *X*.

The trigonometric expression cos *X* is defined as the length of the side adjacent to ∠*X* divided by the length of the hypotenuse. In this diagram, the side adjacent to ∠*X* has length *b*, and the length of the hypotenuse is *c*.

$$\cos X = \frac{b}{c}$$

Note: Problems 12.38–12.40 refer to the diagram in Problem 12.38.

12.40 Verify that tan *Y* is equal to the quotient of sin *Y* and cos *Y*.

Your goal is to prove the following statement true.

$$\tan Y = \sin Y \div \cos Y$$

Evaluate each trigonometric expression based upon the diagram. Note that you should reference ∠*Y*, not ∠*X*, which was the argument of the trigonometric expressions in Problems 12.38–12.39.

$$\frac{\text{side opposite } \angle Y}{\text{side adjacent } \angle Y} = \frac{\text{side opposite } \angle Y}{\text{hypotenuse}} \div \frac{\text{side adjacent } \angle Y}{\text{hypotenuse}}$$

$$\frac{b}{a} = \frac{b}{c} \div \frac{a}{c}$$

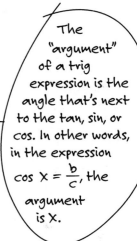

The "argument" of a trig expression is the angle that's next to the tan, sin, or cos. In other words, in the expression $\cos X = \frac{b}{c}$, the argument is X.

Dividing by a rational number is equivalent to multiplying by its reciprocal.

$$\frac{b}{a} = \frac{b}{c} \cdot \frac{c}{a}$$

$$\frac{b}{a} = \frac{b \cdot \cancel{c}}{a \cdot \cancel{c}}$$

$$\frac{b}{a} = \frac{b}{a}$$

See Problem 1.19.

The final statement $\left(\dfrac{b}{a} = \dfrac{b}{a} \right)$ is true; therefore, it is also true that tan *Y* is equal to the quotient of sin *Y* and cos *Y*.

12.41 Problem 12.29 states that the shortest side of a 30°-60°-90° triangle is half as long as the hypotenuse; verify this assertion using the sine ratio and $\triangle JKL$, as illustrated below.

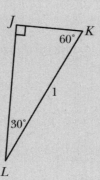

In this diagram, hypotenuse $\overline{LK}$ has length 1. Your goal is to demonstrate that $JK = \dfrac{1}{2}$ by means of the sine ratio, which is defined as the quotient of the lengths of an opposite side and the hypotenuse. Evaluate $\sin \angle L$.

You want to create an equation containing sine to solve for JK. Because sine references an opposite side, you need to use ∠L, as it is opposite JK.

$$\sin \angle L = \frac{JK}{LK}$$

$$\sin 30° = \frac{JK}{1}$$

$$\sin 30° = JK$$

Use a scientific or graphing calculator to evaluate $\sin 30°$. Ensure that the calculator is set to report values in degrees, not radians.

$$0.5 = \frac{1}{2} = JK$$

Therefore, $JK = \sin 30° = \dfrac{1}{2}$, which verifies that $\overline{JK}$ is half as long as $\overline{LK}$.

Note: Problems 12.42–12.44 refer to the following diagram.

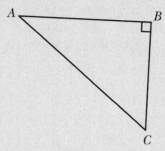

12.42 Given $m\angle C = 35°$ and $BC = 9$, calculate AB. Round the answer to the thousandths place. ◄

> Three places after the decimal point.

You are given the measure of $\angle C$. Note the relationship between this angle and the sides indicated: $\overline{BC}$ is adjacent to $\angle C$, and $\overline{AB}$ is opposite $\angle C$. The tangent ratio relates the opposite and adjacent sides, so evaluate $\tan C$.

$$\tan C = \frac{AB}{BC}$$
$$\tan 35° = \frac{AB}{9}$$

Multiply both sides of the equation by 9 to eliminate the fraction.

$$9(\tan 35°) = AB$$

Use a calculator to evaluate $\tan 35°$.

$$AB \approx 9(0.700207538)$$
$$AB \approx 6.302$$

Note: Problems 12.42–12.44 refer to the diagram in Problem 12.42.

12.43 Given $m\angle A = 40°$ and $AB = 13$, use a cosine ratio to calculate AC. Round the answer to the thousandths place.

> Cosine is the length of the adjacent side divided by the length of the hypotenuse.

Side $\overline{AC}$ is the hypotenuse of the right triangle, and $\overline{AB}$ is adjacent to $\angle A$. Evaluate $\cos 40°$. ◄

$$\cos A = \frac{AB}{AC}$$
$$\cos 40° = \frac{13}{AC}$$

Apply the means-extremes property.

$$\frac{\cos 40°}{1} = \frac{13}{AC}$$
$$(\cos 40°)(AC) = (1)(13)$$
$$(\cos 40°)(AC) = 13$$

Solve for AC.

$$AC = \frac{13}{\cos 40°}$$

Evaluate cos 40° using a calculator and round the quotient to the thousandths place.

$$AC \approx \frac{13}{0.766044443}$$
$$AC \approx 16.970$$

Sine references an opposite side, but you're given information about only ∠A and its adjacent side AB. However, AB is opposite ∠C, and it's easy to calculate m∠C.

Note: Problems 12.42–12.44 refer to the diagram in Problem 12.42.

12.44 Verify the solution to Problem 12.43 using a sine ratio.

The sum of the measures of a triangle is 180°.

$$m\angle A + m\angle B + m\angle C = 180°$$

Substitute $m\angle A = 40°$ and $m\angle B = 90°$ into the equation and solve for $m\angle C$.

$$40° + 90° + m\angle C = 180°$$
$$130° + m\angle C = 180°$$
$$m\angle C = 180° - 130°$$
$$m\angle C = 50°$$

You are given the length of the side opposite $\angle C$: $AB = 13$. Evaluate sin C.

$$\sin C = \frac{AB}{AC}$$
$$\sin 50° = \frac{13}{AC}$$

Apply the means-extremes property and solve for AC.

$$\frac{\sin 50°}{1} = \frac{13}{AC}$$
$$(\sin 50°)(AC) = (1)(13)$$
$$(\sin 50°)(AC) = 13$$
$$AC = \frac{13}{\sin 50°}$$

Use a calculator to evaluate sin 50° and round the final answer to the thousandths place.

$$AC \approx \frac{13}{0.766044443}$$
$$AC \approx 16.970$$

Notice that cos 40° and sin 50° have the same value (approximately 0.766044443), and the sine ratio produces the same value for AC as the cosine ratio generated in Problem 12.43 ($AC \approx 16.970$).

12.45 Calculate $m\angle F$ in the following diagram. Round the answer to the thousandths place.

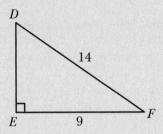

The lengths of two sides are given: $EF = 9$ and $DF = 14$. Side $\overline{EF}$ is adjacent to $\angle F$, and $\overline{DF}$ is the hypotenuse. Therefore, you should apply the cosine ratio.

$$\cos \angle F = \frac{EF}{DF}$$
$$\cos \angle F = \frac{9}{14}$$

Use the inverse cosine function on a scientific or graphing calculator to calculate $m\angle F$.

$$m\angle F = \cos^{-1}\left(\frac{9}{14}\right)$$
$$m\angle F \approx 49.995°$$

> Use cos, sin, and tan to calculate lengths of sides. Use cos⁻¹, sin⁻¹, and tan⁻¹ (the inverse trig functions) to calculate the measures of angles.

> If you got an answer of 0.873, your calculator is in radians mode—change it to degrees mode.

Note: Problems 12.46–12.48 refer to the following diagram.

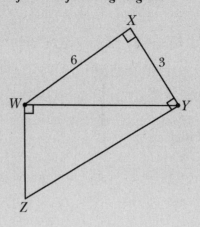

12.46 Calculate $m\angle XWY$. Round the answer to the thousandths place.

Sides $\overline{WX}$ and $\overline{XY}$ are, respectively, adjacent and opposite to $\angle XWY$. Apply the tangent ratio.

$$\tan\angle XWY = \frac{XY}{WX}$$
$$\tan\angle XWY = \frac{3}{6}$$
$$\tan\angle XWY = \frac{1}{2}$$

Use the inverse tangent function on a calculator to identify $m\angle XWY$.

$$m\angle XWY = \tan^{-1}\left(\frac{1}{2}\right)$$
$$m\angle XWY \approx 26.5650511771°$$
$$m\angle XWY \approx 26.565°$$

Note: Problems 12.46–12.48 refer to the diagram in Problem 12.46.

12.47 Calculate YZ.

Apply the Pythagorean Theorem to the sides of $\triangle WXY$ to calculate WY.

$$(WX)^2 + (XY)^2 = (WY)^2$$
$$6^2 + 3^2 = (WY)^2$$
$$36 + 9 = (WY)^2$$
$$45 = (WY)^2$$
$$\sqrt{45} = \sqrt{(WY)^2}$$
$$3\sqrt{5} = WY$$

Notice that $\angle XYW$ and $\angle ZYW$ are complementary angles because $m\angle XYW + m\angle ZYW = 90°$. Furthermore, $\angle XYW$ and $\angle XWY$ are complementary because they are the acute angles of a right triangle. Two angles that are complements of the same angle ($\angle XYW$) are congruent to each other: $m\angle XWY = m\angle ZYW \approx 26.5650511771°$. ←

According to Problem 12.46.

Side $\overline{WY}$ is adjacent to $\angle ZYW$, and you are instructed to calculate the length of hypotenuse $\overline{YZ}$. Apply the cosine ratio.

$$\cos \angle WYZ = \frac{WY}{YZ}$$

$$\cos 26.5650511771° = \frac{3\sqrt{5}}{YZ}$$

Solve for YZ.

$$\left(\cos 26.5650511771°\right)\left(YZ\right) = 3\sqrt{5}$$

$$YZ = \frac{3\sqrt{5}}{\cos 26.5650511771°}$$

$$YZ = \frac{15}{2}$$

Note: Problems 12.46–12.48 refer to the diagram in Problem 12.46.

12.48 Calculate the perimeter of *WXYZ*.

The lengths of three sides of quadrilateral $WXYZ$ are known. The diagram indicates that $WX = 6$ and $XY = 3$; according to Problem 12.47, $YZ = \frac{15}{2}$. Apply the Pythagorean Theorem to the sides of $\triangle WYZ$ to calculate the remaining side of the quadrilateral.

$$(WZ)^2 + (WY)^2 = (YZ)^2$$

$$(WZ)^2 + \left(3\sqrt{5}\right)^2 = \left(\frac{15}{2}\right)^2$$

$$(WZ)^2 + (9)(5) = \frac{225}{4}$$

$$(WZ)^2 = \frac{225}{4} - 45$$

$$(WZ)^2 = \frac{225}{4} - \frac{180}{4}$$

$$(WZ)^2 = \frac{45}{4}$$

$$\sqrt{(WZ)^2} = \frac{\sqrt{45}}{\sqrt{4}}$$

$$WZ = \frac{3\sqrt{5}}{2}$$

Add the lengths of the sides of quadrilateral $WXYZ$.

$$\text{perimeter of } WXYZ = WX + XY + YZ + WZ$$

$$= 6 + 3 + \frac{15}{2} + \frac{3\sqrt{5}}{2}$$

$$= 9 + \frac{15}{2} + \frac{3\sqrt{5}}{2}$$

$$= \frac{18}{2} + \frac{15}{2} + \frac{3\sqrt{5}}{2}$$

$$= \frac{33 + 3\sqrt{5}}{2}$$

Chapter 13

CIRCLES

Round collections of points the same distance from a center

This chapter introduces the circle, the first geometric figure you've encountered thus far that is not constructed of segments. You'll learn about characteristics of circles and their related segments, and you'll explore the notion of tangency. The chapter concludes with the introduction and classification of arcs. This serves as an introduction to Chapter 14, which explores the relationships between lengths of arcs, chords, secants, and tangents in more detail.

The preceding chapters focused primarily on arithmetic, points, lines, planes, and geometric figures you can create with straightedges—figures that are made up of segments and vertices, such as triangles, parallelograms, and trapezoids. Circles are different because, well, they're curvy.

Circles and Related Segments

Including chords and secants

13.1 Identify the key difference between the definitions of a circle and a sphere.

> A circle is basically a cross-section of a sphere.

A circle is defined as a set of *coplanar* points equidistant from a fixed point called the center, whereas a sphere consists of *all* points equidistant from a center. Therefore, a circle is a two-dimensional figure and a sphere is a three-dimensional figure.

Note: Problems 13.2–13.4 refer to the following diagram.

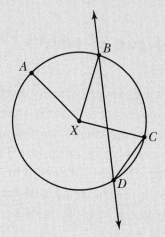

> "Radii" is the plural form of "radius." So in less correct English, this problem is asking you to find the "radiuses."

13.2 Identify the radii in the diagram.

A radius is a segment with one endpoint that is the center of a circle and one endpoint that lies on the circle. In this diagram, the center of the circle is X. Exactly three segments connect X to a point on the circle and are, therefore, radii of $\bigcirc X$: $\overline{AX}$, $\overline{BX}$, and $\overline{CX}$.

Note: Problems 13.2–13.4 refer to the diagram in Problem 13.2.

13.3 Identify the chord(s) in the diagram.

A chord is a segment connecting two points on a circle. This diagram contains exactly two chords: $\overline{BD}$ and $\overline{CD}$.

Note: Problems 13.2–13.4 refer to the diagram in Problem 13.2.

13.4 Identify the secant(s) in the diagram.

A secant is a line that contains a chord. Notice that chord $\overline{BD}$ is a segment of line $\overleftrightarrow{BD}$. Therefore, $\overleftrightarrow{BD}$ is a secant of $\bigcirc X$.

13.5 Determine whether the following statement is true or false. Justify your response.

A radius may also be a chord.

The statement is false. Both endpoints of a chord are on a circle, whereas one endpoint of a radius must be the center of the circle. ←

> If the radius of a circle is 0, then I guess you could say that the points on the circle and the center overlap, but that's because a circle with radius 0 isn't really a circle—it's a point.

13.6 Determine whether the following statement is true or false. Justify your response.

A secant may also be a chord.

The statement is false. A chord is a line segment, whereas a secant is a line that contains a chord. Though the concepts are related, a line does not have endpoints that lie on a circle; a line has no endpoints.

13.7 Determine whether the following statement is true or false. Justify your response.

A chord may also be a diameter.

The statement is true. Any chord that passes through the center of a circle is, by definition, a diameter of the circle. ←

> A diameter is twice as long as a radius. It's basically made up of two collinear radii that are glued together at the center of the circle.

13.8 Determine whether the following statement is true or false. Justify your response.

All radii of a circle are congruent.

The statement is true. A circle is defined as the set of coplanar points equidistant from a center. Thus, every segment connecting a point on the circle to the center of the circle (every radius) is the same length.

13.9 Construct a diagram that contains circle P with chord $\overline{XY}$ and tangent $\overleftrightarrow{YZ}$.

The endpoints of a chord lie on a circle, so X and Y must be points on $\bigodot P$. A tangent line intersects a circle at exactly one point. In this problem, the point of intersection is Y, because Y is explicitly identified as both a point on the circle and a point that lies on tangent line $\overleftrightarrow{YZ}$.

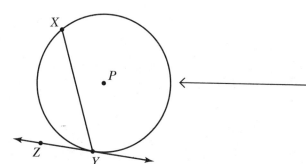

> This isn't the only way to draw this diagram. Just make sure you draw X and Y somewhere on the circle, and draw a tangent line through Y that also has a point Z on it.

13.10 Complete the following statement.

If A is a point on ⊙C, B is a point on ⊙D, and ⊙B ≅ ⊙D, then _____ ≅ _____.

If *A* is a point on ⊙*C*, then $\overline{AC}$ is a radius of the circle. Similarly, $\overline{BD}$ is a radius of ⊙*D*. Congruent circles have congruent radii. Therefore, if *A* is a point on ⊙*C*, *B* is a point on ⊙*D*, and ⊙*B* ≅ ⊙*D*, then $\underline{\overline{AC}} ≅ \underline{\overline{BD}}$.

13.11 Draw a rectangle inscribed in a circle.

Rectangle *ABCD* is inscribed in ⊙*Z* below because each of its vertices lies on the circle.

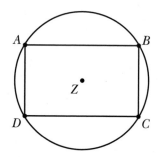

13.12 Draw a circle circumscribed about an isosceles trapezoid.

⊙*M* is circumscribed about isosceles trapezoid *WXYZ* below because each vertex of the trapezoid lies on ⊙*M*.

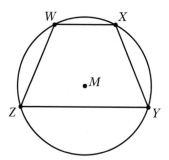

13.13 Draw a regular pentagon circumscribed about a circle.

Regular pentagon *LMNOP* is circumscribed about ○*C* because each of its sides is tangent to the circle. ←

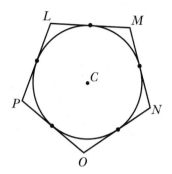

A polygon is regular when all its sides and angles are congruent.

When a polygon is inscribed in a circle, its vertices intersect the circle. When a polygon circumscribes a circle, each of its sides is tangent to the circle. Tangents are covered in Problems 13.17–13.20.

13.14 Draw a circle inscribed in an obtuse triangle.

○*K* is inscribed in Δ*DEF* because it is tangent to each side of the triangle.

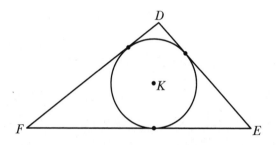

13.15 Given that $\overline{AC}$ and $\overline{BC}$ are radii of ○*C* such that *AC* = *BC* = 9 and *m*∠*ACB* = 90°, calculate the length of chord $\overline{AB}$.

Construct a diagram illustrating ○*C* with perpendicular radii $\overline{AC}$ and $\overline{BC}$. Each radius has length 9, so Δ*ACB* is an isosceles right triangle.

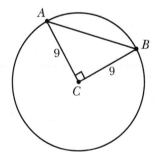

See Problem 12.26.

The hypotenuse of a 45°-45°-90° triangle is $\sqrt{2}$ times the length of a leg. Therefore, *AB* = $9\sqrt{2}$.

13.16 Given that *W*, *X*, *Y*, and *Z* are collinear points and *WZ* = 2(*XZ*) = 6(*YZ*), draw and classify the circles centered at *Z* passing through points *W*, *X*, and *Y*.

Each of the segments $\overline{WZ}$, $\overline{XZ}$, and $\overline{YZ}$ share the common endpoint *Z*, which is also the center of the three circles described here. Additionally, all four points lie along the same line. The longest of the segments is $\overline{WZ}$; it is twice as long as $\overline{XZ}$, so *X* is the midpoint of $\overline{WZ}$. The shortest segment is $\overline{YZ}$, which is one-third as long as $\overline{XZ}$.

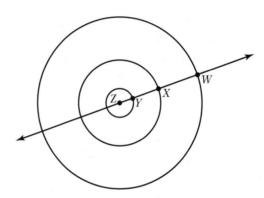

The circles are best classified as "concentric" because they lie on the same plane and share a common center.

Tangent Theorems

Radius-tangent intersection, internal tangents, and external tangents

13.17 Draw two externally tangent circles.

In other words, one of the circles isn't inside the other one; they're completely separate apart from the one point where the circles touch.

Externally tangent circles are tangent to the same line at the same point, such that the regions bounded by the circles do not overlap. In the diagram below, ⊙*A* and ⊙*B* are externally tangent because they are both tangent to line *l* at point *E*.

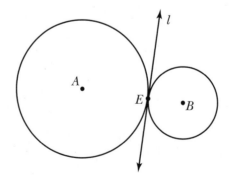

13.18 Draw two internally tangent circles.

As with the externally tangent circles described in Problem 13.17, internally tangent circles are tangent to the same line at the same point; however, the regions bounded by internally tangent circles overlap. In the diagram below, $\bigcirc C$ and $\bigcirc D$ are internally tangent because they are both tangent to line m at the same point, I.

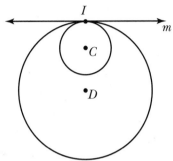

Note: Problems 13.19–13.20 refer to the following diagram.

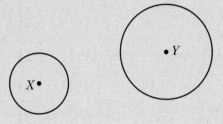

13.19 Draw the internal tangent lines common to $\bigcirc X$ and $\bigcirc Y$.

The internal tangent lines common to a pair of circles are tangent to each of the circles and pass between the centers of the circles. Two such lines can be drawn for circles X and Y.

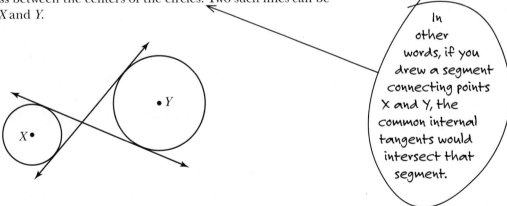

In other words, if you drew a segment connecting points X and Y, the common internal tangents would intersect that segment.

13.20 Draw the external tangent lines common to ⊙*X* and ⊙*Y*.

External tangent lines common to a pair of circles are tangent to each of the circles and do *not* pass between their centers. Two such lines can be drawn for circles *X* and *Y*.

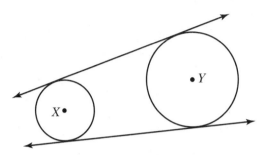

13.21 Complete the following statement.

If line k is tangent to ⊙C at point X, then the radius $\overline{CX}$ is _____ k at X. The converse is true as well:

_____

If line *k* is tangent to ⊙*C* at point *X*, then the radius $\overline{CX}$ is <u>perpendicular to</u> *k* at *X*. The converse is true as well: <u>if the radius $\overline{CX}$ of ⊙*C* is perpendicular to line *k* at point *X*, then *k* is tangent to ⊙*C* at *X*.</u>

> Point *C* will be somewhere else on the line. You probably won't draw it exactly in the right place so that m∠*BAC* = 60°, but that's okay. You'll still get the right answer.

13.22 Construct a diagram in which $\overleftrightarrow{BC}$ is tangent to ⊙*A* at point *B*. Given the radius of ⊙*A* is 5 and m∠*BAC* = 60°, calculate *BC*.

If $\overleftrightarrow{BC}$ is tangent to ⊙*A* at point *B*, then *B* is the intersection point of the line and the circle, as illustrated below.

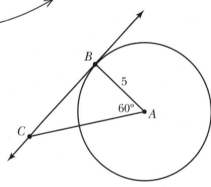

The tangent line to a circle is perpendicular to the radius drawn to the point of tangency. Therefore, $\overleftrightarrow{BC} \perp \overline{AB}$, m∠*CBA* = 90°, and Δ*ABC* is a 30°-60°-90° triangle. According to Problem 12.30, the side opposite the 60° angle is $\sqrt{3}$ times as long as the side opposite the 30° angle.

$$BC = \sqrt{3}\,(AB)$$
$$= \sqrt{3}\,(5)$$
$$= 5\sqrt{3}$$

13.23 Given $\overleftrightarrow{ZM}$ is tangent to $\bigcirc X$ at Z and $\overleftrightarrow{NY} \parallel \overleftrightarrow{ZM}$, prove that $\overleftrightarrow{NY}$ is tangent to $\bigcirc X$ at Y.

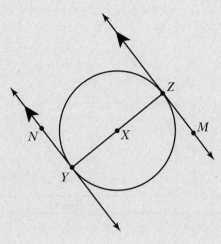

Statement	Reason
1. $\overleftrightarrow{ZM}$ is tangent to $\bigcirc X$ at Z; $\overleftrightarrow{NY} \parallel \overleftrightarrow{ZM}$	1. Given
2. $\overline{XY}$ and $\overline{XZ}$ are radii of $\bigcirc X$	2. A radius extends from the center of a circle to a point on the circle
3. $\overline{XZ} \perp \overleftrightarrow{ZM}$	3. A line tangent to the circle is perpendicular to the radius drawn to the point of tangency
4. $\overline{XY} \perp \overleftrightarrow{NY}$	4. If two parallel lines are intersected by a transversal that is perpendicular to one of the lines, then the other line is perpendicular to the transversal as well
5. $\overleftrightarrow{NY}$ is tangent to $\bigcirc X$ at Y	5. If a line is perpendicular to a radius at a point on the circle, then the line is tangent to the circle at that point

You could also use same-side interior angles or alternate interior angles to prove that $\overline{XY} \perp \overleftrightarrow{NY}$.

13.24 Given $\overrightarrow{JP}$ bisects $\angle LJK$ and $JL = JK$, prove that $\overleftrightarrow{LK}$ is tangent to $\bigcirc J$.

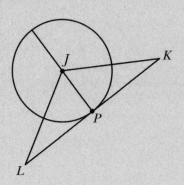

Statement	Reason
1. $\overrightarrow{JP}$ bisects $\angle LJK$; $JL = JK$	1. Given
2. $\angle LJP \cong \angle KJP$	2. An angle bisector divides an angle into two congruent angles
3. $\overline{JP} \cong \overline{JP}$	3. Reflexive property of congruency
4. $\triangle LJP \cong \triangle KJP$	4. SAS postulate
5. $\angle LPJ \cong \angle KPJ$	5. CPCTC
6. $m\angle LPJ + m\angle JPK = 180°$	6. Angle addition postulate
7. $m\angle LPJ = m\angle JPK = 90°$	7. If two angles are congruent and supplementary, then they must be right angles
8. $\overline{JP} \perp \overleftrightarrow{LK}$	8. Perpendicular lines form right angles
9. $\overleftrightarrow{LK}$ is tangent to $\bigcirc J$ at P	9. If a line is perpendicular to a radius at a point on the circle, then the line is tangent to the circle at that point

13.25 Complete the following theorem.

Two segments tangent to a circle that extend from a _____ endpoint are _____.

Two segments tangent to a circle that extend from a <u>common exterior</u> endpoint are <u>congruent</u>. In other words, two segments are congruent if they share one endpoint outside a circle and the remaining endpoints of each segment represent the points at which the segments are tangent to the circle.

13.26 Use the following diagram to prove the theorem completed in Problem 13.25. Assume that $\overline{PX}$ and $\overline{PY}$ are tangent to $\bigcirc A$ at points X and Y, respectively.

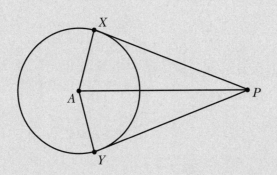

Segments tangent to a circle that extend from a common point outside the circle are congruent, according to the theorem in Problem 13.25. Therefore, your goal is to prove $\overline{PX} \cong \overline{PY}$.

Statement	Reason
1. $\overline{PX}$ is tangent to $\bigcirc A$ at X; $\overline{PY}$ is tangent to $\bigcirc A$ at Y	1. Given
2. $\overline{AX} \cong \overline{AY}$	2. All radii of a circle are congruent
3. $m\angle AXP = m\angle AYP = 90°$	3. The radius drawn to a point of tangency is perpendicular to the tangent line at that point
4. $\overline{AP} \cong \overline{AP}$	4. Reflexive property of congruency
5. $\triangle AXP \cong \triangle AYP$	5. HL theorem ←
6. $\overline{PX} \cong \overline{PY}$	6. CPCTC

Hypotenuse-leg theorem: $\overline{AP}$ is the hypotenuse of both triangles and the congruent legs are the radii (identified in Step 2).

13.27 In the diagram below, $\overline{PM}$ and $\overline{PN}$ are tangent to $\bigcirc C$. Given $MC = 3$ and $PC = 9$, calculate PN.

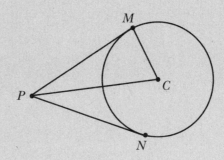

$\overline{PC}$ is the hypotenuse of ΔPMC because it's opposite the right angle. The Pythagorean Theorem is $a^2 + b^2 = c^2$, and c HAS to represent the hypotenuse, so set $c = 9$.

A radius drawn to a point of tangency is perpendicular to the tangent line at that point. Therefore, $\overline{CM} \perp \overline{PM}$ and $m\angle PMC = 90°$. Because ΔPMC is a right triangle, you can apply the Pythagorean Theorem to calculate the length of leg $\overline{PM}$.

$$(PM)^2 + (MC)^2 = (PC)^2$$
$$(PM)^2 + 3^2 = 9^2$$
$$(PM)^2 + 9 = 81$$
$$(PM)^2 = 81 - 9$$
$$\sqrt{(PM)^2} = \sqrt{72}$$
$$PM = 6\sqrt{2}$$

Segments $\overline{PM}$ and $\overline{PN}$ are tangent to $\bigcirc C$ and extend from the common exterior point P. According to the theorem presented in Problem 13.25, $PM = PN$. Therefore, $PN = 6\sqrt{2}$.

Note: Problems 13.28–13.29 refer to the following diagram, in which $\overline{AB}$ and $\overline{AX}$ are tangent to $\bigcirc M$ and $\overline{AC}$ and $\overline{AX}$ are tangent to $\bigcirc N$.

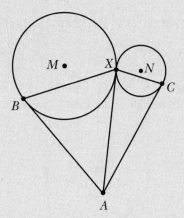

13.28 Prove $\overline{AB} \cong \overline{AC}$.

Statement	Reason
1. $\overline{AB}$ and $\overline{AX}$ are tangent to $\bigcirc M$; $\overline{AC}$ and $\overline{AX}$ are tangent to $\bigcirc N$	1. Given
2. $\overline{AB} \cong \overline{AX}$; $\overline{AX} \cong \overline{AC}$	2. Two segments tangent to a circle that extend from a common exterior endpoint are congruent
3. $\overline{AB} \cong \overline{AC}$	3. Transitive property of congruency

Note: Problems 13.28–13.29 refer to the diagram in Problem 13.28, in which $\overline{AB}$ and $\overline{AX}$ are tangent to $\odot M$, and $\overline{AC}$ and $\overline{AX}$ are tangent to $\odot N$.

13.29 Given $BX = 11$ and $m\angle BAX = 42°$, calculate AB accurate to the thousandths place.

$\overline{AB}$ and $\overline{AX}$ are tangent to $\odot M$ and extend from the common external point A; thus, $AB = AX$ and ABX is an isosceles triangle with vertex angle BAX. Consider the following diagram, which omits $\odot N$ and other extraneous information.

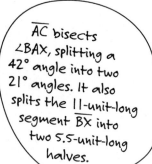

$\overline{AC}$ bisects $\angle BAX$, splitting a 42° angle into two 21° angles. It also splits the 11-unit-long segment $\overline{BX}$ into two 5.5-unit-long halves.

See Problem 12.41 if you need to review sine.

The angle bisector of the vertex angle of an isosceles triangle is the perpendicular bisector of the base. In this diagram, angle bisector $\overline{AC}$ bisects base $\overline{BX}$ at point C, resulting in the congruent segments $\overline{BC}$ and $\overline{CX}$.

Apply the sine ratio to calculate AB.

$$\sin\angle BAC = \frac{BC}{AB}$$

$$\sin 21° = \frac{5.5}{AB}$$

$$(\sin 21°)(AB) = 5.5$$

$$AB = \frac{5.5}{\sin 21°}$$

$$AB \approx \frac{5.5}{0.35836795}$$

$$AB \approx 15.347$$

Note: Problems 13.30–13.31 refer to the following diagram, in which $\overleftrightarrow{WZ}$ and $\overleftrightarrow{XY}$ are common internal tangent lines to $\bigcirc A$ and $\bigcirc B$.

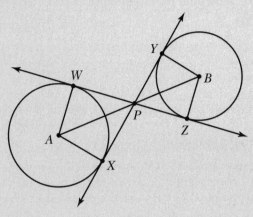

13.30 Prove $\triangle AXP \sim \triangle BYP$.

Statement	Reason
1. $\overleftrightarrow{WZ}$ and $\overleftrightarrow{XY}$ are tangent to $\bigcirc A$ and $\bigcirc B$	1. Given
2. $m\angle AXP = 90°$; $m\angle BYP = 90°$	2. The radius drawn to a point of tangency is perpendicular to the tangent line at that point
3. $m\angle AXP = m\angle BYP$	3. Transitive property of equality
4. $\angle APX \cong \angle BPY$	4. Vertical angles are congruent
5. $\triangle AXP \sim \triangle BYP$	5. AA similarity postulate

Note: Problems 13.30–13.31 refer to the diagram in Problem 13.30, in which $\overleftrightarrow{WZ}$ and $\overleftrightarrow{XY}$ are common internal tangent lines to $\bigcirc A$ and $\bigcirc B$.

13.31 Given $XP = 6$ and $XY = 14$, calculate PZ.

Express XY using the segment addition postulate.

$$XP + PY = XY$$

Substitute $XP = 6$ and $XY = 14$ into the equation to calculate PY.

$$6 + PY = 14$$
$$PY = 14 - 6$$
$$PY = 8$$

$\overline{PY}$ and $\overline{PZ}$ are segments tangent to $\bigcirc B$ that extend from common external point P. Therefore, $\overline{PY} \cong \overline{PZ}$ and $PY = PZ = 8$.

Properties of Arcs
Classifying little pieces of circles

13.32 Identify the differences between a minor arc, a major arc, and a semicircle, including their naming conventions.

> **In other words, what's different about the way you'd name a minor arc versus a major arc, given a diagram?**

A minor arc measures less than 180°, a major arc measures between 180° and 360°, and a semicircle measures exactly 180°. All arcs are named based upon their endpoints. However, two points on a circle define both a major and a minor arc, so a third point is included in the names of major arcs (and in the names of semicircles) for clarification.

> **The endpoints of a semicircle are also the endpoints of a diameter of the circle.**

In the following diagram, $\overline{AC}$ is a diameter of OX, so $\overset{\frown}{ABC}$ and $\overset{\frown}{ADC}$ are semicircles. Note that the middle points in the name of each arc (B and D, respectively) indicate the half of the circle through which each semicircle passes.

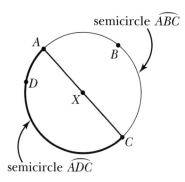

semicircle $\overset{\frown}{ABC}$

semicircle $\overset{\frown}{ADC}$

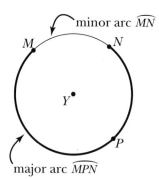

minor arc $\overset{\frown}{MN}$

major arc $\overset{\frown}{MPN}$

Minor arcs (such as $\overset{\frown}{MN}$) need to be identified only by their endpoints. To indicate a major arc (for example, an arc with a measure greater than that of a semicircle), include a point through which the arc passes between the endpoints in its name (for instance, $\overset{\frown}{MPN}$).

Note: Problems 13.33–13.36 refer to the following diagram, in which $\overline{XY}$ and $\overline{MN}$ are diameters of $\angle C$.

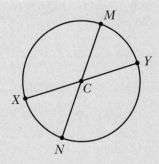

13.33 Identify four semicircles in the diagram.

Semicircles are arcs whose endpoints are also the endpoints of a diameter. The diameter $\overline{XY}$ divides the circle into two semicircles: $\overset{\frown}{XMY}$ and $\overset{\frown}{XNY}$. Diameter $\overline{MN}$ also forms two semicircles: $\overset{\frown}{MXN}$ and $\overset{\frown}{MYN}$.

Note: Problems 13.33–13.36 refer to the diagram in Problem 13.33, in which $\overline{XY}$ and $\overline{MN}$ are diameters of $\angle C$.

13.34 Identify four minor arcs in the diagram.

Minor arcs measure less than 180° and therefore are shorter than semicircles. In this diagram, $\widehat{XM}$ is a minor arc because it is shorter than semicircle $\widehat{XMY}$. Similarly, $\widehat{MY}$, $\widehat{YN}$, and $\widehat{NX}$ are minor arcs.

> The order of the endpoints doesn't matter: $\widehat{XM}$ and $\widehat{MX}$ refer to the same arc, as do $\widehat{XNY}$ and $\widehat{YNX}$.

Note: Problems 13.33–13.36 refer to the diagram in Problem 13.33, in which $\overline{XY}$ and $\overline{MN}$ are diameters of $\angle C$.

13.35 Identify four major arcs in the diagram.

Major arcs measure between 180° and 360°. Therefore, $\widehat{XMN}$ is a major arc because it begins at the same endpoint as semicircle $\widehat{XMY}$ but extends beyond point Y. Similarly, $\widehat{MYX}$, $\widehat{YNM}$, and $\widehat{NXY}$ are major arcs.

> They're longer than a semicircle but not as long as a full circle.

Note: Problems 13.33–13.36 refer to the diagram in Problem 13.33, in which $\overline{XY}$ and $\overline{MN}$ are diameters of $\angle C$.

13.36 Identify four central angles in the diagram.

A central angle is formed by two radii that intersect at the center of a circle. In this diagram, $\angle XCM$, $\angle MCY$, $\angle YCN$, and $\angle NCX$ are central angles. Mathematical convention dictates that central angles measure less than 180°, so straight and reflex angles are not normally classified as central angles.

> Reflex angles measure between 180° and 360°.

Note: Problems 13.37–13.39 refer to $\bigcirc Z$ in the diagram below.

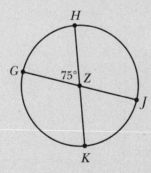

13.37 Calculate $m\widehat{GH}$.

> The m stands for "measure," so you're supposed to calculate the measure of arc GH.

The measure of an arc is defined as the measure of the central angle that intercepts the arc. In this diagram, central angle GZH intercepts $\widehat{GH}$, so $m\widehat{GH} = m\angle GZH = 75°$.

> The sides of the angle intersect the arc's endpoints.

Note: Problems 13.37–13.39 refer to the diagram in Problem 13.37.

13.38 Calculate $m\overset{\frown}{JK}$.

Notice that $\angle GZH$ and $\angle JZK$ are vertical angles; therefore, they have the same measure: $m\angle GZH = m\angle JZK = 75°$. The measure of an arc is defined as the measure of the central angle that intercepts the arc. In this diagram, central angle $\angle JZK$ intercepts $\overset{\frown}{JK}$, so $m\overset{\frown}{JK} = m\angle JZK = 75°$.

Note: Problems 13.37–13.39 refer to the diagram in Problem 13.37.

13.39 Calculate $m\overset{\frown}{HJ}$.

Because chord $\overline{GJ}$ passes through the center of $\bigcirc Z$, it is a diameter of the circle. It shares endpoints with $\overset{\frown}{GHJ}$, so $\overset{\frown}{GHJ}$ is a semicircle. Express the measure of the semicircle using the arc addition postulate.

$$m\overset{\frown}{GH} + m\overset{\frown}{HJ} = m\overset{\frown}{GHJ}$$

According to Problem 13.37, $m\overset{\frown}{GH} = 75°$. The measure of a semicircle is $180°$, so $m\overset{\frown}{GHJ} = 180°$. Substitute these values into the equation and solve for $m\overset{\frown}{HJ}$.

$$75° + m\overset{\frown}{HJ} = 180°$$
$$m\overset{\frown}{HJ} = 180° - 75°$$
$$m\overset{\frown}{HJ} = 105°$$

> The arc addition postulate is basically the segment addition postulate for curvy segments: the measure of a whole thing is equal to the sum of the measures of its parts.

Note: Problems 13.40–13.42 refer to the following diagram.

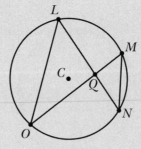

13.40 Given $m\overset{\frown}{LM} = 114°$, calculate $m\angle LNM$.

In this diagram, $\angle LNM$ is best described as an inscribed angle because its sides are chords of the circle and its vertex lies on the circle. The measures of inscribed angles are exactly one-half the measures of the arcs they intercept. Here, $\angle LNM$ intercepts $\overset{\frown}{LM}$.

$$m\angle LNM = \frac{1}{2} m\overset{\frown}{LM}$$

> The angle intercepts the arc because the sides of the angle intersect the endpoints of the arc.

Substitute $m\overarc{LM} = 114°$ into the equation and solve for $m\angle LNM$.

$$m\angle LNM = \frac{1}{2}(114°)$$
$$m\angle LNM = 57°$$

Note: Problems 13.40–13.42 refer to the diagram in Problem 13.40.

13.41 Given $m\overarc{ON} < 180°$, prove that $\angle OMN$ is acute.

Statement	Reason
1. $m\overarc{ON} < 180°$	1. Given
2. $\angle OMN$ is an inscribed angle	2. The vertex of an inscribed angle lies on a circle, and its sides are chords of the circle
3. $\angle OMN$ intercepts $\overarc{ON}$	3. The sides of $\angle OMN$ intersect $\odot C$ at the endpoints of $\overarc{ON}$
4. $m\angle OMN = \frac{1}{2}m\overarc{ON}$	4. The measure of an inscribed angle is half the measure of the intercepted arc
5. $2(m\angle OMN) = m\overarc{ON}$	5. Multiplication property of equality
6. $2(m\angle OMN) < 180°$	6. Substitution property of inequality (Steps 1, 5)
7. $m\angle OMN < 90°$	7. Two unequal quantities divided by the same positive value are unequal in the same order ←
8. $\angle OMN$ is an acute angle	8. Acute angles measure less than 90°

If A is smaller than B, and you divide both in half, half of A is still smaller than half of B.

Note: Problems 13.40–13.42 refer to the diagram in Problem 13.40.

13.42 Prove $m\angle OLN = m\angle OMN$.

Statement	Reason
1. $\angle OLN$ and $\angle OMN$ are inscribed angles	1. The vertex of an inscribed angle lies on a circle, and its sides are chords of the circle
2. $\angle OLN$ intercepts $\overarc{ON}$; $\angle OMN$ intercepts $\overarc{ON}$	2. The sides of $\angle OMN$ and $\angle OLN$ intersect $\bigcirc C$ at the endpoints of $\overarc{ON}$
3. $m\angle OLN = \dfrac{1}{2}m\overarc{ON}$; $m\angle OMN = \dfrac{1}{2}m\overarc{ON}$	3. The measure of an inscribed angle is half the measure of the intercepted arc
4. $m\angle OLN = m\angle OMN$	4. Transitive property of equality

13.43 Given $\overline{DQ} \perp \overline{EF}$ in the diagram below, prove $\dfrac{EQ}{DQ} = \dfrac{DQ}{FQ}$.

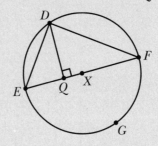

Statement	Reason
1. $\overline{DQ} \perp \overline{EF}$	1. Given
2. $\overline{EF}$ is a diameter of the circle	2. A diameter is a chord that passes through the center of a circle
3. $\overarc{EGF}$ is a semicircle	3. The endpoints of a diameter are also the endpoints of a semicircle
4. $m\overarc{EGF} = 180°$	4. All semicircles measure 180°
5. $\angle EDF$ is an inscribed angle	5. The vertex of an inscribed angle lies on a circle, and its sides are chords of the circle

6. $\angle EDF$ intercepts $\overarc{EGF}$

6. The sides of $\angle EDF$ intersect $\bigcirc X$ at the endpoints of $\overarc{EGF}$

7. $m\angle EDF = \dfrac{1}{2}\, m\overarc{EGF}$

7. The measure of an inscribed angle is half the measure of the intercepted arc

8. $m\angle EDF = \dfrac{1}{2}(180°)$

8. Substitution property of equality (Steps 4, 7)

9. $m\angle EDF = 90°$

9. Reduce $\dfrac{180°}{2}$ to lowest terms

10. $\triangle DEF$ is a right triangle

10. A right triangle contains a right angle

11. $\overline{DQ}$ is an altitude of $\triangle DEF$

11. An altitude is a segment perpendicular to one side of a triangle that extends from the opposite vertex

12. $\dfrac{EQ}{DQ} = \dfrac{DQ}{FQ}$

12. The length of the altitude drawn to a hypotenuse is the geometric mean between the segments of the hypotenuse it creates ←

See Problems 12.6–12.9.

Note: Problems 13.44–13.46 refer to the diagram below, in which $m\angle WCX = 80°$.

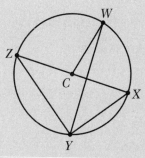

13.44 Calculate $m\overarc{ZW}$.

Notice that chord $\overline{ZX}$ is a diameter of $\bigcirc C$ because it passes through center C. Thus, $\overarc{ZWX}$ is a semicircle because it shares the same endpoints as a diameter. Express the measure of the semicircle using the arc addition postulate.

$$m\overarc{ZW} + m\overarc{WX} = m\overarc{ZWX}$$

You are given the measure of central angle $\angle WCX$. The measure of $\overset{\frown}{WX}$ is defined as the measure of the central angle that intercepts it: $m\overset{\frown}{WX} = m\angle WCX = 80°$. Substitute this value and the measure of the semicircle $\left(m\overset{\frown}{ZWX} = 180°\right)$ into the arc addition equation, and solve for $m\overset{\frown}{ZW}$.

$$m\overset{\frown}{ZW} + 80° = 180°$$
$$m\overset{\frown}{ZW} = 180° - 80°$$
$$m\overset{\frown}{ZW} = 100°$$

Note: Problems 13.44–13.46 refer to the diagram in Problem 13.44, in which m∠WCX = 80°.

13.45 Calculate $m\angle WYX$.

Arc $\overset{\frown}{WX}$ is intercepted by $\angle WCX$, a central angle, so the arc and the angle share the same measure: $m\overset{\frown}{WX} = 80°$. Note that $\overset{\frown}{WX}$ is also intercepted by $\angle WYX$, an inscribed angle. The measure of an inscribed angle is half the measure of the angle it intercepts.

$$m\angle WYX = \frac{1}{2} m\overset{\frown}{WX}$$

Substitute $m\overset{\frown}{WX} = 80°$ into the equation and solve for $m\angle WYX$.

$$m\angle WYX = \frac{1}{2}(80°) = \frac{80°}{2} = 40°$$

Note: Problems 13.44–13.46 refer to the diagram in Problem 13.44, in which m∠WCX = 80°.

13.46 Given $m\angle YZX = \frac{1}{2}(m\angle ZXY) + 15°$, calculate $m\overset{\frown}{XY}$.

Inscribed angle $\angle ZYX$ intercepts semicircle $\overset{\frown}{ZWX}$. Therefore, $m\angle ZYX = 90°$. The measures of the angles of $\triangle ZXY$ must have a sum of 180°.

$$m\angle YZX + m\angle ZYX + m\angle ZXY = 180°$$

Substitute $m\angle ZYX = 90°$ and $m\angle YZX = \frac{1}{2}(m\angle ZXY) + 15°$ into the equation.

$$\left[\frac{1}{2}(m\angle ZXY) + 15°\right] + [90°] + [m\angle ZXY] = 180°$$
$$\left(\frac{1}{2}m\angle ZXY + \frac{2}{2}m\angle ZXY\right) + (15° + 90°) = 180°$$
$$\frac{3}{2}m\angle ZXY + 105° = 180°$$

Any inscribed angle that intercepts a semicircle is a right angle. That's because semicircles measure 180° and inscribed angles have measures half as large as their intercepted arcs: 180° ÷ 2 = 90°.

Write m∠ZXY as $\frac{2}{2}$ m∠ZXY so you can add using common denominators.

Solve for $m\angle ZXY$.

$$\frac{3}{2} m\angle ZXY = 180° - 105°$$

$$\frac{3}{2} m\angle ZXY = 75°$$

$$\frac{2}{3}\left(\frac{3}{2} m\angle ZXY\right) = \frac{2}{3}\left(\frac{75°}{1}\right)$$

$$m\angle ZXY = \frac{150°}{3}$$

$$m\angle ZXY = 50°$$

Recall that $m\angle YZX = \frac{1}{2}(m\angle ZXY) + 15°$. Evaluate this expression given $m\angle ZXY = 50°$.

$$m\angle YZX = \frac{1}{2}(50°) + 15 = 25° + 15 = 40°$$

Because $\angle YZX$ is an inscribed angle, its measure is half the measure of the arc it intercepts $\left(\widehat{XY}\right)$.

$$m\angle YZX = \frac{1}{2} m\widehat{XY}$$

$$40° = \frac{1}{2} m\widehat{XY}$$

$$2(40°) = \frac{2}{1}\left(\frac{1}{2} m\widehat{XY}\right)$$

$$80° = m\widehat{XY}$$

Chapter 14

CHORD, SECANT, AND TANGENT THEOREMS

An avalanche of theorems about circles

Whereas Chapter 13 served primarily to familiarize you with the basic concepts of circles, this chapter investigates more advanced theorems that describe the relationships between those concepts. Thus, you should ensure that you have a firm mastery of Chapter 13 before moving on to explore the theorems in this chapter.

Chapter 13 was a definite change of pace from most of the chapters before it. With the introduction of circles came new ideas. Now that you know what circles are, it's time to get down to the nitty gritty. You may have noticed that the last chapter didn't cover a lot of theorems; you were just getting acclimated to circles. Hopefully you're acclimated now, because you're about to see more circle theorems than you can shake a stick at.

Arc and Chord Theorems

Theorems about segments inside a circle

Note: Problems 14.1–14.2 refer to ○X illustrated below.

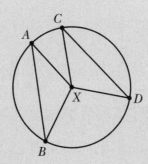

> You're proving an important theorem here: congruent chords intercept congruent arcs.

14.1 Given $\overline{AB} \cong \overline{CD}$, prove $\overset{\frown}{AB} \cong \overset{\frown}{CD}$.

Statement	Reason
1. $\overline{AB} \cong \overline{CD}$	1. Given
2. $\overline{AX} \cong \overline{BX} \cong \overline{CX} \cong \overline{DX}$	2. All radii of a circle are congruent
3. $\triangle AXB \cong \angle CXD$	3. SSS postulate
4. $m\angle AXB = m\angle CXD$	4. CPCTC
5. $m\angle AXB = m\overset{\frown}{AB}$; $m\angle CXD = m\overset{\frown}{CD}$	5. The measure of central angle is equal to the measure of the arc it intercepts
6. $m\overset{\frown}{AB} = m\overset{\frown}{CD}$	6. Substitution property of equality (Steps 4, 5)
7. $\overset{\frown}{AB} \cong \overset{\frown}{CD}$	7. Arcs of the same circle that have equal measures are congruent

Note: Problems 14.1–14.2 refer to the diagram in Problem 14.1.

14.2 Given $m\overset{\frown}{AB} = 3y + 2$, $m\overset{\frown}{CD} = 2 - (8y + 1)$, and $\overline{AB} \cong \overline{CD}$, calculate y.

According to Problem 14.1, congruent chords intercept congruent arcs. If $\overline{AB} \cong \overline{CD}$, then $\overset{\frown}{AB} \cong \overset{\frown}{CD}$ and the measures of the arcs are equal.

$$m\overset{\frown}{AB} = m\overset{\frown}{CD}$$
$$3y + 2 = 2 - (8y + 1)$$
$$3y + 2 = 2 - 8y - 1$$
$$3y + 2 = 1 - 8y$$

Solve for y.

$$3y + 8y = 1 - 2$$
$$11y = -1$$
$$y = -\frac{1}{11}$$

Note: Problems 14.3–14.4 refer to the diagram of $\bigcirc G$ below.

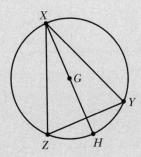

14.3 Given $\widehat{XZ} \cong \widehat{XY}$, prove that $\triangle XYZ$ is isosceles.

Statement	Reason
1. $\widehat{XZ} \cong \widehat{XY}$	1. Given
2. $\overline{XZ}$ intercepts $\widehat{XZ}$; $\overline{XY}$ intercepts $\widehat{XY}$	2. A chord that shares the same endpoints as an arc intercepts that arc
3. $\overline{XZ} \cong \overline{XY}$	3. Chords that intercept congruent arcs are congruent ←
4. $\triangle XYZ$ is isosceles	4. An isosceles triangle has at least two congruent sides

> This is the converse of the theorem in Problem 14.1, so the theorem works both ways. Congruent arcs mean congruent chords, and congruent chords mean congruent arcs. You could also say arcs are congruent iff (if and only if) chords are congruent.

Note: Problems 14.3–14.4 refer to the diagram of $\bigcirc G$ in Problem 14.3.

14.4 Given $\overline{XH}$ bisects $\angle ZXY$, prove that $\overline{XH}$ bisects $\widehat{ZY}$.

Statement	Reason
1. $\overline{XH}$ bisects $\angle ZXY$	1. Given
2. $m\angle ZXH = m\angle YXH$	2. An angle bisector divides an angle into two congruent angles
3. $m\angle ZXH = \frac{1}{2}m\widehat{ZH}$; $m\angle YXH = \frac{1}{2}m\widehat{YH}$	3. The measure of an inscribed angle is half the measure of the arc it intercepts

continues

continued

Statement	Reason
4. $\frac{1}{2}m\widehat{ZH} = \frac{1}{2}m\widehat{YH}$	4. Substitution property of equality (Steps 2, 3)
5. $m\widehat{ZH} = m\widehat{YH}$	5. Multiplication property of equality
6. H is the midpoint of $\widehat{ZY}$	6. A midpoint divides an arc into two congruent arcs
7. $\overline{XH}$ bisects $\widehat{ZY}$	7. An arc bisector intersects an arc at its midpoint

> Multiply both sides of the equation in Step 4 by 2 to get rid of the fractions.

> Arcs can have midpoints, just as segments can.

14.5 Explain why a regular, *n*-sided polygon inscribed in a circle divides the circle into *n* congruent arcs.

Consider the diagram below, in which regular hexagon *ABCDEF* is inscribed in ⊙*T*.

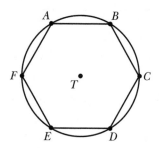

Because the hexagon is inscribed in the circle, each of its vertices lies on the circle. Thus, each side of hexagon *ABCDEF* is a chord of ⊙*T*. The hexagon is regular, so $\overline{AB} \cong \overline{BC} \cong \overline{CD} \cong \overline{DE} \cong \overline{EF} \cong \overline{FA}$. According to Problem 14.1, congruent chords intercept congruent arcs. Therefore, $\widehat{AB} \cong \widehat{BC} \cong \widehat{CD} \cong \widehat{DE} \cong \widehat{EF} \cong \widehat{FA}$.

Any *n*-sided regular polygon will produce a similar result. All *n* sides will be congruent chords of the circle in which the polygon is inscribed. The chords will intercept *n* arcs, which are therefore congruent.

Note: Problems 14.6–14.7 refer to the diagram of ⊙Z below.

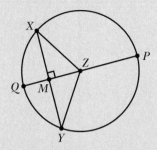

14.6 Use the diagram to prove that a diameter perpendicular to a chord bisects the chord.

Assume that the diameter and chord are perpendicular $\left(\overline{PQ} \perp \overline{XY}\right)$, and prove that the diameter bisects the chord (prove that $\overline{PQ}$ bisects $\overline{XY}$).

Statement	Reason
1. $\overline{PQ} \perp \overline{XY}$	1. Given
2. $\angle ZMX$ and $\angle ZMY$ are right angles	2. Perpendicular lines meet at right angles
3. $\overline{ZX} \cong \overline{ZY}$	3. All radii of a circle are congruent
4. $\overline{ZM} \cong \overline{ZM}$	4. Reflexive property of congruency
5. $\triangle ZMX \cong \triangle ZMY$	5. HL theorem
6. $\overline{XM} \cong \overline{YM}$	6. CPCTC
7. M is the midpoint of $\overline{XY}$	7. A midpoint divides a segment into two congruent segments
8. $\overline{PQ}$ bisects $\overline{XY}$	8. A segment bisector intersects the segment at its midpoint

Note: Problems 14.6–14.7 refer to the diagram of ⊙Z in Problem 14.6.

14.7 Use the diagram to prove that a diameter perpendicular to a chord bisects the arc intercepted by the chord.

Statement	Reason
1. $\overline{PQ} \perp \overline{XY}$	1. Given
2. $\angle ZMX$ and $\angle ZMY$ are right angles	2. Perpendicular lines meet at right angles
3. $\overline{ZX} \cong \overline{ZY}$	3. All radii of a circle are congruent

continues

continued

Statement	Reason
4. $\overline{ZM} \cong \overline{ZM}$	4. Reflexive property of congruency
5. $\triangle ZMX \cong \triangle ZMY$	5. HL theorem
6. $m\angle XZQ = m\angle YZQ$	6. CPCTC
7. $m\overset{\frown}{XQ} = m\angle XZQ$; $m\overset{\frown}{YQ} = m\angle YZQ$	7. The measure of an arc is defined as the measure of the central angle that intercepts the arc
8. $m\overset{\frown}{XQ} = m\overset{\frown}{YQ}$	8. Substitution property of equality (Steps 6, 7)
9. Q is the midpoint of $\overset{\frown}{XY}$	9. A midpoint divides an arc into two congruent arcs
10. $\overline{PQ}$ bisects $\overset{\frown}{XY}$	10. A bisector of an arc intersects the arc at its midpoint

Note: Problems 14.8–14.10 refer to the diagram of $\bigcirc M$ below, in which $m\overset{\frown}{AB} = 65°$, $m\overset{\frown}{DF} = 130°$, and $GC = 8.25$.

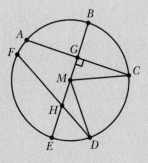

14.8 Calculate $m\angle BMC$.

According to the theorems in Problem 14.6–14.7, a diameter perpendicular to a chord bisects the chord and the arc intercepted by the chord. In this diagram, diameter $\overline{BE}$ is perpendicular to chord $\overline{AC}$. Therefore, $\overline{BE}$ bisects $\overline{AC}$ and $\overset{\frown}{AC}$, producing two congruent segments and two congruent arcs: $\overline{AG} \cong \overline{GC}$ and $\overset{\frown}{AB} \cong \overset{\frown}{BC}$.

You are given $m\overset{\frown}{AB} = 65°$, so $m\overset{\frown}{BC} = 65°$ as well. A central angle has the same measure as the arc it intercepts. In this diagram, $\angle BMC$ intercepts arc $\overset{\frown}{BC}$. Therefore, $m\angle BMC = m\overset{\frown}{BC} = 65°$.

Note: Problems 14.8–14.10 refer to the diagram of ⊙M in Problem 14.8, in which
$m\widehat{AB} = 65°$, $m\widehat{DF} = 130°$, *and* $GC = 8.25$.

14.9 Calculate *DF.*

According to Problem 14.8, $m\widehat{AB} = m\widehat{BC} = 65°$. Apply the arc addition postulate to calculate $m\widehat{AC}$.

$$m\widehat{AB} + m\widehat{BC} = m\widehat{AC}$$
$$65° + 65° = m\widehat{AC}$$
$$130° = m\widehat{AC}$$

Because $m\widehat{DF} = m\widehat{AC} = 130°$, $\widehat{DF} \cong \widehat{AC}$. Because congruent arcs must be intercepted by congruent chords, $\overline{DF} \cong \overline{AC}$. Apply the segment addition postulate to calculate *AC.*

$$AG + GC = AC$$

You are given $GC = 8.25$. Recall that $\overline{AG} \cong \overline{GC}$, so $AG = 8.25$ as well.

$$8.25 + 8.25 = AC$$
$$16.5 = AC$$

Because $\overline{DF} \cong \overline{AC}$, $DF = AC = 16.5$.

> $m\widehat{DF} = 130°$ is one of the given statements in the problem.

Note: Problems 14.8–14.10 refer to the diagram of ⊙M in Problem 14.8, in which
$m\widehat{AB} = 65°$, $m\widehat{DF} = 130°$, *and* $GC = 8.25$.

14.10 Calculate the radius of ⊙M accurate to the thousandths place.

Consider right triangle *MGC.* Problem 14.8 states that $m\angle GMC = 65°$, and you are given the length of the side opposite $\angle GMC$: $GC = 8.25$. The hypotenuse $\overline{MC}$ of $\triangle MGC$ is a radius of ⊙M. Apply the sine ratio to calculate *MC.*

$$\sin \angle GMC = \frac{GC}{MC}$$
$$\sin 65° = \frac{8.25}{MC}$$
$$\sin 65° (MC) = 8.25$$
$$MC = \frac{8.25}{\sin 65°}$$
$$MC \approx \frac{8.25}{0.906307787}$$
$$MC \approx 9.103$$

> $\angle BMC$ and $\angle GMC$ are different names for the same angle.

> See Problem 12.41 if you need to review sine.

Your final answer should be AB = something containing r's.

14.11 Assume that $\overarc{AB}$ is a minor arc intercepted by chord $\overline{AB}$, such that $\overline{AB}$ is 3 units from point C in the diagram below. Express AB in terms of r, the radius of $\bigcirc C$.

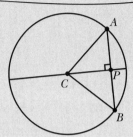

A diameter perpendicular to a chord bisects the chord, so $AP = PB$. Notice that $\overline{AC}$ and $\overline{BC}$ are radii of $\bigcirc C$; the problem states that the radius of the circle is represented by the variable r, so $AC = BC = r$.

The problem also states that chord $\overline{AB}$ is 3 units from the center of the circle. The distance between $\overline{AB}$ and C is CP, the length of the segment connecting $\overline{AB}$ to C that is perpendicular to $\overline{AB}$. Thus, $CP = 3$.

Apply the Pythagorean Theorem to right triangle CAP to calculate AP.

$$(CP)^2 + (AP)^2 = (AC)^2$$
$$3^2 + (AP)^2 = r^2$$
$$9 + (AP)^2 = r^2$$
$$(AP)^2 = r^2 - 9$$
$$\sqrt{(AP)^2} = \sqrt{r^2 - 9}$$
$$AP = \sqrt{r^2 - 9}$$

Multiply both sides of the equation by 2. If AP is half of AB, then AB is twice as large as AP.

Recall that $\overline{CP}$ bisects $\overline{AB}$ at point P, so AP is half of AB.

$$AP = \frac{1}{2}AB$$
$$2(AP) = AB$$
$$2\sqrt{r^2 - 9} = AB$$

Note: Use the diagram of ⊙A below to prove the theorems in Problems 14.12–14.13.

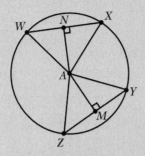

14.12 Prove that chords equidistant from the center of a circle are congruent.

This diagram contains chords $\overline{WX}$ and $\overline{ZY}$. Assume that the chords are equidistant from center A ($AN = AM$), and prove that the chords are congruent ($WX = ZY$).

> To prove chords congruent, show that they have the same length.

Statement	Reason
1. $AN = AM$	1. Given
2. $WN + NX = WX$; $ZM + MY = ZY$	2. Segment addition postulate
3. $AW = AX = AY = AZ$	3. Radii of the same circle are congruent
4. $\triangle ANW \cong \triangle ANX \cong \triangle AMZ \cong \triangle AMY$	4. HL theorem
5. $WN = ZM$; $NX = MY$	5. CPCTC
6. $WN + NX = ZM + MY$	6. Addition property of equality
7. $WX = ZY$	7. Substitution property of equality (Steps 2, 6)

> The hypotenuses of all four triangles are radii (and are therefore congruent). Each triangle also contains either leg $\overline{AN}$ or leg $\overline{AM}$, which are congruent.

> Start with the equation WN = ZM. You can add something to each side of the equation, as long as you're adding equal values. Step 5 says NX = MY, so add NX to the left side and MY to the right side.

Note: Use the diagram of ○A in Problem 14.12 to prove the theorems in Problems 14.12–14.13.

14.13 Prove that congruent chords are equidistant from the center of a circle.

Assume that $\overline{WX} \cong \overline{ZY}$ (and therefore $WX = ZY$) to prove that $AN = AM$.

Statement	Reason
1. $WX = ZY$	1. Given
2. $\frac{1}{2}(WX) = \frac{1}{2}(ZY)$	2. Multiplication property of equality
3. $NX = \frac{1}{2}(WX);\ MY = \frac{1}{2}(ZY)$	3. A diameter perpendicular to a chord bisects the chord
4. $NX = MY$	4. Substitution property of equality (Steps 2, 3)
5. $AX = AY$	5. All radii of the same circle are congruent
6. $\triangle ANX \cong \triangle AMY$	6. HL theorem
7. $AN = AM$	7. CPCTC
8. $\overline{WX}$ and $\overline{ZY}$ are equidistant from A	8. The distance between a point and a segment is the length of the perpendicular segment connecting them

It doesn't really matter that $\overline{AN}$ and $\overline{AM}$ aren't diameters. If you extend the segments through the center, you could make them diameters if you felt like it, but either way, chords $\overline{WX}$ and $\overline{ZY}$ get bisected.

Note: Problems 14.14–14.17 refer to the diagram below: kite JKLM inscribed in ○Q. Assume $m\widehat{KJ} = (2x-5)°$, $m\widehat{KL} = (5y)°$, and $m\widehat{MR} = 52.5°$.

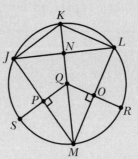

14.14 Calculate $m\widehat{JM}$.

A kite has exactly two pairs of congruent, adjacent sides. In this diagram, $\overline{KJ} \cong \overline{KL}$ and $\overline{JM} \cong \overline{ML}$. Congruent chords intercept congruent arcs, so $\widehat{JM} \cong \widehat{ML}$. Radius $\overline{QR}$ is perpendicular to $\overline{ML}$, so $\overline{QR}$ bisects $\widehat{ML}$ at R and $m\widehat{RL} = m\widehat{MR} = 52.5°$. Apply the arc addition postulate to calculate $m\widehat{ML}$.

$$m\,\overarc{MR} + m\,\overarc{RL} = m\,\overarc{ML}$$

$$52.5° + 52.5° = m\,\overarc{ML}$$

$$105° = m\,\overarc{ML}$$

Recall that $\overarc{JM} \cong \overarc{ML}$, so $m\,\overarc{JM} = m\,\overarc{ML} = 105°$.

Note: Problems 14.14–14.17 refer to the diagram in Problem 14.14: kite JKLM inscribed in ⊙Q. Assume $m\,\overarc{KJ} = (2x - 5)°$, $m\,\overarc{KL} = (5y)°$, and $m\,\overarc{MR} = 52.5°$.

14.15 Is the following statement true or false: $QP = QO$? Justify your answer.

The statement is true. As stated in Problem 14.14, $\overline{JM}$ and $\overline{ML}$ are congruent, adjacent sides of kite $JKLM$. The segments are also chords of ⊙Q. Congruent chords are equidistant from the center of a circle. The distance between Q and $\overline{JM}$ is defined as QP; the distance between Q and $\overline{ML}$ is defined as QO. Therefore, $QP = QO$.

Note: Problems 14.14–14.17 refer to the diagram in Problem 14.14: kite JKLM inscribed in ⊙Q. Assume $m\,\overarc{KJ} = (2x - 5)°$, $m\,\overarc{KL} = (5y)°$, and $m\,\overarc{MR} = 52.5°$.

14.16 Calculate x.

Chord $\overline{KM}$ is a diameter of the circle because it passes through center Q. Express $m\,\overarc{KJM}$ using the arc addition postulate.

$$m\,\overarc{KJ} + m\,\overarc{JM} = m\,\overarc{KJM}$$

Diameters of a circle intercept semicircles, so $m\,\overarc{KJM} = m\,\overarc{KLM} = 180°$. According to Problem 14.14, $m\,\overarc{JM} = 105°$. Substitute these values into the arc addition equation and solve for $m\,\overarc{KJ}$.

$$m\,\overarc{KJ} + 105° = 180°$$

$$m\,\overarc{KJ} = 180° - 105°$$

$$m\,\overarc{KJ} = 75°$$

You are given $m\,\overarc{KJ} = (2x - 5)°$.

$$2x - 5 = 75$$

$$2x = 75 + 5$$

$$2x = 80$$

$$x = \frac{80}{2}$$

$$x = 40$$

Note: Problems 14.14–14.17 refer to the diagram in Problem 14.14: kite JKLM inscribed in ⊙Q. Assume $m\widehat{KJ} = (2x - 5)°$, $m\widehat{KL} = (5y)°$, and $m\widehat{MR} = 52.5°$.

14.17 Calculate y.

As stated in Problem 14.14, $\overline{KJ}$ and $\overline{KL}$ are congruent, adjacent sides of kite JKLM. Congruent arcs intercept congruent chords: $\widehat{KJ} \cong \widehat{KL}$. Set the measures of the chords equal.

$$m\widehat{KJ} = m\widehat{KL}$$

According to Problem 14.16, $m\widehat{KJ} = 75°$. You are given $m\widehat{KL} = (5y)°$. Substitute the measures of each arc into the above equation and solve for y.

$$75 = 5y$$
$$\frac{75}{5} = y$$
$$15 = y$$

14.18 Complete the following statement.

The measure of an angle formed by two _____ intersecting inside a circle is _____ the sum of the _____.

The measure of an angle formed by two <u>chords</u> intersecting inside a circle is <u>half</u> the sum of the measures of the <u>intercepted arcs</u>.

14.19 Draw a conclusion about $m\angle AZB$ in the diagram below, based on the statement completed in Problem 14.18.

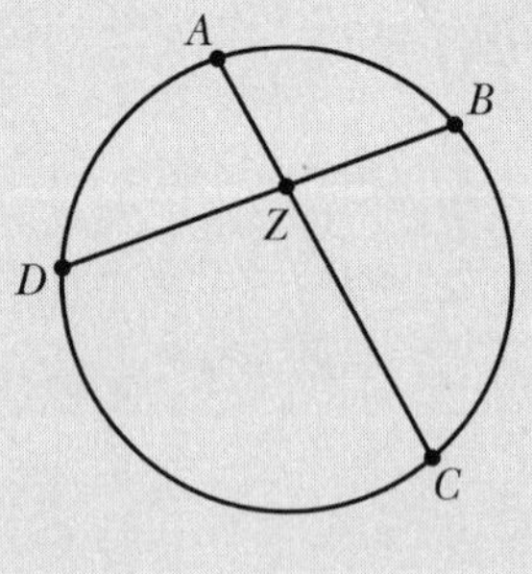

These aren't central angles, so the measure of each one is not equal to the measure of the arc it intercepts.

Notice that $\angle AZB$ and $\angle DZC$ are vertical angles and are therefore congruent. Each of the congruent angles intercepts an arc: $\angle AZB$ intercepts $\widehat{AB}$, and $\angle DZC$ intercepts $\widehat{DC}$. Add the measures of the intercepted arcs and multiply by one-half to calculate the measures of the vertical angles.

$$m\angle AZB = m\angle DZC = \frac{1}{2}\left(m\widehat{AB} + m\widehat{DC}\right)$$

14.20 Use the diagram below to prove the conclusion generated in Problem 14.19.

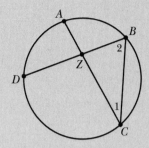

Given the additional chord $\overline{BC}$ included in this diagram (when compared to the diagram in Problem 14.19), your goal is to prove $m\angle AZB = \frac{1}{2}\left(m\widehat{AB} + m\widehat{DC}\right)$.

Statement	Reason
1. $\overline{AC}$, $\overline{BC}$, and $\overline{DB}$ are chords of $\bigodot Z$	1. Given
2. $m\angle AZB = m\angle 1 + m\angle 2$	2. The measure of an exterior angle of a triangle is equal to the sum of the measures of the remote interior angles
3. $m\angle 1 = \frac{1}{2}m\widehat{AB}$; $m\angle 2 = \frac{1}{2}m\widehat{DC}$	3. The measures of inscribed angles are half the measures of the arcs they intercept
4. $m\angle AZB = \frac{1}{2}m\widehat{AB} + \frac{1}{2}m\widehat{DC}$	4. Substitution property of equality (Steps 2, 3)
5. $m\angle AZB = \frac{1}{2}\left(m\widehat{AB} + m\widehat{DC}\right)$	5. Distributive property of equality

This is the hardest step of the proof. You have to remember the theorem from Problem 7.29.

According to the distributive property, $a(b + c) = ab + ac$. In this case, you're sort of using it in reverse. You start with a common multiple, like a in the expression $ab + ac$, and factor it out: $a(b + c)$.

Note: Problems 14.21–14.23 refer to the diagram below.

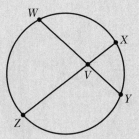

14.21 Is the following statement true or false? Justify your answer.

If $m\widehat{WX} = 90$, then $\angle WVX$ is a right angle.

The statement is false. If $\angle WVX$ were a central angle (that is, if V were the center of the circle), then $\widehat{WX}$ and $\angle WVX$ would have the same measure, 90°. However, V is not the center of the circle.

Note: Problems 14.21–14.23 refer to the diagram in Problem 14.21.

14.22 Given $m\widehat{WZ} = 115°$ and $m\widehat{XY} = 45°$, calculate $m\angle XVY$.

Chords $\overline{WY}$ and $\overline{ZX}$ intersect inside the circle, so the measures of the angles formed by the chords are equal to half the sum of the measures of the intercepted arcs. In this diagram, vertical angles XVY and WVZ intercept arcs $\widehat{XY}$ and $\widehat{WZ}$.

$$m\angle XVY = m\angle WVZ = \frac{1}{2}\left(m\widehat{XY} + m\widehat{WZ}\right)$$

Substitute the given arc measurements into the equation.

$$m\angle XVY = m\angle WVZ = \frac{1}{2}(45° + 115°)$$
$$= \frac{1}{2}(160°)$$
$$= 80°$$

Thus, $m\angle XVY = 80°$.

Note: Problems 14.21–14.23 refer to the diagram in Problem 14.21.

14.23 Given $m\angle ZVY = 120°$ and $m\widehat{ZY} = 170°$, calculate $m\widehat{WX}$. ←

Vertical (and therefore congruent) angles ZVY and WVX intercept $\widehat{ZY}$ and $\widehat{WX}$. Therefore, the angle measures are equal to half the sum of the measures of the intercepted arcs.

$$m\angle ZVY = \frac{1}{2}\left(m\widehat{ZY} + m\widehat{WX}\right)$$

Substitute the given measurements into the equation.

$$120° = \frac{1}{2}\left(170° + m\widehat{WX}\right)$$

Multiply both sides of the equation by the denominator, 2, to eliminate the fraction.

$$2(120°) = \frac{2}{1}\left(\frac{1}{2}\right)\left(170° + m\widehat{WX}\right)$$
$$240° = 170° + m\widehat{WX}$$
$$240° - 170° = m\widehat{WX}$$
$$70° = m\widehat{WX}$$

Note: Problems 14.24-14.25 refer to the diagram below, in which $\overline{AD} \cong \overline{AB}$, $m\widehat{CD} = 158°$, and $m\widehat{BC} = 88°$.

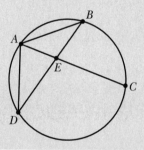

14.24 Calculate $m\angle AED$.

The problem states that $\overline{AD} \cong \overline{AB}$. Congruent chords of a circle intercept congruent arcs. Therefore, $\widehat{AB} \cong \widehat{AD}$. Let x represent the measure of the congruent arcs and apply the arc addition postulate.

$$m\widehat{AB} + m\widehat{BC} + m\widehat{CD} + m\widehat{DA} = 360°$$

Substitute the known measurements into the equation and solve for x.

> Read this problem carefully— the book gives you one angle measurement and one arc measurement this time.

> All the arcs added together form a full circle, which always measures 360°.

$$x + 88° + 158° + x = 360°$$
$$2x + 246° = 360°$$
$$2x = 360° - 246°$$
$$2x = 114°$$
$$x = \frac{114°}{2}$$
$$x = 57°$$

Thus, $m\widehat{AD} = m\widehat{AB} = 57°$. To calculate $m\angle AED$, recall that the angles formed by chords intersecting inside a circle have measures equal to one-half the sum of the measures of the intercepted arcs.

$$m\angle AED = \frac{1}{2}\left(m\widehat{AD} + m\widehat{BC}\right)$$
$$= \frac{1}{2}(57° + 88°)$$
$$= \frac{1}{2}(145°)$$
$$= 72.5°$$

Note: Problems 14.24-14.25 refer to the diagram in Problem 14.24, in which $\overline{AD} \cong \overline{AB}$, $m\widehat{CD} = 158°$, and $m\widehat{BC} = 88°$.

14.25 Calculate $m\angle AEB$.

> Although this technique is quicker, you could add the measures of the intercepted arcs ($\widehat{AB}$ and $\widehat{CD}$) and divide by 2. You'd get the same answer.

When combined, angles AED and AEB form the straight angle DEB.

$$m\angle AED + m\angle AEB = 180°$$

According to Problem 14.24, $m\angle AED = 72.5°$.

$$72.5° + m\angle AEB = 180°$$
$$m\angle AEB = 180° - 72.5°$$
$$m\angle AEB = 107.5°$$

14.26 Complete the following statement.

If two chords intersect inside a circle, then the _____ of the lengths of one chord are equal to the _____ of the lengths of the other chord.

If two chords intersect inside a circle, then the <u>product</u> of the lengths of one chord are equal to the <u>product</u> of the lengths of the other chord.

Note: Problems 14.27–14.28 refer to the diagram below.

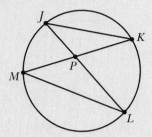

14.27 Draw a conclusion about the diagram based on the statement completed in Problem 14.26.

Chords $\overline{JL}$ and $\overline{MK}$ intersect inside the circle, so the lengths of the segments of each chord produce the same value when multiplied. ←

$$JP \cdot PL = MP \cdot PK$$

> The intersection point P breaks both chords into two pieces. When you multiply the lengths of the pieces of each chord, you get the same number.

Note: Problems 14.27–14.28 refer to the diagram in Problem 14.27.

14.28 Prove the conclusion generated in Problem 14.27.

Your goal is to prove $JP \cdot PL = MP \cdot PK$, given the diagram in Problem 14.27.

Statement	Reason
1. $\overline{JL}$ and $\overline{MK}$ are chords of the circle that intersect at P	1. Given
2. $\angle JPK \cong \angle MPL$	2. Vertical angles are congruent
3. $m\angle KJL = \frac{1}{2} m\widehat{KL}$; $m\angle KML = \frac{1}{2} m\widehat{KL}$	3. The measure of an inscribed angle is half the measure of the arc it intercepts
4. $m\angle KJL = m\angle KML$	4. Transitive property of equality
5. $\triangle JPK \sim \triangle MPL$	5. AA similarity postulate
6. $\dfrac{JP}{MP} = \dfrac{PK}{PL}$	6. CSSTP
7. $JP \cdot PL = MP \cdot PK$	7. Means-extremes property of proportions

Note: Problems 14.29–14.30 refer to the diagram below.

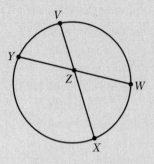

14.29 Given $VZ = 4$, $ZW = 6$, and $ZX = 8$, calculate YZ.

If chords intersect within a circle, the products of the lengths of the segments of the chords are equal. In this diagram, $\overline{VX}$ and $\overline{YW}$ intersect at point Z within the circle.

$$VZ \cdot ZX = YZ \cdot ZW$$

Substitute the given values into the equation and solve for YZ.

$$4 \cdot 8 = YZ \cdot 6$$
$$32 = 6(YZ)$$
$$\frac{32}{6} = YZ$$
$$\frac{32 \div 2}{6 \div 2} = YZ$$
$$\frac{16}{3} = YZ$$

Note: Problems 14.29–14.30 refer to the diagram in Problem 14.29.

14.30 Assume that Z is the midpoint of $\overline{YW}$. Given $VZ = 6$ and $ZX = 10$, calculate ZW.

If Z is the midpoint of $\overline{YW}$, then the point divides the segment into two congruent segments, each half as long as $\overline{YW}$. Thus, $YZ = ZW = \frac{1}{2}(YW)$.

Chords $\overline{VX}$ and $\overline{YW}$ intersect at point Z within the circle, so the products of the lengths of the segments formed by their intersection are equal.

$$VZ \cdot ZX = YZ \cdot ZW$$

Let x represent the length of congruent segments $\overline{YZ}$ and $\overline{ZW}$. Substitute all known values into the above equation and solve for x.

$$6 \cdot 10 = x \cdot x$$
$$60 = x^2$$
$$\sqrt{60} = \sqrt{x^2}$$
$$\sqrt{4 \cdot 15} = x$$
$$2\sqrt{15} = x$$

Therefore, $ZW = YZ = 2\sqrt{15}$.

Secant and Tangent Theorems
Now featuring things outside the circle

14.31 Given $\overleftrightarrow{NO}$ is tangent to $\bigcirc X$ and $m\overparen{NM} = 144°$ in the diagram below, calculate $m\angle MNO$.

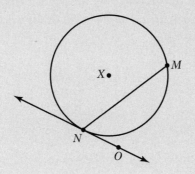

The measure of an angle formed by a chord and a tangent is half the measure of the arc it intercepts. Angle MNO is formed by chord $\overline{MN}$ and tangent $\overline{NO}$, so its measure is half the measure of $\overparen{NM}$, the arc it intercepts.

$$m\angle MNO = \frac{1}{2} m\overparen{NM}$$

Substitute $m\overparen{NM} = 144°$ into the equation.

$$m\angle MNO = \frac{1}{2}(144°) = 72°$$

This is similar to the measure of an inscribed angle, which is also half the measure of the arc it intercepts.

Note: Problems 14.32–14.34 refer to the diagram below, in which $\overrightarrow{MN}$ is tangent to the circle, $\overline{JK} \parallel \overline{ML}$, $m\widehat{JK} = 156°$, and $m\angle LMN = 44°$.

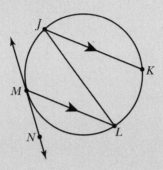

14.32 Calculate $m\widehat{ML}$.

The measure of an angle formed by a chord and a tangent is half the measure of the arc it intercepts. In this diagram, $\angle LMN$ is formed by chord $\overline{ML}$ and tangent $\overrightarrow{MN}$, so its measure is half the measure of $\widehat{ML}$, the arc it intercepts. You are given $m\angle LMN = 44°$.

$$m\angle LMN = \frac{1}{2} m\widehat{ML}$$

$$44° = \frac{1}{2} m\widehat{ML}$$

Multiply both sides of the equation by 2 to solve for $m\widehat{ML}$.

$$2(44°) = \frac{2}{1}\left(\frac{1}{2}\right) m\widehat{ML}$$

$$88° = m\widehat{ML}$$

Note: Problems 14.32–14.34 refer to the diagram in Problem 14.31, in which $\overline{JK} \parallel \overline{ML}$, $m\widehat{JK} = 156°$, and $m\angle LMN = 44°$.

14.33 Calculate $m\widehat{MJ}$.

Parallel chords $\overline{JK}$ and $\overline{ML}$ are intersected by transversal $\overline{JL}$, forming congruent, alternate interior angles $\angle KJL$ and $\angle MLJ$. Note that $\angle KJL$ and $\angle MLJ$ are inscribed angles, so their measures are each half the measures of the arcs they intercept.

$$m\angle KJL = \frac{1}{2} m\widehat{KL} \qquad m\angle MLJ = \frac{1}{2} m\widehat{MJ}$$

Note that $\widehat{KL} \cong \widehat{MJ}$. Apply the arc addition property.

If the angles are congruent, then the arcs they intercept must be congruent as well—both arcs have a measure exactly twice as large as the measures of the congruent angles.

$$m\,\widehat{JK} + m\,\widehat{KL} + m\,\widehat{ML} + m\,\widehat{MJ} = 360°$$

You are given $m\,\widehat{JK} = 156°$; according to Problem 14.32, $m\,\widehat{ML} = 88°$. Let x represent the measures of the congruent arcs.

$$156° + x + 88° + x = 360°$$
$$2x + 244° = 360°$$
$$2x = 360° - 244°$$
$$2x = 116°$$
$$x = \frac{116°}{2}$$
$$x = 58°$$

Therefore, $m\,\widehat{MJ} = m\,\widehat{KL} = 58°$.

> *Note: Problems 14.32–14.34 refer to the diagram in Problem 14.31, in which $\overline{JK} \parallel \overline{ML}$,*
> *$m\,\widehat{JK} = 156°$, and $m\angle LMN = 44°$.*

14.34 Calculate $m\angle KJL$.

According to Problem 14.33, $m\,\widehat{KL} = 58°$. Notice that $\angle KJL$ is an inscribed angle that intercepts $\widehat{KL}$, so the measure of the angle is half the measure of the arc.

$$m\angle KJL = \frac{1}{2}\,m\,\widehat{KL} = \frac{1}{2}(58°) = 29°$$

14.35 Complete the following statement.

> *An angle formed by _____ or _____ that extend from a common point outside*
> *a circle has a measure equal to _____ of the measures of the intercepted arcs.*

An angle formed by <u>tangents</u> or <u>secants</u> that extend from a common point outside a circle has a measure equal to <u>half the difference</u> of the measures of the intercepted arcs. ←

If you combine all four arcs, you get the entire circle, so the measures add up to 360°.

Compare this to the theorem in Problem 14.18: an angle formed by two chords intersecting INSIDE a circle has a measure equal to half the SUM of the intercepted arcs.

14.36 Draw a conclusion about the diagram below based on the theorem completed in Problem 14.35.

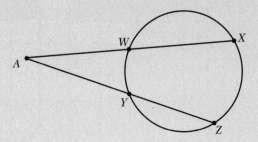

They're secant lines because they contain chords of the circle.

Secant lines $\overleftrightarrow{AX}$ and $\overleftrightarrow{AZ}$ extend from common point A, which lies outside the circle. Therefore, the measure of the angle formed is half the difference of the measures of the intercepted arcs.

$$m\angle A = \frac{1}{2}\left(m\widehat{XZ} - m\widehat{WY}\right)$$

$(m\widehat{WY} < m\widehat{XY})$

Note that the order in which you subtract is crucial. An angle measurement must be positive, so you must subtract the smaller arc measurement from the larger arc measurement.

14.37 Given $m\angle V = 35°$, $m\widehat{RS} = 75°$, and $\overrightarrow{VR}$ is tangent to the circle in the diagram below, calculate $m\widehat{RT}$.

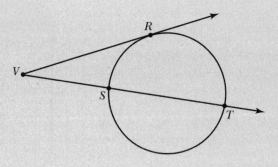

Tangent $\overrightarrow{VR}$ and secant $\overrightarrow{VT}$ extend from common point V outside the circle, forming $\angle V$. Therefore, the measure of $\angle V$ is half the difference of the intercepted arcs.

$$m\angle V = \frac{1}{2}\left(m\widehat{RT} - m\widehat{RS}\right)$$

Substitute $m\angle V = 35°$ and $m\widehat{RS} = 75°$ into the equation, multiply both sides of the equation by 2 to eliminate the fraction, and solve for $m\widehat{RT}$.

$$35° = \frac{1}{2}\left(m\widehat{RT} - 75°\right)$$

$$2(35°) = \frac{2}{1}\left(\frac{1}{2}\right)\left(m\widehat{RT} - 75°\right)$$

$$70° = m\widehat{RT} - 75°$$

$$70° + 75° = m\widehat{RT}$$

$$145° = m\widehat{RT}$$

14.38 Given $\overline{ZG}$ and $\overline{ZH}$ are tangent to the circle in the diagram below, and $m\widehat{GPH}$ is 40° larger than three times $m\widehat{GH}$, calculate $m\angle GZH$. Note that the figure may not be drawn to scale.

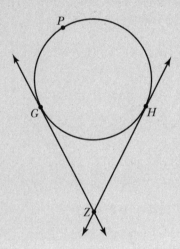

Of the two named arcs, less is known about the measure of $\widehat{GH}$. Thus, let $x = m\widehat{GH}$. You are given that $m\widehat{GPH}$ is 40° larger than three times $m\widehat{GH}$, so $m\widehat{GPH} = 3x + 40$. When an exterior angle (such as $\angle GZH$) is formed by two tangents that extend from a common point, the sum of the intercepted arcs is 360°. ◄

$$m\widehat{GPH} + m\widehat{GH} = 360°$$

Substitute $x = m\widehat{GH}$ and $m\widehat{GPH} = 3x + 40$ into the equation and solve for x.

$$(3x + 40) + x = 360$$

$$4x + 40 = 360$$

$$4x = 360 - 40$$

$$4x = 320$$

$$x = \frac{320}{4}$$

$$x = 80$$

Both arcs in the diagram connect G to H in a different direction around the circle. The arcs have the same endpoints, so when you combine them, you get the whole circle.

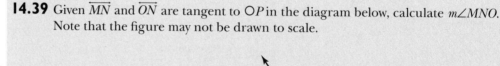

Thus, $m\overset{\frown}{GH} = x = 80°$ and $m\overset{\frown}{GPH} = 3(80) + 40 = 280°$. Because $\angle GZH$ is formed by tangents $\overleftrightarrow{GZ}$ and $\overleftrightarrow{HZ}$ that extend from the common exterior point Z, its measure is half the difference of the measures of the intercepted arcs.

$$m\angle GZH = \frac{1}{2}\left(m\overset{\frown}{GPH} - m\overset{\frown}{GH}\right)$$

Substitute the measures of the arcs into the equation to evaluate $m\angle GZH$.

$$m\angle GZH = \frac{1}{2}(280° - 80°) = \frac{1}{2}(200°) = 100°$$

14.39 Given $\overleftrightarrow{MN}$ and $\overleftrightarrow{ON}$ are tangent to $\odot P$ in the diagram below, calculate $m\angle MNO$. Note that the figure may not be drawn to scale.

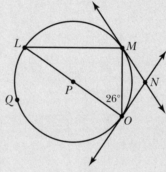

This solution reads a lot like a proof. Every statement made and step taken is justified by a reason. Even when you're not proving something, you need to explain how you're doing each step and what theorems allow you to do what you're doing.

Notice that chord $\overline{LO}$ is a diameter because it passes through P, the center of the circle. Diameters intercept semicircles, so $m\overset{\frown}{LQO} = 180°$. That semicircle is also intercepted by $\angle LMO$, an inscribed angle. The measure of an inscribed angle is half the measure of the arc it intercepts.

$$m\angle LMO = \frac{1}{2}m\overset{\frown}{LQO} = \frac{1}{2}(180°) = 90°$$

Now consider $\triangle LMO$. The sum of the angles of a triangle is $180°$, and the measures of two angles of the triangle are known: $m\angle LMO = 90°$ and $m\angle LOM = 26°$ (according to the diagram). Calculate $m\angle MLO$, the third angle of the triangle.

$$m\angle LOM + m\angle MLO + m\angle LMO = 180°$$
$$26° + m\angle MLO + 90° = 180°$$
$$116° + m\angle MLO = 180°$$
$$m\angle MLO = 180° - 116°$$
$$m\angle MLO = 64°$$

Because $\angle MLO$ is an inscribed angle that intercepts $\overarc{MO}$, the measure of the angle is half the measure of the arc. Calculate $m\overarc{MO}$.

$$m\angle MLO = \frac{1}{2}m\overarc{MO}$$

$$64° = \frac{1}{2}m\overarc{MO}$$

$$2(64) = \frac{2}{1}\left(\frac{1}{2}\right)m\overarc{MO}$$

$$128° = m\overarc{MO}$$

Combining $\overarc{MO}$ and $\overarc{MQO}$ produces $\bigcirc P$ because the arcs share the same endpoints. Apply the arc addition postulate, recalling that circles measure 360°, and calculate $m\overarc{MQO}$.

$$m\overarc{MO} + m\overarc{MQO} = 360°$$

$$128° + m\overarc{MQO} = 360°$$

$$m\overarc{MQO} = 360° - 128°$$

$$m\overarc{MQO} = 232°$$

Angle *MNO* is formed by tangents $\overline{MN}$ and $\overline{ON}$ that extend from the common exterior point *N*, so its measure is half the difference of the measures of the intercepted arcs.

$$m\angle MNO = \frac{1}{2}\left(m\overarc{MQO} - m\overarc{MO}\right)$$

$$= \frac{1}{2}(232° - 128°)$$

$$= \frac{1}{2}(104°)$$

$$= 52°$$

> Look at my note in Problem 14.38 for another example of two arcs (one major arc and one minor arc) with the same endpoints combining to form the entire circle.

14.40 Complete the following statement.

If two secants to a circle extend from a common point outside the circle, the _____ of the lengths of each secant and its _____ are equal.

If two secants to a circle extend from a common point outside the circle, the <u>products</u> of the lengths of each secant and its <u>external segment</u> are equal.

Note: Problems 14.41–14.42 refer to the diagram below.

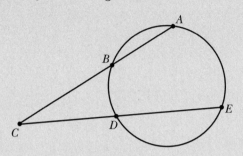

14.41 Given $CB = 6$, $BA = 5$, and $CE = 14$, calculate DE.

Secants $\overline{CA}$ and $\overline{CE}$ extend from the common exterior point C, so the products of the lengths of each secant and the portion of each secant that lies outside the circle are equal. For instance, the length of one secant in the diagram is CA, and the portion of that secant lying outside the circle is CB.

> The other secant has length CE, and the portion of it that's outside the circle is CD. Multiply each secant's length by the part of it that's not in the circle.

$$CA \cdot CB = CE \cdot CD$$

According to the segment addition postulate, $CB + BA = CA$. You are given $CB = 6$ and $BA = 5$, so $6 + 5 = 11 = CA$. Substitute all known lengths into the equation above and solve for CD.

$$11 \cdot 6 = 14 \cdot CD$$
$$66 = 14 \cdot CD$$
$$\frac{66}{14} = CD$$
$$\frac{33}{7} = CD$$

The problem instructs you to calculate DE. You are given $CE = 14$, and the above work demonstrates that $CD = \dfrac{33}{7}$. Apply the segment addition postulate to calculate DE.

$$CD + DE = CE$$
$$\frac{33}{7} + DE = 14$$
$$DE = 14 - \frac{33}{7}$$

Multiply the numerator and denominator of $\dfrac{14}{1}$ by 7 to express the fractions with a common denominator.

$$DE = \frac{14}{1}\left(\frac{7}{7}\right) - \frac{33}{7}$$
$$= \frac{98}{7} - \frac{33}{7}$$
$$= \frac{65}{7}$$

Note: Problems 14.41–14.42 refer to the diagram in Problem 14.41.

14.42 Given $CA = 3.6$, $CE = 4.5$, and $\overline{CB}$ is $\frac{1}{2}$ a unit longer than $\overline{CD}$, calculate CB.

The problem indicates that $\overline{CB}$ is $\frac{1}{2}$ a unit longer than $\overline{CD}$. You are presented with more information about $\overline{CB}$, so let x represent the length of the segment about which you know the least: $x = CD$. If CB is $\frac{1}{2}$ a unit greater than CD, then $CB = x + \frac{1}{2}$.

Secants $\overline{CA}$ and $\overline{CE}$ extend from the common exterior point C, so the products of the lengths of each secant and the portion of each secant that lies outside the circle are equal.

$$CA \cdot CB = CE \cdot CD$$

Substitute the given values of CA and CE into the equation, as well as the variable expressions defined above for CD and CB.

$$(3.6)\left(x + \frac{1}{2}\right) = (4.5)(x)$$

$$3.6x + 3.6\left(\frac{1}{2}\right) = 4.5x$$

$$3.6x + 1.8 = 4.5x$$

$$1.8 = 4.5x - 3.6x$$

$$1.8 = 0.9x$$

$$\frac{1.8}{0.9} = x$$

$$2 = x$$

Therefore, $x = CD = 2$. Evaluate CB.

$$CB = x + \frac{1}{2}$$

$$= 2 + \frac{1}{2}$$

$$= \frac{2}{1}\left(\frac{2}{2}\right) + \frac{1}{2}$$

$$= \frac{4}{2} + \frac{1}{2}$$

$$= \frac{5}{2}$$

When this theorem mentions the "length of the tangent," it means the length of the SEGMENT along the tangent line that connects the common external point to the circle.

14.43 Complete the following statement.

If a secant and a tangent to a circle extend from a common point outside the circle, the _____ of the lengths of the secant segment and _____ equals _____.

If a secant and a tangent to a circle extend from a common point outside the circle, the <u>product</u> of the lengths of the secant segment and <u>its external segment</u> equals <u>the square of the length of the tangent</u>.

Note: Problems 14.44–14.46 refer to the diagram below, in which $\overrightarrow{WX}$ is tangent to the circle.

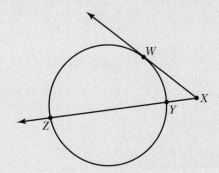

14.44 Draw a conclusion about the diagram based on the theorem completed in Problem 14.43.

Secant $\overleftrightarrow{ZX}$ and tangent $\overrightarrow{WX}$ extend from common external point X. Therefore, the product of the lengths of secant segment $\overline{ZX}$ and the portion of the secant segment that lies outside the circle $\left(\overline{YX}\right)$ is equal to the square of the length of the tangent segment $\left(\overline{WX}\right)$.

$$ZX \cdot YX = (WX)^2$$

Note: Problems 14.44–14.46 refer to the diagram in Problem 14.44, in which $\overrightarrow{WX}$ is tangent to the circle.

14.45 Given $WX = 6$ and $YX = 3$, calculate ZX.

According to Problem 14.44, $ZX \cdot YX = (WX)^2$. Substitute the values of WX and YX into the equation and solve for ZX.

$$ZX \cdot 3 = 6^2$$
$$ZX \cdot 3 = 36$$
$$ZX = \frac{36}{3}$$
$$ZX = 12$$

Note: Problems 14.44–14.46 refer to the diagram in Problem 14.44, in which $\overrightarrow{WX}$ is tangent to the circle.

14.46 Given $ZX = 15$ and $ZY = 10$, calculate WX.

Apply the segment addition postulate to $\overline{ZX}$ to calculate YX, the length of the external portion of the secant segment.

$$ZY + YX = ZX$$
$$10 + YX = 15$$
$$YX = 15 - 10$$
$$YX = 5$$

The product of the lengths of the secant segment and the portion of the secant segment that lies outside the circle is equal to the square of the length of the tangent segment.

$$ZX \cdot YX = (WX)^2$$
$$15 \cdot 5 = (WX)^2$$
$$\sqrt{75} = \sqrt{(WX)^2}$$
$$\sqrt{25 \cdot 3} = WX$$
$$5\sqrt{3} = WX$$

Chapter 15
BASIC GEOMETRIC CONSTRUCTIONS

Break out the compass and straightedge

Compass and straightedge constructions have been an integral component of Euclidean geometry since its inception. Although modern tools can provide more exact diagrams, they do not supplant the importance of traditional, hand-drawn geometric constructions. For example, the physical act of drawing angle bisectors using the simplest of tools provides insight into geometric figures that any other means cannot replicate.

Note that all constructions in Chapters 15 and 16 should be completed using only a compass and a straightedge.

Geometry is a unique mathematics course, blending the pure theory of proofs with its polar opposite, hands-on (and hand-drawn!) diagrams. You know a lot of theorems about parallel and perpendicular lines, for example, but now you get to sit down and construct them by hand with a compass and a ruler or something that can draw straight lines. By the way, when I say "compass," I don't mean something that indicates which direction is north. A compass is a small metal tool with two arms, one that ends in a pencil and one that ends in a sharp point.

Here's a checklist of the constructions you should master by the end of the chapter:

- ☐ Congruent segment
- ☐ Congruent angle
- ☐ Angle bisector
- ☐ Perpendicular bisector
- ☐ Equilateral triangle
- ☐ Perpendicular through a point on a line
- ☐ Perpendicular through a point not on a line
- ☐ Parallel through a point not on a line

Angle and Segment Constructions
Copy and bisect segments and angles

15.1 Traditional geometric constructions require the use of a straightedge. Describe the difference between a straightedge and a ruler, and explain why the distinction is made.

A ruler has a straight edge, allowing you to draw straight lines and segments easily. However, by definition, a ruler features demarcations used for measurement. These lines may not be used to complete a geometric construction. For instance, Problems 15.5–15.6 investigate the construction of congruent segments. Using a ruler makes this procedure trivial but arguably no more accurate. You may use a ruler as a straightedge (in fact, it is encouraged), but disregard the units on the surface of the ruler.

> Rulers and compasses are only as accurate as the person who uses them, and it's nearly impossible to construct things with robotic accuracy. Don't worry, though: you aren't using these constructions to build a skyscraper—you're using them to approach geometry from a different angle (pun intended).

15.2 Describe the primary use of a compass in a geometric construction.

Compasses have one major purpose: to draw arcs. Arcs, of course, are portions of circles, and circles are defined as the set of coplanar points equidistant from a fixed point called the center. Practically speaking, all points on the arc drawn by the pencil of a compass are equidistant from the spike of the compass, the sharp, stationary arm of the compass that marks the center of the circle.

> In other words, the distance between the point of the pencil and the sharp point that could poke your eye out if you're not careful.

In the following constructions, you are occasionally asked to draw arcs of any radius. In these cases, the distance between the points of the compass is not important. Usually, these arcs are drawn to identify points of intersection from which more precise arcs are drawn later.

Other constructions require you to draw arcs of a specific radius. The points of a compass may be placed at the endpoints of a segment, for instance, to draw an arc with a radius congruent to that segment.

15.3 Construct OP with radius $\overline{PX}$.

Draw a point and label it P. Place the spike of the compass on P and draw a circle of any radius.

> The sharp end.

> It helps if you hold only the top of the compass and keep constant, light pressure on the spike to keep the compass from jumping around on the page. The technique is all in the wrist. Twist your fingers so that the pencil glides in a circle.

Draw a point X somewhere on $\bigcirc P$. By definition, the radius of a circle connects the center to a point on the circle. Use a straightedge to draw radius $\overline{PX}$ by connecting points P and X.

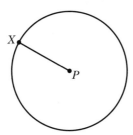

15.4 Construct $\bigcirc Q$ with diameter $\overline{YZ}$. Justify the steps you take to complete the construction.

A diameter is a chord that passes through the center of the circle. Begin by drawing point Q and a circle of any radius with center Q. Next, draw a point Y (one endpoint of the diameter) somewhere on the circle.

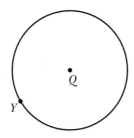

Use the straightedge to draw a chord that begins at Y and passes through center Q. The point at which the line segment next intersects the circle is point Z, the other endpoint of the diameter.

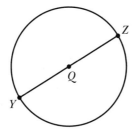

A "unique point" means "a point that's not already labeled on the circle." In other words, make a new point.

15.5 Inscribe a right triangle within ⊙Q, the circle constructed in Problem 15.4.

Draw a unique point X on ⊙Q and use your straightedge to construct chords $\overline{XZ}$ and $\overline{XY}$.

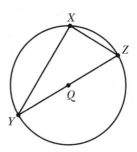

Notice that ∠YXZ is an inscribed angle that intercepts a semicircle. The measure of an inscribed angle is half the measure of the arc it intercepts. All semicircles measure 180°, so m∠YXZ = 90°. Therefore, ΔXYZ is a right triangle, and ΔXYZ is inscribed in ⊙Q because each of its vertices are points on the circle.

15.6 Construct a segment $\overline{MN}$ that is congruent to $\overline{AB}$, illustrated below.

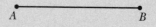

Use your straightedge to draw a line k that is visibly longer than $\overline{AB}$, and draw a point M on line k.

Don't worry about drawing arrowheads at the ends of line k. Basically, one section of line k will become $\overline{MN}$. After that, you can erase the leftover parts of k if you want to.

Place the points of your compass at points A and B to set a radius equal to the distance between the points. Then place the spike of your compass at point M and draw an arc with that radius that intersects k. The point at which the arc intersects k is N.

When you use the compass to measure the distance between A and B, don't move the arms of the compass again until you're finished with the problem.

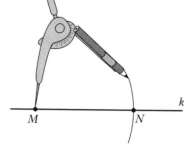

Line k now contains $\overline{MN}$, which is the same length as (and thus congruent to) $\overline{AB}$.

15.7   Construct a segment $\overline{PQ}$ that is congruent to $\overline{GH}$, illustrated below.

Use your straightedge to draw a line j that is visibly longer than $\overline{GH}$. Draw a point P on the line. It is not necessary to draw j at the same angle as $\overline{GH}$. In the following diagrams, j is drawn horizontally, and the final segment $\overline{PQ}$ that is constructed along j is still congruent to $\overline{GH}$. Congruent segments need only be the same length; the orientation of the segments is irrelevant.

Place the points of your compass on the endpoints of $\overline{GH}$ to set a radius equal to the distance between the points. Then place the spike of your compass on P and construct an arc of that radius that intersects j. The point at which the arc intersects j is Q.

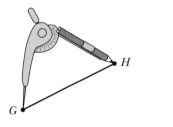
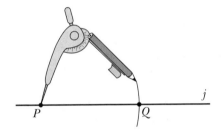

Some constructions have a lot of steps, and it's easy to get them confused. To keep things as clear as possible, the book breaks important new constructions into separate problems that represent the individual steps.

Note: Problems 15.8–15.9 represent the two steps necessary to construct $\angle XYZ$ that is congruent to $\angle ABC$ in the following diagram.

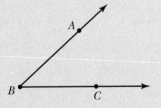

15.8   Step 1: Draw line $\overleftrightarrow{YZ}$ and construct two arcs of the same radius, one with center B and one with center Y.

The vertex of $\angle ABC$ is the second letter of its name, point B. Similarly, the vertex of $\angle XYZ$ is point Y. Draw a line using your straightedge and mark two points on the line, Y and Z.

Put Y at the left end of the line you just drew, just as B is on the left side of $\angle ABC$.

Place the spike of your compass at point B and draw an arc of any radius that intersects $\overline{BA}$ and $\overline{BC}$. Then draw an arc of the same radius centered at point Y that intersects $\overrightarrow{YZ}$.

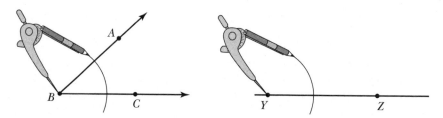

Make note of the points at which the congruent arcs intersect the sides of $\angle ABC$ and $\overrightarrow{YZ}$. These points play an important role in the next step of the construction.

Note: Problems 15.8–15.9 represent the two steps necessary to construct $\angle XYZ$ that is congruent to $\angle ABC$, illustrated in Problem 15.8.

15.9 Step 2: Set a radius equal to the distance between the points of intersection of $\angle ABC$ and the arc constructed in Step 1. Use that radius to construct X.

Let M and N represent the points of intersection of $\angle ABC$ and the arc drawn in Problem 15.8. Let P represent the point of intersection of $\overrightarrow{YZ}$ and the arc also drawn in Problem 15.8.

Place the points of your compass at M and N. Then draw an arc with that radius centered at P. Let X represent the point at which the arcs intersect.

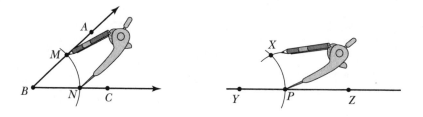

Use your straightedge to draw $\overrightarrow{YX}$. This ray is the remaining side of $\angle XYZ$, which is congruent to $\angle ABC$.

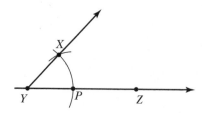

15.10 Explain why ∠XYZ, as constructed in Problems 15.8–15.9, is congruent to ∠ABC.

Consider △MBN and △XYP in the following diagram, which is based on the arcs and points constructed in Problems 15.8–15.9.

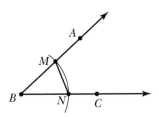 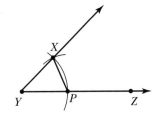

Recall that $\overarc{MN}$ and $\overarc{XP}$ were drawn with the same radius. Thus, M and N are equidistant from B, X and P are equidistant from Y, and $\overline{BM} \cong \overline{BN} \cong \overline{YX} \cong \overline{YP}$. Thus, two sides of △MBN are congruent to two sides of △XYP.

Point X was constructed by drawing an arc centered at P that has the same radius as $\overarc{MN}$. Therefore, $\overline{MN} \cong \overline{XP}$, and △MBN ≅ △XYP by the SSS triangle congruency postulate. Corresponding parts of congruent triangles are congruent, so ∠MBN ≅ ∠XYP. Therefore, ∠ABC ≅ ∠XYZ. ←

∠ABC is another name for ∠MBN, and ∠XYZ is another name for ∠XYP.

15.11 Construct ∠JKL congruent to ∠RST in the following diagram.

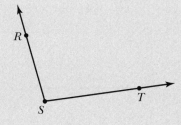

Use your straightedge to construct line $\overleftrightarrow{KL}$. Draw an arc of any radius that is centered at S and intersects $\overrightarrow{SR}$ and $\overrightarrow{ST}$. Draw an arc of the same radius centered at K. Let Z represent the point of intersection of the arc and $\overleftrightarrow{KL}$.

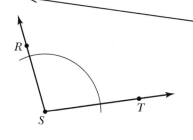 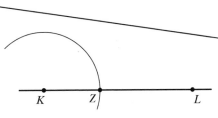

The arc you draw centered at K needs to extend left of point K because ∠RST is obtuse.

Place the points of your compass at the points of intersection of $\angle RST$ and the arc you constructed. Use this radius to draw an arc centered at Z. Let J represent the point at which the arcs intersect. Draw $\overrightarrow{KJ}$ to complete congruent $\angle JKL$.

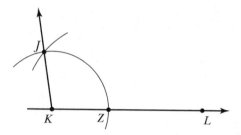

An equilateral triangle has three congruent sides.

15.12 Construct equilateral triangle ABC with sides congruent to $\overline{AB}$, illustrated below.

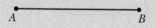

The third vertex of the equilateral triangle will be somewhere above and between A and B, so draw your arcs there. (You could also place the arcs below $\overline{AB}$, and you'd end up with the upside-down version of the same triangle.)

Place the points of your compass at A and B to set a radius equal to the distance between the points. Then construct two arcs of that radius, one centered at A and one centered at B.

Let C represent the point of intersection of the arcs. Draw segments $\overline{AC}$ and $\overline{AB}$ to complete equilateral triangle ABC.

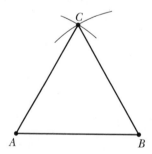

15.13 Explain why △*ABC*, constructed in Problem 15.12, is equilateral.

Placing the points of your compass on *A* and *B* prepares you to draw arcs of radius *AB*. You then proceed to construct two arcs, one centered at each endpoint of $\overline{AB}$. Every point on the arc centered at *A* is exactly *AB* units away from *A*; similarly, every point on the arc centered at *B* is exactly *AB* units away from *B*.

Point *C* belongs to both arcs and is therefore exactly *AB* units away from points *A* and *B*. Therefore, *CA* = *CB* = *AB* and △*ABC* is equilateral.

15.14 Construct △*XYZ* given ∠*X*, $\overline{XY}$, and $\overline{XZ}$ as illustrated below.

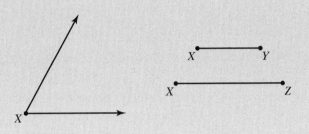

Use your straightedge to draw a line *l* and mark point *X* on line *l*. Construct a congruent ∠*X* on line *l* using the technique described in Problems 15.8–15.9.

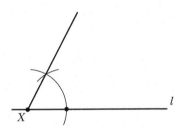

Use the technique described in Problem 15.6 to construct a segment that is congruent to either $\overline{XY}$ or $\overline{XZ}$ along line *l*. In the diagram below, $\overline{XZ}$ is constructed along *l*. Note that the arcs drawn during the construction of ∠*X* on line *l* have been erased to avoid cluttering the diagram.

It doesn't matter which one you draw on line l. The final result will be a triangle with the same three sides, although it might be rotated a little differently.

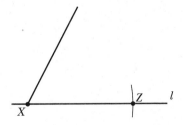

Construct a segment congruent to $\overline{XY}$ along the remaining side of $\angle X$, and connect points Y and Z to complete $\triangle XYZ$.

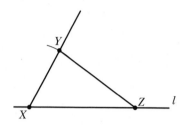

15.15 Construct acute isosceles triangle DEF with vertex $\angle F$, given $\overline{EF}$ as illustrated below.

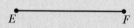

If $\angle F$ is the vertex angle of $\triangle DEF$, then the base of the triangle is $\overline{DE}$ and the legs of the triangle are $\overline{DF}$ and $\overline{EF}$. Place the points of your compass at E and F in the diagram to set a radius equal to the length of a leg of the triangle. Draw a point F and construct an arc of that radius centered at F.

$F_{\bullet}$

In other words, don't draw them too far apart from each other along the arc.

Draw two points D and E on the arc such that the segments that join those two points to F form an acute angle. Connect the three points to complete $\triangle DEF$.

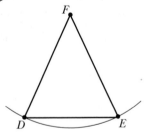

Note: Problems 15.16–15.17 represent the two steps necessary to construct the angle bisector of ∠RST, illustrated below.

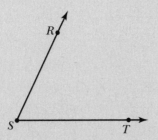

15.16 Step 1: Draw an arc of any radius centered at the vertex of ∠RST that intersects both sides of the angle.

Place the spike of your compass at *S*, the vertex of the angle, and construct an arc that intersects $\overline{SR}$ and $\overline{ST}$. Let the points of intersection be *G* and *H*, respectively.

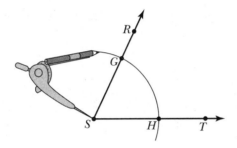

Note: Problems 15.16–15.17 represent the two steps necessary to construct the angle bisector of ∠RST, illustrated in Problem 15.16.

15.17 Step 2: Construct two arcs with congruent radii, one centered at *G* and one centered at *H*. The intersection point of the arcs lies on the angle bisector of ∠RST.

Place the spike of your compass at *H* and draw an arc within ∠RST. Next, place the spike of your compass at *G* and draw an intersecting arc of the same radius.

> You don't have to use the radius from Problem 15.16; just make sure you use the same radius when you draw the two arcs in this problem.

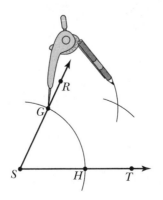

Use your straightedge to connect *S* and the point at which the arcs intersect. This ray is the angle bisector of ∠*RST*.

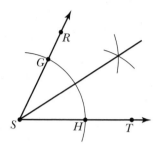

15.18 Explain why the ray constructed in Problems 15.16–15.17 bisects ∠*RST*.

In Problem 15.17, two intersecting arcs are constructed, centered at points equidistant from the vertex of the angle. All points on the arcs are the same distance from their respective centers, located on the sides of the angle.

Consider the converse of the theorem presented in Problem 6.41: every point on an angle bisector is equidistant from the sides of the angle it bisects. The point at which the arcs intersect is the same distance from the sides of ∠*RST* and therefore lies on the angle bisector.

15.19 Construct the angle bisector of ∠*K*, illustrated below.

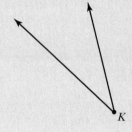

Draw an arc of any radius centered at *K*.

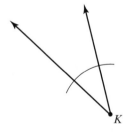

Draw two congruent arcs, each centered at an intersection point of ∠*K* and the arc drawn in the previous diagram.

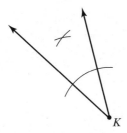

Use your straightedge to draw the ray with endpoint K that passes through the point at which the two newly drawn arcs intersect.

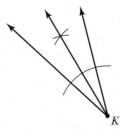

You can't use a protractor to construct the angle; you can use only a compass and a straightedge.

15.20 Construct a 30° angle.

Use your straightedge to draw a segment, and follow the procedure outlined in Problem 15.12 to construct an equilateral triangle with sides congruent to that segment.

The endpoints of the segment you drew will be two vertices of the triangle. Draw an arc at each endpoint of the segment using the segment's length as the radius. The third vertex of the triangle is the point at which the arcs intersect.

Each angle of an equilateral triangle measures 60°. Construct the angle bisector of one of the angles of the triangle.

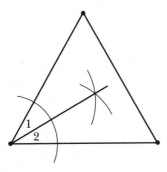

An angle bisector divides an angle into two congruent halves. Therefore, ∠1 and ∠2 in the above diagram each measure 60° ÷ 2 = 30°.

15.21 Construct the perpendicular bisector of $\overline{AB}$, illustrated below.

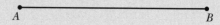

Each point on the perpendicular bisector of a segment is equidistant from the segment's endpoints. Draw two congruent arcs, each centered at an endpoint of $\overline{AB}$. In the diagram below, the arcs intersect above the segment at point M and below the segment at N.

> Make sure the radius of your arcs is larger than half the segment. Otherwise, the arcs won't intersect.

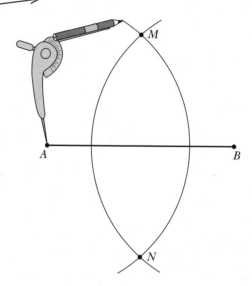

> Technically, you just constructed the angle bisector of the straight angle lying along $\overleftrightarrow{AB}$.

Use your straightedge to connect M and N; $\overline{MN}$ is the perpendicular bisector of $\overline{AB}$. In the diagram below, note that the entire lengths of the arcs centered at A and B need not be drawn; only the portions of the arcs containing the points of intersection are required.

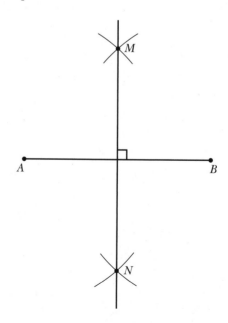

15.22 Construct the perpendicular bisector of $\overline{XY}$, illustrated below.

Draw arcs of the same radius centered at X and Y. Use a straightedge to connect the points at which the arcs intersect. ←

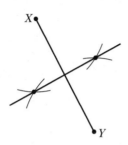

Follow the steps outlined in Problem 15.21. They'll work whether the segment is horizontal (like $\overline{AB}$ in that problem) or slanty (like $\overline{XY}$ in this problem).

15.23 Construct a segment that is one-fourth the length of $\overline{CD}$, illustrated below.

Construct the perpendicular bisector of $\overline{CD}$. Let E be the midpoint of $\overline{CD}$. ←

A perpendicular bisector intersects a segment at its midpoint.

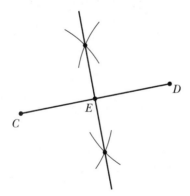

A bisector divides a segment into two congruent segments, each half as long as the original segment.

$$CE = ED = \frac{1}{2}(CD)$$

Now construct the perpendicular bisector of $\overline{EF}$. Let F be the midpoint of the segment.

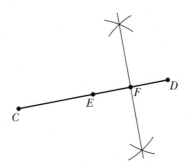

Because F is the midpoint of $\overline{EF}$, $EF = FD = \frac{1}{2}(ED)$. Recall that $ED = \frac{1}{2}(CD)$; substitute that value into the equality statement.

$$EF = FD = \frac{1}{2}(ED)$$

$$EF = FD = \frac{1}{2}\left[\frac{1}{2}(CD)\right]$$

$$EF = FD = \frac{1}{4}(CD)$$

Both $\overline{EF}$ and $\overline{FD}$ are one-fourth as long as $\overline{CD}$.

15.24 Construct a 135° angle.

Use your straightedge to draw a segment and then construct its perpendicular bisector. In the diagram below, $\overleftrightarrow{AB}$ is the perpendicular bisector of $\overline{YZ}$.

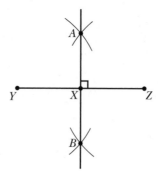

Construct the angle bisector of a right angle. In the following diagram, $\overline{XW}$ is the angle bisector of $\angle AXY$.

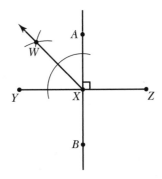

An angle bisector divides an angle into two congruent angles, each half the measure of the original angle.

$$m\angle AXW = m\angle WXY = \frac{1}{2}m\angle AXY$$

$$m\angle AXW = m\angle WXY = \frac{1}{2}(90°)$$

$$m\angle AXW = m\angle WXY = 45°$$

Apply the angle addition postulate.

$$m\angle WXZ = m\angle WXA + m\angle AXZ$$
$$= 45° + 90°$$
$$= 135°$$

Note that $\angle WXB$ also measures 135° because it also consists of a 45° angle and a right angle.

Point–Line Constructions

Create parallel and perpendicular lines

Note: Given point P on line l, as illustrated below, Problems 15.25–15.26 represent the steps necessary to construct a line perpendicular to l at P.

l

P

15.25 Step 1: Draw an arc of any radius, centered at *P*, that intersects *l* twice. ←

> The arc should intersect *l* once to the left of point *P* and once to the right.

Place the spike of your compass at *P* and mark two points on *l* that are equidistant from *P*. Note that only the portions of the arc that intersect *l* are drawn in subsequent steps. Let *X* and *Y* represent the points of intersection. ←

> You end up with $\overline{XY}$, which has midpoint *P*. In Step 2, you construct the perpendicular bisector of the segment, like you do in Problem 15.21.

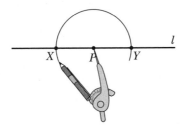

Note: Problems 15.25–15.26 represent the steps necessary to construct a line perpendicular to l at P, given the illustration in Problem 15.25.

15.26 Step 2: Draw two congruent arcs, one centered at each of the intersection points of *l* and the arc drawn in Step 1.

You can swing the arcs below *l* if you want to, like you did in Problem 15.21, but you don't have to. That point will lie on $\overleftrightarrow{ZP}$.

Choose a radius that's larger than *XP* and draw congruent arcs centered at *X* and *Y* (the intersection points created in Step 1). Let *Z* represent the point at which the arcs intersect above *l*. Two points are required to draw a line, and you already know one of those points; the line passes through point *P*. Therefore, you need not locate the intersection point of the arcs below *l* (as was necessary when constructing the perpendicular bisector of a segment).

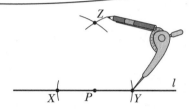

Use your straightedge to connect *Z* and *P* to complete the construction; $\overleftrightarrow{ZP} \perp l$ at point *P*.

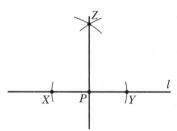

15.27 Given point *A* on line *k* illustrated below, construct line *q* perpendicular to *k* at point *A*.

Draw an arc of any radius, centered at *A*, that intersects *k* twice.

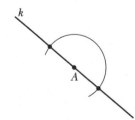

Draw two congruent arcs, centered at the intersection points of *k* and the arc drawn in the previous diagram. Use a straightedge to draw line *q*, which connects *A* to the point at which the two newly drawn arcs intersect.

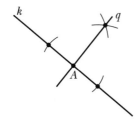

Note: Given line m and point X (which is not on line m), as illustrated below, Problems 15.28–15.29 represent the steps necessary to construct a line through X that is perpendicular to m.

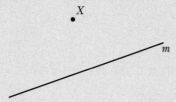

15.28 Step 1: Locate two points on *m* that are equidistant from *X*.

Place the spike of your compass on *X* and draw an arc that intersects *m* twice. Let *A* and *B* represent the points of intersection. To avoid cluttering your diagram, do not draw the entire arc. Instead, draw only the portions of the arc near line *m*.

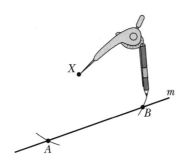

Note: Problems 15.28–15.29 represent the steps necessary to construct a line perpendicular to m through X, given the diagram in Problem 15.28.

15.29 Step 2: Draw two congruent arcs, one centered at each of the intersection points of *m* and the arc drawn in Step 1.

Draw congruent arcs centered at the intersection points *A* and *B* described in Problem 15.28. Let *Y* represent the point at which the congruent arcs intersect. Use your straightedge to connect *X* and *Y* to complete the construction: $\overleftrightarrow{XY} \perp m$.

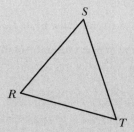

Note: Problems 15.30–15.33 refer to △RST, illustrated below.

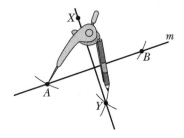

15.30 Construct $\overleftrightarrow{AB}$, the perpendicular bisector of $\overline{RS}$.

Draw congruent arcs centered at the endpoints of $\overline{RS}$, and use a straightedge to connect the two points, *A* and *B*, at which the arcs intersect.

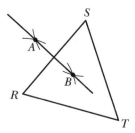

Note: Problems 15.30–15.33 refer to △RST, as illustrated in Problem 15.30.

15.31 Construct median $\overline{TX}$.

A median is a segment that connects a vertex of a triangle to the midpoint of the opposite side. In this example, median $\overline{TX}$ extends from vertex T to the midpoint of $\overline{RS}$, the side of △RST that is opposite ∠T.

Problem 15.30 constructs the perpendicular bisector of $\overline{RS}$. Let X represent the point at which the perpendicular bisector intersects $\overline{RS}$. Draw a line segment connecting T and X to complete the median.

> A perpendicular bisector intersects a segment at its midpoint, so X is the midpoint of $\overline{RS}$.

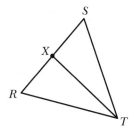

Note: Problems 15.30–15.33 refer to △RST, as illustrated in Problem 15.30.

15.32 Construct angle bisector $\overrightarrow{SQ}$.

The endpoint of $\overrightarrow{SQ}$ is point S, so $\overrightarrow{SQ}$ bisects ∠S in △RST. Draw an arc of any radius centered at S. Then construct congruent arcs centered at the intersection points on $\overline{RS}$ and $\overline{ST}$. Let Q represent the point at which the congruent arcs intersect. Draw a ray from S through Q to complete the construction.

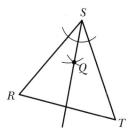

Note: Problems 15.30–15.33 refer to △RST, as illustrated in Problem 15.30.

15.33 Construct altitude $\overline{RY}$.

An altitude passes through a vertex of a triangle and is perpendicular to the line containing the opposite side. In this diagram, R is a vertex of △RST, so altitude $\overline{RY}$ extends from point R to opposite side $\overline{ST}$.

Draw an arc centered at R that intersects $\overline{ST}$ twice. Construct two congruent arcs at those intersection points. Connect R and the point at which the new, congruent arcs intersect.

Construct a line perpendicular to $\overline{ST}$ that passes through R using the technique outlined in Problems 15.28–15.29.

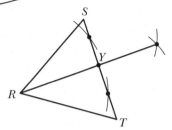

Let Y represent the point at which the perpendicular line intersects $\overline{ST}$.

Note: Given line l and point X (which is not on line l), as illustrated below, Problems 15.34–15.36 represent the steps necessary to construct a line through X that is parallel to l.

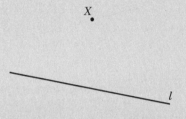

15.34 Step 1: Draw a transversal that intersects l and X.

The transversal is simply a line that passes through l. Let A represent the intersection point of the transversal and line l. Let B represent another point on l.

Point B is introduced to make naming $\angle XAB$ easier.

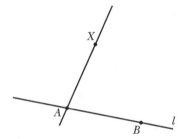

Note: Problems 15.34–15.36 represent the steps necessary to construct a line parallel to l through X, given the illustration in Problem 15.34.

15.35 Step 2: Draw an arc centered at X and a congruent arc centered at the intersection point of l and the transversal.

Parallel lines intersected by a transversal form congruent corresponding angles. Your goal in Steps 2 and 3 is to construct an angle with vertex X that corresponds to, and is congruent to, $\angle XAB$. Draw congruent arcs centered at X and A.

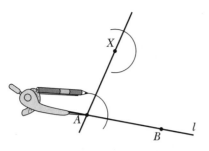

Note: Problems 15.34–15.36 represent the steps necessary to construct a line parallel to l through X, given the illustration in Problem 15.34.

15.36 Step 3: Construct an angle congruent to the angle formed by *l* and the transversal that has vertex *X*.

Place the points of your compass at the intersection points of ∠*XAB* and the arc drawn in Step 2, as illustrated below.

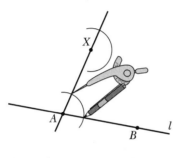

Use this radius to construct a corresponding arc centered at the intersection point near *X*. Let *P* represent the point at which the newly drawn arc intersects the arc drawn in the preceding step.

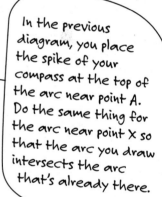

In the previous diagram, you place the spike of your compass at the top of the arc near point A. Do the same thing for the arc near point X so that the arc you draw intersects the arc that's already there.

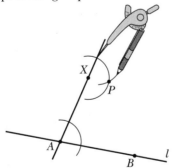

Use your straightedge to draw the line through points *X* and *P* to complete the construction: $\overrightarrow{XP} \parallel l$.

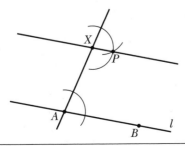

15.37 Construct a right triangle.

The following steps create right triangle *JKL*. Begin by drawing a line *p* that contains a segment $\overline{JK}$.

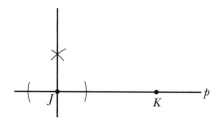

Construct a segment perpendicular to *p* that passes through *J*. ◄

Problems 15.25 and 15.26 explain this construction.

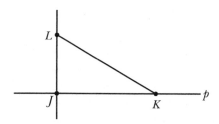

Plot a point *L* on the line perpendicular to *p*, and connect points *J*, *K*, and *L* to complete the construction.

L
J   K   p

15.38 Construct a rectangle.

The following steps create rectangle *WXYZ*. Begin by drawing a line *l* that contains a segment $\overline{WX}$.

W   X   l

Construct a line perpendicular to *l* that passes through *W*. Let *Z* be a point on the new line.

If you call this point Y instead, the final rectangle's name will be WXZY.

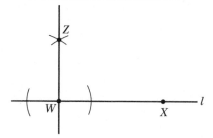

Construct a line perpendicular to $\overline{ZW}$ that passes through Z.

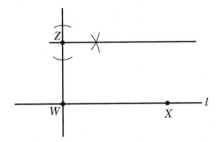

The newly constructed line and $\overleftrightarrow{WX}$ are both perpendicular to $\overline{ZW}$ and are therefore parallel to each other. Construct a segment $\overline{ZY}$ along the new line that is congruent to $\overline{WX}$.

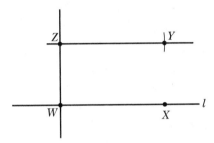

Draw a line connecting Y and X to complete the construction.

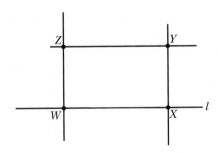

$WXYZ$ is a rectangle because opposite sides are parallel and adjacent sides intersect at right angles.

15.39 Construct a parallelogram.

The following steps create parallelogram *LMNO*. Begin by drawing a line *k* that contains a segment $\overline{LM}$, and then draw a transversal that intersects *k* at *L*. Let *O* represent another point on the transversal.

Problems 15.34–15.36 explain this construction.

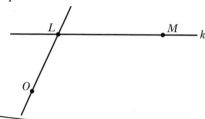

Construct a line parallel to *k* that passes through *O*.

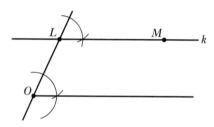

Construct a line parallel to $\overline{LO}$ that passes through *M*.

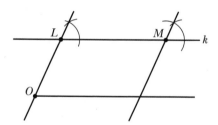

Let *N* represent the vertex of the angle opposite ∠*OLM*. The construction is complete; quadrilateral *LMNO* is a parallelogram because its opposite sides are parallel.

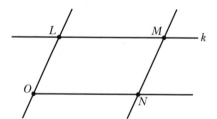

Note: Problems 15.40–15.41 refer to $\overleftrightarrow{RS}$ and $\overrightarrow{RT}$, as illustrated in the following.

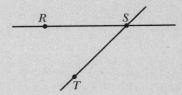

15.40 Construct $\angle STV$ such that $\angle STV$ and $\angle RST$ are congruent alternate interior angles.

Draw congruent arcs centered at S and T. Place the points of your compass at the intersection points of the arc centered at S and the sides of $\angle RST$. Use this radius to construct a congruent angle with vertex T, as demonstrated in the following illustration.

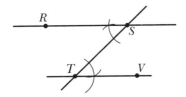

Let V represent a point on the newly constructed line; $\angle RST \cong \angle STV$.

Note: Problems 15.40–15.41 refer to $\overleftrightarrow{RS}$, $\overleftrightarrow{ST}$, and $\overrightarrow{TV}$, as described in Problem 15.40.

15.41 What conclusion can you draw about $\overleftrightarrow{RS}$ and $\overrightarrow{TV}$?

Consider the diagram constructed in Problem 15.40; $\angle RST$ and $\angle STV$ are congruent alternate interior angles formed by transversal $\overleftrightarrow{ST}$ intersecting lines $\overleftrightarrow{RS}$ and $\overrightarrow{TV}$. If alternate interior angles formed by two lines and a transversal are congruent, then the lines are parallel: $\overleftrightarrow{RS} \parallel \overrightarrow{TV}$.

Chapter 16
ADVANCED GEOMETRIC CONSTRUCTIONS

Let's move past bisecting everything in sight

When you master the constructions in Chapter 15, you can apply them to more advanced constructions, such as identifying the orthocenter of a triangle or inscribing a triangle within a circle. Therefore, ensure that you have a firm understanding of the constructions in the preceding chapter before you attempt the more advanced techniques outlined in this chapter.

The first 18 problems in this chapter are based directly on constructions you learn in Chapter 15, so the truly new material (including circles and proportional segments) doesn't start until after that.

Here's a checklist of the constructions you should be able to handle at the end of this chapter:

- ☐ Orthocenter of a triangle
- ☐ Incenter of a triangle
- ☐ Centroid of a triangle
- ☐ Circumcenter of a triangle
- ☐ Circle inscribed in a triangle
- ☐ Circle circumscribed about a triangle
- ☐ Tangent line at a point on a circle
- ☐ Tangent line from a point outside a circle
- ☐ Segment divided into n congruent segments ←
- ☐ Segments that are in proportion
- ☐ Geometric mean

In other words, you divide a segment into n = 3, n = 4, n = 5, and so on congruent segments.

This is the last time this book mentions constructions. In Chapter 17, it's back to compassless geometry, so enjoy constructions while you can.

Constructions Involving Concurrent Lines
Angle and perpendicular bisectors, medians, and altitudes

Note: Problems 16.1–16.4 refer to △ABC below.

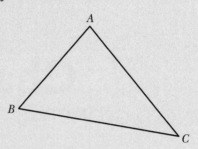

16.1 Construct the perpendicular bisector of $\overline{AB}$.

Draw congruent arcs centered at A and B that intersect on both sides of $\overline{AB}$. The line connecting the intersection points of the arcs is the perpendicular bisector of $\overline{AB}$.

> See Problems 15.21–15.22 to practice constructing perpendicular bisectors.

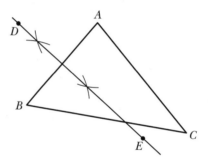

In the above diagram, $\overline{DE}$ is the perpendicular bisector of $\overline{AB}$.

Note: Problems 16.1–16.4 refer to △ABC, illustrated in Problem 16.1.

16.2 Construct the perpendicular bisector of $\overline{BC}$.

Repeat the procedure described in Problem 16.1, drawing the line connecting the intersection points of two congruent arcs drawn at the endpoints of $\overline{BC}$.

> Draw this right on top of the diagram you created in Problem 16.1. By the end of Problem 16.3, you'll have constructed the perpendicular bisectors of all three sides of the triangle.

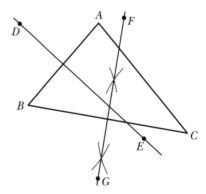

In this diagram, $\overline{FG}$ bisects $\overline{BC}$. Note that perpendicular bisector $\overline{DE}$, constructed in Problem 16.1, is included in the diagram as well (although the arcs used to construct $\overline{DE}$ have been removed for clarity).

Note: Problems 16.1–16.4 refer to △ABC, illustrated in Problem 16.1.

16.3 Construct the perpendicular bisector of $\overline{AC}$.

Repeat the procedure described in Problems 16.1 and 16.2, drawing the line connecting the intersection points of two congruent arcs drawn at the endpoints of $\overline{AC}$.

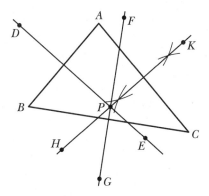

In the diagram above, $\overline{HK}$ is the perpendicular bisector of $\overline{AC}$. Note that all three perpendicular bisectors intersect at a single point. Let P represent the point of intersection.

Two or more lines that intersect at a single point are called "concurrent." That's why this section is titled "Constructions Involving Concurrent Lines."

Note: Problems 16.1–16.4 refer to △ABC, illustrated in Problem 16.1.

16.4 Complete the following statement and verify your answer with a construction.

The point at which the perpendicular bisectors of a triangle intersect is called the _____, and it _____ from the _____ of the triangle.

The point at which the perpendicular bisectors of a triangle intersect is called the <u>circumcenter</u>, and it is <u>equidistant</u> from the <u>vertices</u> of the triangle. All coplanar points equidistant from a fixed point belong to a circle. In the diagram generated by Problems 16.1–16.3, points A, B, and C belong to the circle with radius $PA = PB = PC$.

This comes straight from Problems 7.22–7.24.

Place the points of your compass on *P* and one of the vertices of the triangle. Draw a circle centered at *P* with that radius; it intersects all three vertices of △*ABC*.

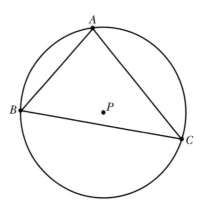

Note: Problems 16.5–16.8 refer to △XYZ below.

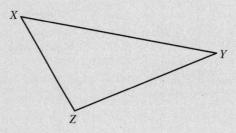

16.5 Construct angle bisector $\overline{XM}$.

The endpoint of $\overrightarrow{XM}$ is *X*, so construct the angle bisector of ∠*X* using the technique described in Problems 15.16–15.17. Let *M* be a point on the angle bisector.

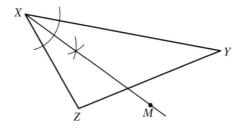

Note: Problems 16.5–16.8 refer to △XYZ, illustrated in Problem 16.5.

16.6 Construct angle bisector $\overrightarrow{YN}$.

Repeat the procedure outlined in Problem 16.5 to construct the angle bisector of ∠Y. Let *N* be a point on the angle bisector. Note that the diagram below also contains $\overrightarrow{XM}$, the angle bisector constructed in Problem 16.5, but it omits the arcs used to create $\overrightarrow{XM}$ for clarity.

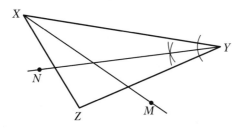

Note: Problems 16.5–16.8 refer to △XYZ, illustrated in Problem 16.5.

16.7 Construct angle bisector $\overrightarrow{ZO}$.

Repeat the procedure described in Problems 16.5 and 16.6 to construct the angle bisector of ∠Z. Let *O* be a point on the bisector.

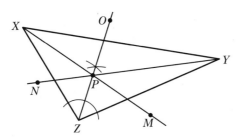

Notice that the angle bisectors of ∠X, ∠Y, and ∠Z intersect at a single point. Let *P* represent the point of intersection. ←

Rule of Thumb: The angle bisectors of a triangle intersect at one point, just as the perpendicular bisectors of the sides of a triangle intersect at one point. Medians and altitudes work the same way, as you'll see in the next few problems.

Note: Problems 16.5–16.8 refer to △XYZ, illustrated in Problem 16.5.

16.8 Complete the following statement and justify your answer using a construction.

The point at which the angle bisectors of a triangle intersect is called the _____, and it is _____ from the _____ of the triangle.

The point at which the angle bisectors of a triangle intersect is called the <u>incenter</u>, and it is <u>equidistant</u> from the <u>sides</u> of the triangle. Thus, you can construct a circle with center *P* (the incenter in Problem 16.7) inscribed in △XYZ such that each side of the triangle is tangent to the circle.

> The radius of the circle is the common distance between P and the sides of the triangle.

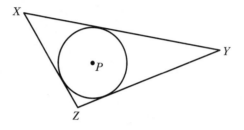

Note: Problems 16.9–16.12 refer to △LMN below.

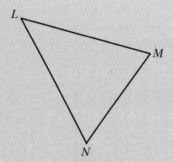

16.9 Construct median $\overline{LA}$.

A median is a segment that extends from the vertex of a triangle to the midpoint of the opposite side. Median $\overline{LA}$ extends from vertex *L* to point *A*; therefore, *A* must be the midpoint of $\overline{MN}$, the side opposite ∠*L*. To identify *A*, construct the perpendicular bisector of $\overline{MN}$.

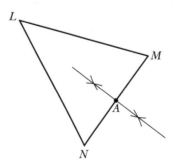

Let A represent the point at which $\overline{MN}$ and its perpendicular bisector intersect. Note that the perpendicular bisector is *not* a median of the triangle. Instead, median $\overline{LA}$ is the segment connecting points L and A, as illustrated below.

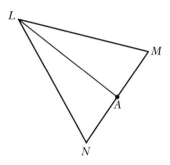

Note: Problems 16.9–16.12 refer to △LMN, illustrated in Problem 16.9.

16.10 Construct median $\overline{MB}$.

Median $\overline{MB}$ extends from vertex M to the midpoint of $\overline{LN}$, the side opposite $\angle M$. Repeat the procedure described in Problem 16.9, constructing the perpendicular bisector of $\overline{LN}$ to identify its midpoint, B.

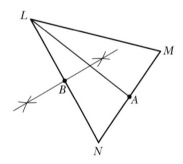

Use a straightedge to draw the segment connecting M and B to complete the construction of median $\overline{MB}$.

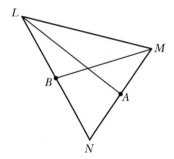

Note: Problems 16.9–16.12 refer to △LMN, illustrated in Problem 16.9.

16.11 Construct median $\overline{NC}$.

Median $\overline{NC}$ extends from vertex N to the midpoint of $\overline{LM}$, the side opposite $\angle N$. Repeat the procedure described in Problems 16.9 and 16.10, constructing the perpendicular bisector of $\overline{LM}$ to identify its midpoint, C.

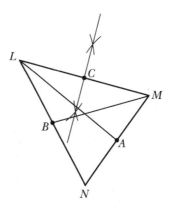

Use a straightedge to draw the segment connecting N and C to complete the construction of median $\overline{NC}$.

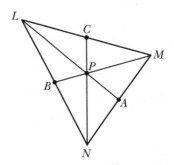

Note that all three medians of △LMN intersect at a single point; let P represent that point of intersection.

Note: Problems 16.9–16.12 refer to △LMN, illustrated in Problem 16.9.

16.12 Let P be the centroid of △LMN, as defined in Problem 16.11. Given $MP = x - 3$, express MB in terms of x.

According to Problem 7.15, the segment connecting a vertex to a centroid is twice as long the segment connecting the centroid to the midpoint of the side opposite that vertex. In this problem, $MP = 2(PB)$. Hence, MP is two-thirds the length of the median: $MP = \dfrac{2}{3}(MB)$. Substitute $MP = x - 3$ into the equation and solve for MB.

The point at which a triangle's medians intersect is called the centroid.

You can split $\overline{MB}$ into thirds; each segment will have length PB. Because MP = 2(PB), MP makes up two of the segment's thirds.

$$MP = \frac{2}{3}(MB)$$

$$x - 3 = \frac{2}{3}(MB)$$

Multiply both sides of the equation by $\frac{3}{2}$ to eliminate the fraction.

$$\frac{3}{2}(x - 3) = \frac{3}{2}\left(\frac{2}{3}\right)(MB)$$

$$\frac{3}{2}(x) + \left(\frac{3}{2}\right)(-3) = MB$$

$$\frac{3}{2}x - \frac{9}{2} = MB$$

$$\frac{3x - 9}{2} = MB$$

Note: Problems 16.13–16.16 refer to △RST below.

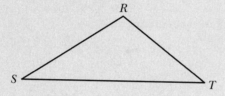

16.13 Construct altitude $\overline{RX}$.

The altitude of a triangle is a segment that is perpendicular to a line containing one side of a triangle and extends from the vertex of the opposite angle. Its endpoints are the vertex and the point at which the altitude intersects that side. Thus, altitude $\overline{RX}$ connects vertex R to a point X on $\overline{ST}$ and $\overline{RX} \perp \overline{ST}$.

Problems 15.28–15.29 explain how to construct a line perpendicular to a given line that passes through a point not on the line. To construct $\overleftrightarrow{RX}$, a line perpendicular to $\overline{ST}$ that passes through R, draw an arc centered at R that intersects $\overline{ST}$ twice. Draw congruent arcs centered at those points of intersection. Finally, draw a line connecting R and the point at which the congruent arcs intersect.

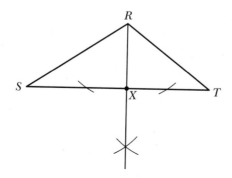

Let X represent the point at which the perpendicular line through R intersects $\overline{ST}$; segment $\overline{RX}$ is an altitude of $\triangle RST$.

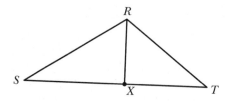

Note: Problems 16.13–16.16 refer to $\triangle RST$, illustrated in Problem 16.13.

16.14 Construct altitude $\overline{SY}$.

Altitude $\overline{SY}$ is the segment perpendicular to $\overrightarrow{RT}$ (the side opposite $\angle S$) that passes through S. However, to construct the altitude, you must first extend $\overrightarrow{RT}$ by drawing the line containing the segment.

The altitudes AND the perpendicular bisectors of an obtuse triangle intersect outside the triangle. Therefore, you need to extend the sides of the triangle here and in Problem 16.15.

Repeat the procedure described in Problem 16.13, drawing an arc centered at S that intersects $\overrightarrow{RT}$ twice and then drawing congruent arcs centered at those points of intersection. Use a straightedge to draw a line connecting S and the point at which the congruent arcs intersect.

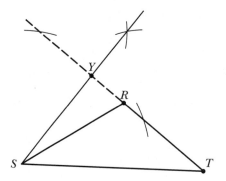

Let Y represent the point at which the perpendicular line through S intersects $\overrightarrow{RT}$; segment $\overline{SY}$ is an altitude of $\triangle RST$.

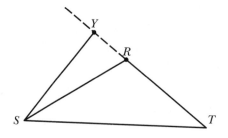

Note: Problems 16.13–16.16 refer to △RST, illustrated in Problem 16.13.

16.15 Construct altitude $\overline{TZ}$.

Altitude $\overline{TZ}$ is the segment perpendicular to $\overleftrightarrow{RT}$ (the side opposite ∠T) that passes through *T*. Repeat the procedure described in Problems 16.13 and 16.14, drawing an arc centered at *T* that intersects $\overleftrightarrow{RS}$ twice and then drawing congruent arcs centered at those points of intersection. Use a straightedge to draw a line connecting *T* and the point at which the congruent arcs intersect.

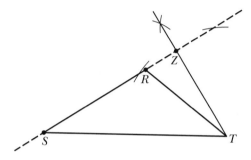

Let *Z* represent the point at which the perpendicular line through *T* intersects $\overleftrightarrow{RS}$; $\overline{TZ}$ is an altitude of △RST.

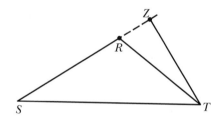

Note: Problems 16.13–16.16 refer to △RST, illustrated in Problem 16.13.

16.16 Complete the following statement and verify it using the diagrams generated in Problems 16.13–16.15.

The point at which the altitudes of a triangle intersect is called the _____, and it always lies outside a triangle that is _____.

Fun fact: The orthocenter of a right triangle is the vertex of the right angle.

The point at which the altitudes of a triangle intersect is called the <u>orthocenter</u>, and it always lies outside a triangle that is <u>obtuse</u>.

Each altitude is extended above the triangle.

In the diagram below, lines are drawn that contain the altitudes constructed in Problems 16.13–16.15; the lines intersect at a single point, *P*.

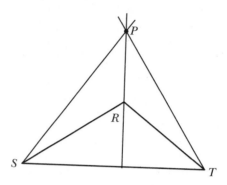

16.17 Construct a circle circumscribed about △*ABC*, illustrated below.

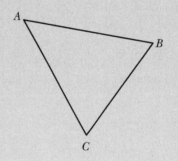

Even though all three bisectors will intersect at the same point, there's no need to draw all three.

According to Problem 16.4, the circumcenter of a triangle is equidistant from the vertices of the triangle. Construct the perpendicular bisectors of any two sides of the triangle; the point at which they intersect is the circumcenter of △*ABC*. In the diagram below, *m* (the perpendicular bisector of $\overline{AB}$) and *n* (the perpendicular bisector of $\overline{BC}$) intersect at *P*, the circumcenter.

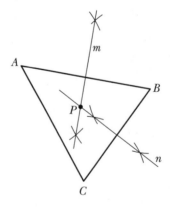

Place the spike of your compass at point *P* and place the pencil of your compass at *A*, *B*, or *C*. Use this radius to draw a circle centered at *P*.

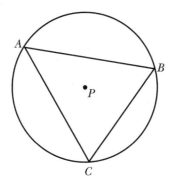

⊙P is circumscribed about $\triangle ABC$ because ⊙P passes through the vertices of the triangle.

16.18 Construct a circle inscribed in $\triangle XYZ$, illustrated below.

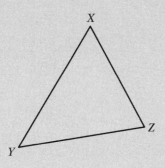

According to Problem 16.8, the incenter of a triangle is the point equidistant from the sides of the triangle. Construct the angle bisectors of any two angles of $\triangle XYZ$; the point at which they intersect is the incenter of the triangle. In the following diagram, j bisects $\angle X$, k bisects $\angle Y$, and the lines intersect at incenter P.

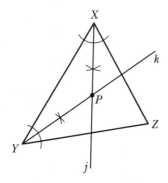

The distance between a line and a point not on that line is defined as the length of the segment perpendicular to the line that passes through the point. Construct line l such that it is perpendicular to $\overline{YZ}$ and passes through P. Let W represent the point at which l and $\overline{YZ}$ intersect.

Place the points of the compass at P and W to set your radius. Then place the spike at P and draw a circle. It'll touch each side of ΔXYZ at exactly one point if you've done everything right.

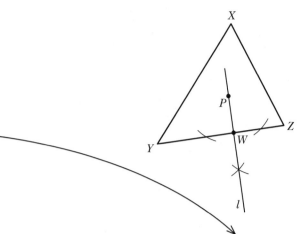

Draw a circle centered at *P* with radius *PW*.

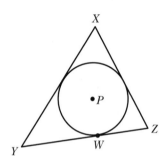

⊙*P* is inscribed in ΔXYZ, and each side of the triangle is tangent to ⊙*P*.

Circle and Tangent Constructions

Draw tangent lines at points on and outside a circle

*Note: Problems 16.19–16.20 refer to the diagram below, in which **Q** is a point on ⊙C.*

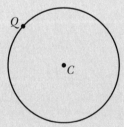

16.19 Construct the line tangent to ⊙*C* at point *Q*.

Use a straightedge to draw $\overrightarrow{CQ}$, the ray extending from the center of the circle that passes through *Q* and contains radius $\overline{CQ}$.

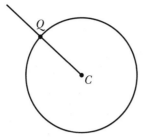

Your next goal is to construct a line perpendicular to $\overleftrightarrow{CQ}$ that passes through Q. Begin by drawing an arc of any radius centered at Q that intersects $\overleftrightarrow{CQ}$ twice.

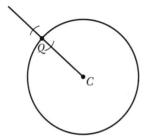

Draw two congruent arcs, each centered at an intersection point of the newly drawn arcs. Let line j represent the line connecting Q and the point at which the congruent arcs intersect.

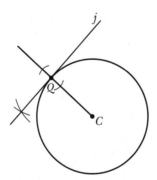

Line j is tangent to $\bigcirc C$, and Q is the point of tangency.

Note: Problems 16.19–16.20 refer to the diagram in Problem 16.19, in which Q is a point on $\bigcirc C$.

16.20 Explain why line j, constructed in Problem 16.19, is tangent to $\bigcirc C$.

If a line is perpendicular to the radius of a circle at a point on the circle, then the line is tangent to the circle at that point. Line j is perpendicular to radius $\overline{CQ}$ at point Q, so j is tangent to $\overline{CQ}$.

See
Problem 13.21.

Problems 16.21–16.22 refer to a circle with center X and diameter $\overline{YZ}$.

16.21 Draw ⊙X with diameter $\overline{YZ}$ and construct lines tangent to ⊙X at Y and Z.

In other words, draw the secant that contains/ overlaps the diameter.

The diameter of a circle is a chord that passes through the center of the circle; thus, Y and Z must be points on the circle and $\overline{YZ}$ must pass through X. Use a straightedge to extend the line containing the diameter beyond endpoints Y and Z.

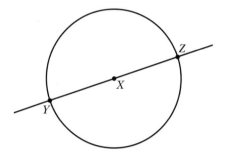

Draw two pairs of arcs, one pair centered at Y and one pair centered at Z, such that each intersects $\overleftrightarrow{YZ}$ twice.

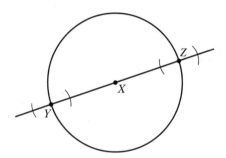

Draw congruent arcs centered at the intersection points around Y, and then do the same thing at the intersection points near Z. The points at which the congruent arcs intersect—M and N in the diagram below—represent points on lines tangent to ⊙X. Use a straightedge to draw tangent lines $\overleftrightarrow{MY}$ and $\overleftrightarrow{NZ}$.

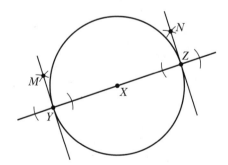

Problems 16.21–16.22 refer to a circle with center X and diameter $\overline{YZ}$.

16.22 Describe the relationship between the tangent lines constructed in Problem 16.21, and justify your answer.

Recall that two lines perpendicular to the same line are parallel to each other. In this problem, $\overleftrightarrow{MY} \perp \overline{YZ}$ and $\overleftrightarrow{NZ} \perp \overline{YZ}$, so $\overleftrightarrow{MY} \parallel \overleftrightarrow{NZ}$.

Note: Problems 16.23–16.26 represent the four steps necessary to construct a line tangent to ⊙A that passes through point B, which lies outside ⊙A.

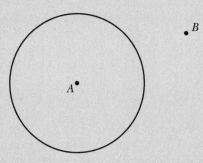

16.23 Step 1: Draw segment $\overline{AB}$.

Use your straightedge to connect the center of ⊙A to point B. This segment will serve as the diameter of a new circle, so the second step of the construction is to identify the center of the new circle, the midpoint of $\overline{AB}$.

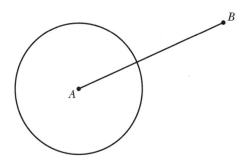

Note: Problems 16.23–16.26 represent the four steps necessary to construct a line tangent to ⊙A that passes through point B, which lies outside ⊙A.

16.24 Step 2: Construct the perpendicular bisector of $\overline{AB}$.

Draw congruent arcs centered at *A* and *B* that intersect each other twice. The line connecting the points of intersection is the perpendicular bisector of $\overline{AB}$ and intersects the segment at its midpoint *M*.

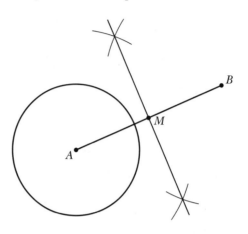

M is the midpoint of $\overline{AB}$ and AM = MB, so you could also say that the radius of the new circle equals MB.

Note: Problems 16.23–16.26 represent the four steps necessary to construct a line tangent to ⊙A that passes through point B, which lies outside ⊙A.

16.25 Step 3: Draw a circle centered at *M* with radius $\overline{AM}$.

Place the points of your compass at *M* and *A* to set a radius equal to the distance between the points. Next, place the spike of your compass at *M* and draw the circle with that radius. The circle has diameter $\overline{AB}$ and intersects ⊙A twice.

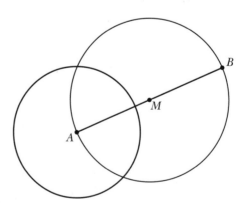

Note: Problems 16.23–16.26 represent the four steps necessary to construct a line tangent to ⊙A that passes through point B, which lies outside ⊙A.

16.26 Step 4: Draw the lines connecting B to the points at which the circles intersect.

Let S and T represent the points at which the circles intersect; $\overleftrightarrow{BS}$ and $\overleftrightarrow{BT}$ are tangent to ⊙A at points S and T, respectively.

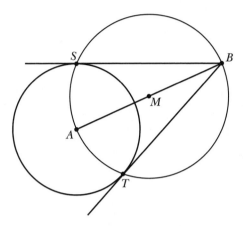

16.27 Explain why $\overleftrightarrow{BS}$ and $\overleftrightarrow{BT}$ (constructed in Problems 16.23–16.26) are tangent to ⊙A.

Consider the diagram in Problem 16.26. Notice that ∠ASB and ∠ATB are inscribed angles on ⊙M, and each intercepts a semicircle: ∠ASB intercepts semicircle $\overarc{ATB}$, and ∠ATB intercepts $\overarc{ASB}$. Inscribed angles that intercept semicircles are right angles, so $\overline{AS} \perp \overline{SB}$ and $\overline{AT} \perp \overline{TB}$. If a line is perpendicular to the radius of a circle at a point on the circle, then the line is tangent to the circle. Thus, $\overleftrightarrow{BS}$ and $\overleftrightarrow{BT}$ are tangent to ∠A.

16.28 Given *l* and $\overline{CD}$, as illustrated below, construct a circle tangent to *l* with radius $\overline{CD}$.

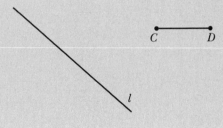

Draw point C on *l* and construct the line perpendicular to *l* at C.

Place the points of your compass at the endpoints of the given segment $\overline{CD}$. Then place the spike of your compass at the new point C (on line l) and draw an arc with that radius to plot a new point D on the perpendicular bisector.

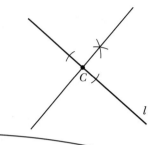

Construct $\overline{CD}$ on the line perpendicular to l.

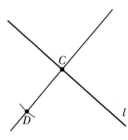

Draw a circle centered at D with radius CD.

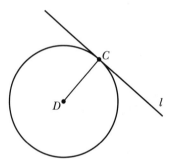

16.29 Given $\bigcirc Z$ and points F and G illustrated below, construct inscribed quadrilateral $ABCD$ such that the vertices are the points at which the tangent lines to $\bigcirc Z$ passing through F and G intersect $\bigcirc Z$.

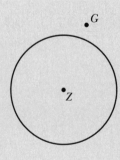

Use a straightedge to draw segments $\overline{FZ}$ and $\overline{GZ}$.

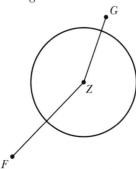

Construct the perpendicular bisectors of the segments. Let M and N represent the midpoints of the segments.

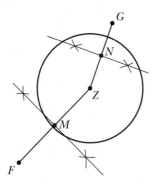

Draw $\bigcirc M$ with radius FM and $\bigcirc N$ with radius GN. Let A, B, C, and D represent the points at which $\bigcirc M$ and $\bigcirc N$ intersect $\bigcirc Z$. ←

Not the points where $\bigcirc F$ and $\bigcirc G$ intersect each other.

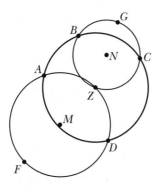

Use a straightedge to connect consecutive vertices of quadrilateral $ABCD$.

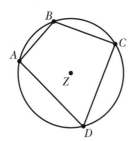

Proportional Segment Constructions

Split a segment into congruent segments or draw a geometric mean

Note: Problems 16.30–16.33 represent the four steps necessary to trisect $\overline{XY}$, illustrated below.

16.30 Step 1: Draw an arbitrary line through X and create an arbitrary segment $\overline{XA}$ on that line.

Use a straightedge to draw a line passing through X. Place point A rather close to X so that $\overline{XA}$ is much shorter than $\overline{XY}$, to aid in later steps of the construction.

Arbitrary means "however you want to draw it." You could draw an arbitrary line through A that slants right, slants left, or even is perpendicular to $\overline{XY}$.

Make $\overline{XA}$ about a third as long as $\overline{XY}$.

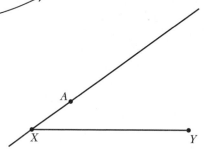

Note: Problems 16.30–16.33 represent the four steps necessary to trisect $\overline{XY}$, illustrated in Problem 16.30.

16.31 Step 2: Construct two segments congruent and adjacent to $\overline{XA}$.

Construct $\overline{AB}$ and $\overline{BC}$, segments congruent to $\overline{XA}$, along the arbitrary line drawn in Step 1.

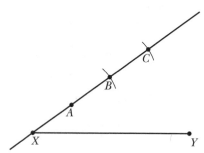

Note: Problems 16.30–16.33 represent the four steps necessary to trisect $\overline{XY}$, illustrated in Problem 16.30.

16.32 Step 3: Draw the segment connecting Y and the endpoint of the segment congruent to $\overline{XA}$ that is farthest from X.

Use a straightedge to connect points C and Y.

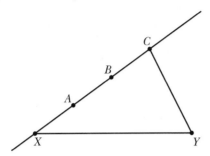

Note: Problems 16.30–16.33 represent the four steps necessary to trisect $\overline{XY}$, illustrated in Problem 16.30.

16.33 Step 4: Construct segments parallel to the segment constructed in Step 3 at the two remaining endpoints of the congruent segments constructed in Step 2.

Construct two lines parallel to $\overleftrightarrow{CY}$, one passing through A and one passing through B. Let L and M represent the points at which the newly drawn parallel lines intersect $\overline{XY}$.

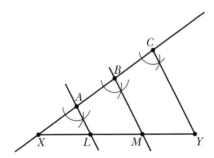

The construction is complete: $\overline{XL} \cong \overline{LM} \cong \overline{MY}$.

16.34 Explain why the construction described in Problems 16.30–16.33 trisects $\overline{XY}$.

Consider the diagram in Problem 16.33. Parallel lines $\overleftrightarrow{AL}$, $\overleftrightarrow{BM}$, and $\overleftrightarrow{CY}$ are intersected by two transversals: $\overleftrightarrow{XC}$ and $\overleftrightarrow{XY}$. According to Problem 9.26, if a group of parallel lines divides one transversal into congruent segments, then those parallel lines divide any transversal into congruent segments.

One of the first steps of the construction was to create three congruent segments on $\overleftrightarrow{XC}$: $\overline{XA} \cong \overline{AB} \cong \overline{BC}$. The parallel lines constructed in Problem 16.33 divide $\overleftrightarrow{XC}$ into those three congruent segments, so the parallel lines must divide $\overline{XY}$ into three congruent segments as well: $\overline{XL} \cong \overline{LM} \cong \overline{MY}$.

The "fourth proportional" is the last number in a proportion. In this problem, you are given three values (lengths a, b, and c) and asked to find the value d. You want a and b to be in the same proportion as c and d.

Note: Problems 16.35–16.39 represent the five steps necessary to construct the fourth proportional d, given three segments with lengths a, b, and c (illustrated below), such that $\frac{a}{b} = \frac{c}{d}$.

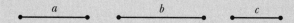

16.35 Step 1: Draw an arbitrary line *l* and construct $\overline{VW}$ with length *a* on *l*.

Draw a line *l* and place point *V* on the line. Construct a segment with endpoint *V* on *l* that is congruent to the given segment with length *a*. Let *W* be the other endpoint of the segment.

Note: Problems 16.35–16.39 represent the five steps necessary to construct the fourth proportional d, given three segments with lengths a, b, and c (illustrated in Problem 16.35), such that $\frac{a}{b} = \frac{c}{d}$.

16.36 Step 2: Construct $\overline{WX}$, a segment with length *b*, on *l*.

Notice that $\overline{VW}$ and $\overline{WX}$ share endpoint *W*, so the segments are adjacent on *l*.

Note: Problems 16.35–16.39 represent the five steps necessary to construct the fourth proportional d, given three segments with lengths a, b, and c (illustrated in Problem 16.35), such that $\frac{a}{b} = \frac{c}{d}$.

16.37 Step 3: Draw an arbitrary line *m* that intersects *l* at *V*, and construct $\overline{VY}$, a segment with length *c* on *m*.

Line *m* must intersect *l* at point *V*. Draw $\overline{VY}$ on *m*, noting that it shares endpoint *V*.

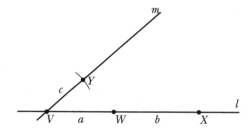

Note: Problems 16.35–16.39 represent the five steps necessary to construct the fourth proportional d, given three segments with lengths a, b, and c (illustrated in Problem 16.35), such that $\dfrac{a}{b} = \dfrac{c}{d}$.

16.38 Step 4: Draw $\overline{YW}$, the segment connecting the endpoints of the segments that share endpoint *V*.

The segment drawn in the final step of the construction will be parallel to $\overline{YW}$.

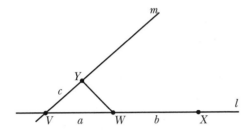

Note: Problems 16.35–16.39 represent the five steps necessary to construct the fourth proportional d, given three segments with lengths a, b, and c (illustrated in Problem 16.35), such that $\dfrac{a}{b} = \dfrac{c}{d}$.

16.39 Step 5: Construct $\overline{ZX}$, a segment parallel to $\overline{YW}$ that passes through *X*.

Let *Z* be the intersection point of *m* and the line parallel to $\overline{YW}$ that passes through *X*. *YZ* is the fourth proportional, *d*.

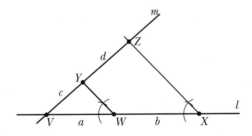

16.40 Explain why the construction described in Problems 16.35–16.39 results in the fourth proportional.

Consider the diagram generated in Problem 16.39. Lines *l* and *m* are divided into two segments by point *V* and parallel lines $\overrightarrow{YW}$ and $\overrightarrow{XZ}$. If parallel lines divide one transversal (in this case, *l*) into segments of a specific ratio, then they divide all transversals in the same ratio. Thus, $\dfrac{a}{b} = \dfrac{c}{d}$.

Note: Problems 16.41–16.45 represent the five steps necessary to construct the geometric mean
f, given two segments with lengths c and d (illustrated below), such that $\dfrac{c}{f} = \dfrac{f}{d}$.

16.41 Step 1: Draw line *l* that contains $\overline{VW}$, a segment with length *c*.

Draw line *l* with a straightedge and plot an arbitrary point *V*. Use the points of your compass to measure length *c*, and plot point *W*.

Note: Problems 16.41–16.45 represent the five steps necessary to construct the geometric mean
f, given two segments with lengths c and d (illustrated in Problem 16.41), such that $\dfrac{c}{f} = \dfrac{f}{d}$.

16.42 Step 2: Construct $\overline{WX}$, a segment with length *d*, on line *l* adjacent to the segment constructed in Step 1.

Segments $\overline{VW}$ and $\overline{WX}$ are adjacent because they share an endpoint (*W*) and belong to the same line (*l*).

Note: Problems 16.41–16.45 represent the five steps necessary to construct the geometric mean
f, given two segments with lengths c and d (illustrated in Problem 16.41), such that $\dfrac{c}{f} = \dfrac{f}{d}$.

16.43 Step 3: Identify the midpoint *Y* of $\overline{VX}$.

Construct the perpendicular bisector of $\overline{VX}$ and let *Y* represent the point at which it intersects $\overline{VX}$.

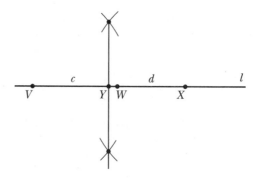

Note: Problems 16.41–16.45 represent the five steps necessary to construct the geometric mean f, given two segments with lengths c and d (illustrated in Problem 16.41), such that $\frac{c}{f} = \frac{f}{d}$.

16.44 Step 4: Draw a semicircle centered at Y with radius VY.

Y is the midpoint of $\overline{VX}$, so $VY = YX$ and the semicircle centered at Y with radius VY will intersect V and X.

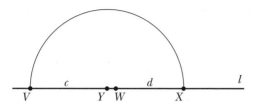

Note: Problems 16.41–16.45 represent the five steps necessary to construct the geometric mean f, given two segments with lengths c and d (illustrated in Problem 16.41), such that $\frac{c}{f} = \frac{f}{d}$.

16.45 Step 5: Construct the line perpendicular to l at W. Let Z be the point at which this line intersects the semicircle constructed in Step 4.

The length of $\overline{ZW}$ is equal to f, the geometric mean of c and d.

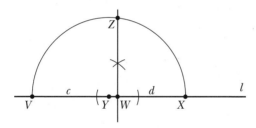

16.46 Explain why the construction described in Problems 16.41–16.45 produces the geometric mean.

Consider the diagram generated in Problem 16.45. Notice that $\angle VZX$ is an inscribed angle that intercepts the diameter of a semicircle. Thus, $\angle VZX$ is a right angle and $\triangle VZX$ is a right triangle.

Just like in Problem 16.27.

Because $\overline{ZW}$ extends from a vertex of the triangle and is perpendicular to the opposite side, it is an altitude of the triangle and splits hypotenuse $\overline{VX}$ into two segments at point W. According to Problem 12.6, the length of the altitude drawn to the hypotenuse of a right triangle is the geometric mean between the lengths of the segments of the hypotenuse. Therefore, ZW is the geometric mean of VW and WX.

Chapter 17
AREA AND PERIMETER

The space inside and the distance around a two-dimensional shape

This chapter outlines the means by which to calculate the areas and perimeters of the core geometric figures discussed thus far. As you progress through the chapter, make a concerted effort to memorize the formulas and the values represented by each variable.

By the end of this chapter, you should know how to calculate all of the following:

☐ Area/perimeter of a rectangle
☐ Area/perimeter of a square
☐ Area/perimeter of a parallelogram
☐ Area/perimeter of a rhombus
☐ Area/perimeter of a triangle
☐ Area/perimeter of a trapezoid
☐ Area/perimeter of a regular polygon
☐ Area/circumference of a circle
☐ Area/arc length of a sector
☐ Area/perimeter of similar figures

The last item in the list is a little different than the others. You'll be given two similar figures and a lot of information about one of them. From that, you need to find the area or perimeter of the other.

Rectangles and Squares

Area = length · width

Note: Problems 17.1–17.2 refer to rectangle ABCD below.

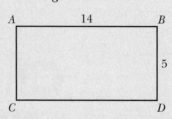

> Technically, it doesn't matter which measurement is the length and which is the width—you get the same area either way. That said, length is usually defined as the larger of the two values.

17.1 Calculate the area of *ABCD*.

The area of a rectangle is defined as the product of its length l and width w. The length of *ABCD* is 14 and the width is 5.

$$A = l \cdot w = 14 \cdot 5 = 70$$

Note: Problems 17.1–17.2 refer to rectangle ABCD, illustrated in Problem 17.1.

17.2 Calculate the perimeter of *ABCD*.

> A rectangle has four sides, two that represent the length (in this case, 14) and two that represent the width (in this case, 5).

The perimeter *P* of a figure is the sum of the lengths of its sides. The opposite sides of rectangle *ABCD*, like the opposite sides of all parallelograms, are congruent. Therefore, *ABCD* has two sides with length 14 and two sides with length 5.

$$P = 2l + 2w = 2(14) + 2(5) = 28 + 10 = 38$$

17.3 Calculate the area and perimeter of square *QRST* with side length $2\sqrt{3}$ feet.

All four sides of a square are congruent, so each side of *QRST* has length $s = 2\sqrt{3}$ ft. Multiply the side length by 4 to calculate the perimeter.

$$\text{perimeter} = 4s = 4\left(2\sqrt{3}\right) = 8\sqrt{3} \text{ ft}$$

> Side length and perimeter use the same units (in this case, feet) because the perimeter is just the sum of the lengths of the sides.

Units are provided in this problem (side length is measured in feet), so perimeter and area measurements should also include units. Note that the perimeter of a figure shares the same units as the length of its sides.

The area of a square is equal to the square of the length of one side. As in any rectangle, the area of a square is equal to the product of its length and width, but because all four sides are congruent, both the length and the width are equal to *s*, the length of a side.

> It makes sense: you're squaring a side length that contains a unit (ft), so you're squaring the unit as well (ft²).

$$\text{area} = s^2 = \left(2\sqrt{3}\right)^2 = (2)^2 \left(\sqrt{3}\right)^2 = 4 \cdot 3 = 12 \text{ ft}^2$$

All area is reported in square units, so the correct response is 12 ft², not 12 ft.

17.4 Calculate the dimensions of a rectangle with area 24, assuming that its length is 3 times its width.

Let w represent the width of the rectangle. The problem states that the length of the rectangle is 3 times the width, so $l = 3w$. Substitute $A = 24$ and $l = 3w$ into the formula for the area of a rectangle.

$$A = l \cdot w$$
$$24 = 3w \cdot w$$
$$24 = 3w^2$$

Solve the equation for w.

$$\frac{24}{3} = w^2$$
$$8 = w^2$$
$$\sqrt{8} = \sqrt{w^2}$$
$$\sqrt{4 \cdot 2} = w$$
$$2\sqrt{2} = w$$

Recall that the length is 3 times the width.

$$l = 3w = 3\left(2\sqrt{2}\right) = 6\sqrt{2}$$

If a rectangle has area 24 and its length is 3 times its width, the dimensions of the triangle are $w = 2\sqrt{2}$ and $l = 6\sqrt{2}$.

17.5 Calculate the area and perimeter of the following figure. Assume that adjacent sides are perpendicular.

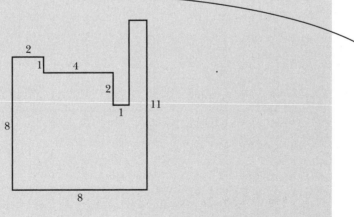

In other words, all the angles are right angles.

If adjacent sides are perpendicular, then the diagram consists of rectangles. In the following diagram, the figure is divided into four rectangular regions: A, B, C, and D.

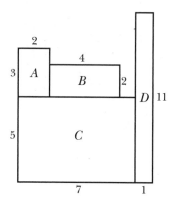

Calculate the area of each region.

$$\text{Area} \left(\text{Region } A\right): \ 3 \cdot 2 = 6$$
$$\text{Area} \left(\text{Region } B\right): \ 4 \cdot 2 = 8$$
$$\text{Area} \left(\text{Region } C\right): \ 7 \cdot 5 = 35$$
$$\text{Area} \left(\text{Region } D\right): \ 11 \cdot 1 = 11$$

According to the area addition postulate, the area of the entire figure is equal to the sum of the areas of its parts.

$$\text{area} = 6 + 8 + 35 + 11 = 60$$

The perimeter is the sum of the lengths of the sides of the figure, indicated in the diagram below.

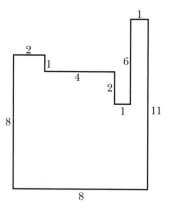

$$\text{perimeter} = 8 + 2 + 1 + 4 + 2 + 1 + 6 + 1 + 11 + 8 = 44$$

> You don't have to use a compass and straightedge—don't take the word construct too literally.

17.6 Calculate the area of the square that can be inscribed in a circle of radius 8 inches.

Construct a diagram that illustrates the given information; square $ABCD$ is inscribed in $\bigodot O$.

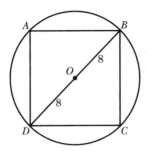

The diagonals of *ABCD* are diameters of the circle; thus, they are twice as long as the radius: *BD* = *AC* = 16. Notice that △*ABD* is an isosceles right triangle with hypotenuse 16. Apply the Pythagorean Theorem to calculate *s*, the length of a side of *ABCD*.

> The sides of ABCD are congruent. Let s represent the length of the congruent sides: s = AB = BC = CD = DA.

$$(AB)^2 + (AD)^2 = (BD)^2$$
$$s^2 + s^2 = (16)^2$$
$$2s^2 = 256$$
$$s^2 = \frac{256}{2}$$
$$s^2 = 128$$

According to Problem 17.3, the area of a square with side length *s* is *s*². Thus, *s*² = 128 is the area of *ABCD*.

Triangles

$A = \frac{1}{2}bh$ and Heron's Formula

Note: Problems 17.7–17.8 refer to △RST, illustrated below.

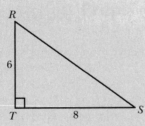

17.7 Calculate the area of △*RST*.

The area of a triangle is equal to half the product of its base and height: $A = \frac{1}{2}bh$. Any side of a triangle may serve as the base; the height of the triangle is the altitude that is perpendicular to the base.

In this diagram, sides $\overline{RT}$ and $\overline{TS}$ are perpendicular. Therefore, either may serve as the base of the triangle, and the other represents the height. Let *b* = 8 and *h* = 6, and apply the triangle area formula.

$$A = \frac{1}{2}bh = \frac{1}{2}(8)(6) = \frac{1}{2}(48) = 24$$

The base and height MUST be perpendicular segments, and the base MUST be a side of the triangle. Apart from those conditions, you can be flexible when choosing the base and height of a triangle.

Note that the area of $\triangle RST$ is the same if you choose to set $b = 6$ and $h = 8$.

$$A = \frac{1}{2}bh = \frac{1}{2}(6)(8) = \frac{1}{2}(48) = 24$$

Note: Problems 17.7–17.8 refer to $\triangle RST$, illustrated in Problem 17.7.

17.8 Calculate the perimeter of $\triangle RST$.

Because $\triangle RST$ is a right triangle, you can apply the Pythagorean Theorem to calculate the length of side $\overline{RS}$.

$$(RT)^2 + (TS)^2 = (RS)^2$$
$$6^2 + 8^2 = (RS)^2$$
$$36 + 64 = (RS)^2$$
$$100 = (RS)^2$$
$$\sqrt{100} = \sqrt{(RS)^2}$$
$$10 = RS$$

The perimeter P of a triangle is the sum of the lengths of its three sides.

$$P = RT + TS + RS = 6 + 8 + 10 = 24$$

Note: Problems 17.9–17.11 refer to the following diagram, in which $\triangle FGH$ has altitude $\overline{FJ}$.

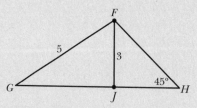

17.9 Calculate the area and perimeter of $\triangle GFJ$.

An altitude extends from the vertex of a triangle (in this case, F) and is perpendicular to the opposite side (in this case, $\overline{GH}$).

If $\overline{FJ}$ is an altitude of $\triangle FGH$, then $\overline{FJ} \perp \overline{GH}$ and $\triangle GFJ$ is a right triangle. Apply the Pythagorean Theorem to calculate GJ.

$$\left(GJ\right)^2 + \left(FJ\right)^2 = \left(GF\right)^2$$
$$\left(GJ\right)^2 + 3^2 = 5^2$$
$$\left(GJ\right)^2 + 9 = 25$$
$$\left(GJ\right)^2 = 16$$
$$GJ = 4$$

Because $\overline{FJ} \perp \overline{GH}$, let one side represent the base of $\triangle GFJ$ and the other represent the height: $b = FJ = 3$ and $h = GJ = \sqrt{34}$. Apply the triangle area formula.

$$A = \frac{1}{2}bh = \frac{1}{2}(3)(4) = \frac{12}{2} = 6$$

The perimeter P of $\triangle GFJ$ is the sum of the lengths of its sides.

$$P = GF + FJ + GJ = 5 + 3 + 4 = 12$$

Note: Problems 17.9–17.11 refer to the diagram in Problem 17.9, in which $\triangle FGH$ has altitude $\overline{FJ}$.

17.10 Calculate the area and perimeter of $\triangle FJH$.

> Problem 17.9 explains why they're perpendicular.

$\triangle FJH$ and $\triangle GFJ$ are right triangles for the same reason: $\overline{FJ} \perp \overline{GH}$. Thus, $m\angle FJH = 90°$. Recall that the sum of the measures of the angles of a triangle is $180°$ and calculate $m\angle JFH$.

$$m\angle JFH + m\angle FHJ + m\angle HJF = 180°$$
$$m\angle JFH + 45° + 90° = 180°$$
$$m\angle JFH + 135° = 180°$$
$$m\angle JFH = 180° - 135°$$
$$m\angle JFH = 45°$$

Because $m\angle JFH = m\angle FHJ$, $\triangle FJH$ is an isosceles right triangle according to the inverse of the isosceles triangle theorem and $\overline{FJ} \cong \overline{JH}$. The congruent sides of the triangle are perpendicular, so let one side represent the base and the other represent the height in the triangle area formula.

$$A = \frac{1}{2}bh = \frac{1}{2}(3)(3) = \frac{1}{2}(9) = \frac{9}{2}$$

Notice that $\triangle FJH$ is a 45°-45°-90° triangle. Therefore, its hypotenuse is $\sqrt{2}$ times as long as the legs.

> See Problems 12.26–12.28.

$$FH = \sqrt{2}\left(FJ\right) = \sqrt{2}(3) = 3\sqrt{2}$$

Now that you know the lengths of all three sides of the triangle, you can calculate the perimeter P of $\triangle FJH$.

$$P = FJ + JH + FH = 3 + 3 + 3\sqrt{2} = 6 + 3\sqrt{2}$$

Note: Problems 17.9–17.11 refer to the diagram in Problem 17.9, in which $\triangle FGH$ has altitude $\overline{FJ}$.

17.11 Calculate the area and perimeter of $\triangle GFH$.

Begin by determining the lengths of the sides of the triangle. According to the diagram, $GF = 5$. Problem 17.10 states that $FH = 3\sqrt{2}$. To calculate the length of the third side, apply the segment addition postulate, noting that $GJ = 4$ (according to Problem 17.9) and $JH = 3$ (according to Problem 17.10).

$$GH = GJ + JH = 4 + 3 = 7$$

Calculate the perimeter P of $\triangle GFH$.

$$P = GF + FH + GH$$
$$= 5 + 3\sqrt{2} + 4 + 3$$
$$= 12 + 3\sqrt{2}$$

Recall that $\overline{FJ} \perp \overline{GH}$. Substitute $b = GH = \sqrt{34} + 3$ and $h = FJ = 3$ into the triangle area formula.

$$A = \frac{1}{2}bh$$
$$= \frac{1}{2}(7)(3)$$
$$= \frac{1}{2}(21)$$
$$= \frac{21}{2}$$

You may also apply the area addition postulate to calculate the area of $\triangle FGH$. Combining $\triangle FGJ$ and $\triangle FJH$ produces $\triangle FGH$; thus, adding the areas of $\triangle FGJ$ and $\triangle FJH$ (as calculated in Problems 17.9 and 17.10) generates the same answer.

$$\text{area}(\triangle FGH) = \text{area}(\triangle FGJ) + \text{area}(\triangle FJH)$$
$$= 6 + \frac{9}{2}$$
$$= \frac{12}{2} + \frac{9}{2}$$
$$= \frac{21}{2}$$

17.12 Calculate the area of the following triangle.

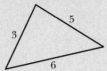

The diagram does not clearly state that any sides of the triangle are perpendicular, so you cannot assign values for the base and the height. Furthermore, applying the Pythagorean Theorem proves that the figure is not a right triangle.

$$3^2 + 5^2 \overset{?}{=} 6^2$$

$$9 + 25 \overset{?}{=} 36$$

$$34 \neq 36$$

Heron's Formula allows you to calculate the area A of any triangle, given only the lengths of its sides (a, b, and c): $A = \sqrt{s(s-a)(s-b)(s-c)}$ when $s = \dfrac{1}{2}(a+b+c)$. To apply Heron's Formula in this problem, set $a = 3$, $b = 5$, and $c = 6$, and calculate the corresponding value of s.

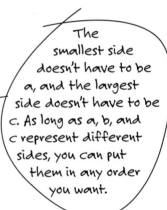

The smallest side doesn't have to be a, and the largest side doesn't have to be c. As long as a, b, and c represent different sides, you can put them in any order you want.

$$s = \frac{1}{2}(a+b+c) = \frac{1}{2}(3+5+6) = \frac{1}{2}(14) = 7$$

Substitute $s = 7$ into Heron's Formula.

$$A = \sqrt{s(s-a)(s-b)(s-c)}$$
$$= \sqrt{7(7-3)(7-5)(7-6)}$$
$$= \sqrt{7(4)(2)(1)}$$
$$= \sqrt{4 \cdot (7 \cdot 2 \cdot 1)}$$
$$= 2\sqrt{14}$$

17.13 Calculate the area of a triangle with sides 5 cm, 7 cm, and 10 cm long.

Apply the procedure described in Problem 7.12. Let $a = 5$, $b = 7$, and $c = 10$, and calculate s.

$$s = \frac{1}{2}(a+b+c) = \frac{1}{2}(5+7+10) = \frac{1}{2}(22) = 11$$

Substitute $s = 11$ into Heron's Formula.

$$A = \sqrt{s(s-a)(s-b)(s-c)}$$
$$= \sqrt{11(11-5)(11-7)(11-10)}$$
$$= \sqrt{11(6)(4)(1)}$$
$$= \sqrt{4 \cdot (11 \cdot 6 \cdot 1)}$$
$$= 2\sqrt{66} \text{ cm}^2$$

Parallelograms and Rhombuses

$A = b \cdot h, \quad A = \frac{1}{2}d_1 d_2$

Note: Problems 17.14–17.15 refer to □ WXYZ, illustrated below.

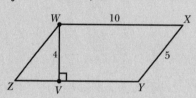

17.14 Calculate the perimeter of □ WXYZ.

The perimeter P of a figure is the sum of the lengths of its sides. Recall that the opposite sides of a parallelogram are congruent, so $\overline{WX} \cong \overline{YZ}$ and $\overline{ZW} \cong \overline{XY}$.

$$P = WX + XY + YZ + ZW$$
$$= 10 + 5 + 10 + 5$$
$$= 30$$

Note: Problems 17.14–17.15 refer to □ WXYZ, illustrated in Problem 17.14.

17.15 Calculate the area of □ WXYZ.

The area of a parallelogram is the product of its base and height. The height is defined by the length of a segment connecting opposite sides of the parallelogram that is perpendicular to both sides. In this problem, $\overline{WV}$ has endpoints on opposite sides of the parallelogram (W lies on $\overline{WX}$, and V lies on $\overline{YZ}$) and is perpendicular to those sides: $\overline{WV} \perp \overline{WX}$ and $\overline{WV} \perp \overline{YZ}$.

> Bases of the parallelogram are the sides perpendicular to $\overline{WV}$ (the segment representing height), so $\overline{WX}$ and $\overline{YZ}$ are both bases of □ WXYZ.

Let h represent the height of the parallelogram: $h = WV = 4$. Let b represent a base of the parallelogram: $b = WX = YZ = 10$. Apply the formula for the area of a parallelogram.

$$A = b \cdot h = 10 \cdot 4 = 40$$

Note: Problems 17.16–17.17 refer to □JKLM, illustrated below.

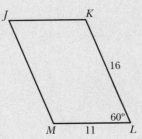

17.16 Calculate the perimeter of □JKLM.

Opposite sides of a parallelogram are congruent, so $JM = KL = 16$ and $JK = ML = 11$. The perimeter P of □JKLM is the sum of the lengths of its sides.

$$P = JK + KL + ML + JM$$
$$= 11 + 16 + 11 + 16$$
$$= 54$$

Note: Problems 17.16–17.17 refer to □JKLM, illustrated in Problem 17.16.

17.17 Calculate the area of □JKLM.

The area of a parallelogram is the product of its base and height. To determine the height of □JKLM, draw $\overline{KN}$, the segment with endpoint K that is perpendicular to $\overline{ML}$, as illustrated below.

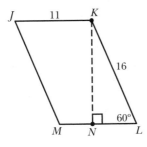

> Technically, the area is the product of the height and the LENGTH of the base, but I'm nitpicking.

Notice that $\triangle KNL$ is a 30°-60°-90° right triangle. Therefore, the side opposite the 30° angle $\left(\overline{NL}\right)$ is half as long as the hypotenuse, and the side opposite the 60° angle $\left(\overline{KN}\right)$ is $\sqrt{3}$ times as long as the shortest side.

> See Problems 12.29–12.31.

$$NL = \frac{1}{2}KL = \frac{1}{2}(16) = 8 \quad \text{and} \quad KN = \sqrt{3}\,(NL) = \sqrt{3}\,(8) = 8\sqrt{3}$$

Let $b = ML = 11$ represent the base and $h = KN = 8\sqrt{3}$ represent the height of □JKLM in the formula for the area of a parallelogram.

$$A = bh = (11)\left(8\sqrt{3}\right) = 88\sqrt{3}$$

Note: Problems 17.18–17.19 refer to the following diagram, in which the diagonals of rhombus WXYZ intersect at point Q.

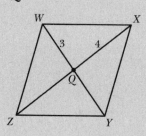

17.18 Calculate the area of *WXYZ*.

> d_1 and d_2 represent the lengths of the diagonals, so in this problem, $d_1 = WY$ and $d_2 = ZX$ (or vice versa).

The diagonals of a parallelogram bisect each other, so *Q* is the midpoint of $\overline{WY}$ and $\overline{ZX}$. Calculate the lengths of the diagonals.

$$WY = 2(WQ) = 2(3) = 6 \quad \text{and} \quad ZX = 2(QX) = 2(4) = 8$$

The area of a rhombus is half the product of its diagonals: $A = \dfrac{1}{2} d_1 d_2$.

$$A = \frac{1}{2}(WY)(ZX) = \frac{1}{2}(6)(8) = \frac{1}{2}(48) = 24$$

Note: Problems 17.18–17.19 refer to the diagram in Problem 17.18, in which the diagonals of rhombus WXYZ intersect at point Q.

17.19 Calculate the perimeter of *WXYZ*.

> $\triangle XQY$, $\triangle ZQY$, and $\triangle WQZ$ are also right triangles.

Recall that the diagonals of a rhombus are perpendicular. Thus, $\overline{WY} \perp \overline{ZX}$ and $\triangle WQX$ is a right triangle. Apply the Pythagorean Theorem to calculate *WX*.

$$\left(WQ\right)^2 + \left(QX\right)^2 = \left(WX\right)^2$$
$$3^2 + 4^2 = \left(WX\right)^2$$
$$9 + 16 = \left(WX\right)^2$$
$$25 = \left(WX\right)^2$$
$$\sqrt{25} = \sqrt{\left(WX\right)^2}$$
$$5 = WX$$

> Problem 17.3 states that the perimeter of a square is 4s, where s is the length of a side. That formula works for all rhombuses, not just squares.

A rhombus has four congruent sides, so the perimeter of *WXYZ* is 4(5) = 20.

Trapezoids

$A = \dfrac{h}{2}(b_1 + b_2)$

Note: Problems 17.20–17.21 refer to trapezoid ABCD, illustrated below.

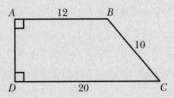

17.20 Calculate the perimeter of *ABCD*.

Divide the trapezoid into a rectangle and a right triangle by drawing $\overline{BF}$, the segment perpendicular to $\overline{DC}$ illustrated in the diagram below.

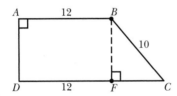

Opposite sides of parallelograms are congruent, so $DF = AB = 12$. Apply the segment addition postulate to calculate *FC*.

$$DC = DF + FC$$
$$20 = 12 + FC$$
$$20 - 12 = FC$$
$$8 = FC$$

Notice that $\triangle BFC$ is a right triangle. Apply the Pythagorean Theorem to calculate *BF*.

$$(BF)^2 + (FC)^2 = (BC)^2$$
$$(BF)^2 + 8^2 = 10^2$$
$$(BF)^2 + 64 = 100$$
$$(BF)^2 = 100 - 64$$
$$(BF)^2 = 36$$
$$\sqrt{(BF)^2} = \sqrt{36}$$
$$BF = 6$$

Opposite sides of *ABCD* are congruent, so $AD = BF = 6$. The perimeter *P* of a trapezoid is the sum of the lengths of its sides.

$$P = AB + BC + CD + DA = 12 + 10 + 20 + 6 = 48$$

> The bases of a trapezoid are the parallel sides, and the height is the length of a segment that is perpendicular to the bases—in this case, AD or BF.

Note: Problems 17.20–17.21 refer to trapezoid ABCD, illustrated in Problem 17.20.

17.21 Calculate the area of *ABCD*.

The area A of a trapezoid with bases b_1 and b_2 and height h is $A = \frac{1}{2}h(b_1 + b_2)$. Substitute $b_1 = 12$, $b_2 = 20$, and $h = 6$ into the formula to calculate the area.

$$A = \frac{1}{2}h(b_1 + b_2) = \frac{1}{2}(6)(12 + 20) = \frac{6}{2}(32) = 3(32) = 96$$

17.22 Calculate the area of a trapezoid with one 7 cm base, a 5.25 cm median, and a 4 cm height.

Recall that the length (m) of the median is equal to half the sum of the bases (b_1 and b_2) of the trapezoid: $m = \frac{1}{2}(b_1 + b_2)$. Substitute $b_1 = 7$ and $m = 5.25$ into the equation and calculate b_2, the length of the remaining base.

$$5.25 = \frac{1}{2}(7 + b_2)$$

$$5.25 = \frac{1}{2}(7) + \frac{1}{2}b_2$$

$$5.25 = 3.5 + \frac{1}{2}b_2$$

$$5.25 - 3.5 = \frac{1}{2}b_2$$

$$1.75 = \frac{1}{2}b_2$$

$$2(1.75) = \frac{2}{1}\left(\frac{1}{2}b_2\right)$$

$$3.5 = b_2$$

Substitute $b_1 = 7$, $b_2 = 3.5$, and $h = 4$ into the formula for the area of a trapezoid.

$$A = \frac{1}{2}h(b_1 + b_2) = \frac{1}{2}(4)(7 + 3.5) = \frac{4}{2}(10.5) = 2(10.5) = 21 \text{ cm}^2$$

> Area is reported in square units. The lengths are measured in centimeters, so the area is measured in cm².

17.23 Prove that the area of a trapezoid is equal to the product of its height and the length of its median.

Given an arbitrary trapezoid with bases of length b_1 and b_2, height h, and median with length m, prove that the area A of the trapezoid is equal to mh.

Statement	Reason
1. An arbitrary trapezoid has bases b_1 and b_2, median m, and height h	1. Given
2. $m = \dfrac{1}{2}\left(b_1 + b_2\right)$	2. The length of the median is half the sum of the lengths of the bases
3. $A = \dfrac{1}{2}h\left(b_1 + b_2\right)$	3. The formula for the area of a trapezoid
4. $A = mh$ ←	4. Substitution property of equality (Steps 2, 3)

Everything on the right side of the Step 3 equation EXCEPT h is equal to m, according to Step 2.

17.24 Calculate the area of trapezoid *WXYZ* illustrated below. Report your answer accurate to the tenths place.

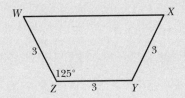

A trapezoid is a quadrilateral with exactly one pair of parallel sides—in this figure, $\overline{WX} \parallel \overline{ZY}$. Same-side interior angles formed by parallel lines and a transversal are supplementary, so $\angle W$ and $\angle Z$ are supplementary. Calculate $m\angle W$.

$$m\angle Z + m\angle W = 180°$$
$$125° + m\angle W = 180°$$
$$m\angle W = 180° - 125°$$
$$m\angle W = 55°$$

Notice that *WXYZ* is isosceles, so both pairs of base angles are congruent: $m\angle Z = m\angle Y = 125°$ and $m\angle W = m\angle X = 55°$. To determine the height of the trapezoid, draw $\overline{ZA}$ and $\overline{YB}$—the segments through *Z* and *Y* that are perpendicular to $\overline{WX}$.

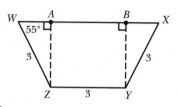

The hypotenuse-angle theorem; see Problem 8.26.

According to the HA theorem, $\triangle WAZ$ and $\triangle XBY$ are congruent right triangles. Apply a trigonometric ratio to calculate AZ, the height of the trapezoid.

$$\sin 55° = \frac{AZ}{3}$$
$$3(\sin 55°) = AZ$$
$$3(0.819152044) \approx AZ$$
$$2.457456133 \approx AZ$$

Corresponding parts of congruent triangles are congruent, so $AW = BX$ and $BY = AZ$. Apply the cosine ratio to calculate AW.

$$\cos 55° = \frac{AW}{3}$$
$$3(\cos 55°) = AW$$
$$3(0.573576436) \approx AW$$
$$1.72072931 \approx AW$$

Use the segment addition postulate to determine the length of $\overline{WX}$.

$$WX = WA + AB + BX \approx 1.72072931 + 3 + 1.72072931 \approx 6.44145862$$

Substitute $b_1 = 3$, $b_2 = 6.44145862$, and $h = 2.457456133$ into the formula for the area of a trapezoid.

$$A = \frac{1}{2}h\left(b_1 + b_2\right)$$
$$\approx \frac{1}{2}(2.457456133)(3 + 6.44145862)$$
$$\approx \frac{1}{2}(2.457456133)(9.44145862)$$
$$\approx 11.6$$

Regular Polygons

$A = \frac{1}{2}ap$

17.25 Draw a regular octagon and identify its radius and apothem.

A regular octagon is an eight-sided polygon with eight congruent sides and eight congruent interior angles, as illustrated below.

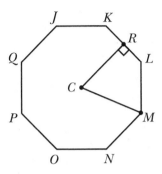

The center of regular octagon $JKLMNOPQ$ is C; it is the unique point inside the polygon that is equidistant from all eight vertices. A segment connecting the center of a regular polygon and one of the vertices is called the radius of the polygon. In the above diagram, $\overline{CM}$ is a radius of $JKLMNOPQ$. ←

An apothem of a regular polygon is a line segment that extends from the center and is perpendicular to one of the sides. In the above diagram, $\overline{CR}$ is an apothem of the octagon. Note that an apothem of a regular polygon always intersects a side at its midpoint, so $KR = RL$.

> This is just one radius of the polygon. Seven other segments connect C to one of the other vertices.

Note: Problems 17.26–17.29 refer to an equilateral triangle ABC, which has three sides with length $4\sqrt{3}$.

17.26 Calculate the length of the apothem.

In the following diagram, Z is the center of regular triangle ABC. By definition, the distances between Z and the vertices are equal; hence, $AZ = CZ$. Because it is perpendicular to side $\overline{AC}$ and extends from the center Z, $\overline{ZD}$ is an apothem of the triangle. Radii of a regular figure bisect interior angles, so

$$m\angle ZAD = m\angle ZCD = \frac{1}{2}(60°) = 30°. \leftarrow$$

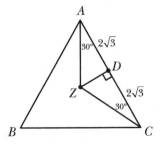

> Each interior angle of a regular polygon measures $180(N-2)°$ divided by N, where N is the number of sides. Each angle of a regular (equilateral) triangle measures 60°.

Recall that an apothem intersects the side of a regular polygon at its midpoint, so D is the midpoint of $\overline{AC}$.

$$AD = DC = \frac{1}{2}(AC) = \frac{1}{2}\left(4\sqrt{3}\right) = 2\sqrt{3}$$

Consider right triangle AZD, a 30°-60°-90° triangle. The side opposite the 60° angle $\left(\overline{AD}\right)$ is $\sqrt{3}$ times as long as the side opposite the 30° angle $\left(\overline{ZD}\right)$.

$$\sqrt{3}\,(ZD) = AD$$
$$\sqrt{3}\,(ZD) = 2\sqrt{3}$$
$$ZD = \frac{2\sqrt{3}}{\sqrt{3}}$$
$$ZD = 2$$

The length of apothem $\overline{ZD}$ is 2.

Note: Problems 17.26–17.29 refer to an equilateral triangle ABC, which has three sides with length $4\sqrt{3}$.

17.27 Calculate the length of the radius of ABC.

The radius of a regular polygon connects the center to one of the vertices. Consider the diagram in Problem 17.26, which contains right triangle AZD with $AD = 2\sqrt{3}$ and $ZD = 2$. The hypotenuse of a 30°-60°-90° triangle is twice the length of the side opposite the 30° angle.

$$AZ = 2(ZD) = 2(2) = 4$$

The length of radius $\overline{AZ}$ is 4.

Note: Problems 17.26–17.29 refer to an equilateral triangle ABC, which has three sides with length $4\sqrt{3}$.

17.28 Calculate the area of $\triangle ABC$ using the formula area $= \frac{1}{2}bh$.

In the diagram below, equilateral triangle ABC has base $\overline{BC}$ and height $\overline{AE}$.

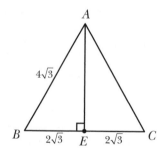

Use the Pythagorean Theorem to calculate AE.

$$(AE)^2 + (BE)^2 = (AB)^2$$
$$(AE)^2 + \left(2\sqrt{3}\right)^2 = \left(4\sqrt{3}\right)^2$$
$$(AE)^2 + 4 \cdot 3 = 16 \cdot 3$$
$$(AE)^2 + 12 = 48$$
$$(AE)^2 = 48 - 12$$
$$(AE)^2 = 36$$
$$\sqrt{(AE)^2} = \sqrt{36}$$
$$AE = 6$$

Substitute $b = BC = 4\sqrt{3}$ and $h = AE = 6$ into the formula for the area of a triangle.

$$\text{area} = \frac{1}{2}bh = \frac{1}{2}\left(4\sqrt{3}\right)(6) = \frac{6}{2}\left(4\sqrt{3}\right) = 3\left(4\sqrt{3}\right) = 12\sqrt{3}$$

> The area of a regular polygon is equal to half the product of its apothem a and its perimeter p.

Note: Problems 17.26–17.29 refer to an equilateral triangle ABC, which has three sides with length $4\sqrt{3}$.

17.29 Verify the area calculated in Problem 17.28 using the formula area $= \dfrac{1}{2}ap$.

According to Problem 17.26, the apothem of ABC has length 2. The perimeter p of a regular polygon is the side length s multiplied by the number of sides n.

$$p = sn = \left(4\sqrt{3}\right)(3) = 12\sqrt{3}$$

Substitute $a = 2$ and $p = 12\sqrt{3}$ into the formula for the area of a regular polygon.

$$\text{area} = \frac{1}{2}ap = \frac{1}{2}(2)\left(12\sqrt{3}\right) = \frac{2}{2}\left(12\sqrt{3}\right) = 12\sqrt{3}$$

The area of $\triangle ABC$ is $12\sqrt{3}$, which verifies the area calculated in Problem 17.28.

Note: Problems 17.30–17.31 refer to a regular nonagon with sides that measure 5 inches.

17.30 Calculate the central angle and report the radius accurate to the tenths place.

The sum of the interior angles of a nonagon is $180(n - 2) = 180(7) = 1{,}260°$. Therefore, each interior angle measures $1{,}260° \div 9 = 140°$. Consider the following figure, which contains a nonagon and $\triangle XYZ$, which is formed by one side of the polygon and the radii that extend to the endpoints of that side.

> A nonagon is a polygon with n = 9 sides.

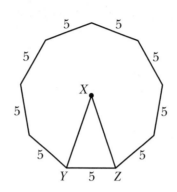
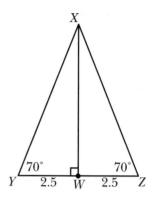

Recall that radii of a regular polygon bisect the interior angles, so $m\angle XYZ = m\angle XZY = \frac{1}{2}(140°) = 70°$. The sum of the measures of a triangle is 180°, so $\angle YXZ$ measures 40°.

You can also calculate the central angle by dividing 360° by the number of sides. In this case, $360° \div 9 = 40°$.

Apothem $\overline{XW}$ bisects $\overline{YZ}$, forming segments $\overline{YW}$ and $\overline{WZ}$ (which measure 2.5 in) and right triangles XYW and XZW. Apply a trigonometric ratio to calculate the radius XY of the nonagon.

$$\cos 70° = \frac{YW}{XY}$$
$$\frac{\cos 70°}{1} = \frac{2.5}{XY}$$
$$XY(\cos 70°) = 2.5$$
$$XY = \frac{2.5}{\cos 70°}$$
$$XY \approx \frac{2.5}{0.342020143}$$
$$XY \approx 7.3$$

The radius of the nonagon is approximately 7.3 inches.

Note: Problems 17.30–17.31 refer to a regular nonagon with sides that measure 5 inches.

17.31 Calculate the area of the nonagon and report the answer accurate to the tenths place.

Consider the diagram of $\triangle XYZ$ in Problem 17.30, which has apothem $\overline{XW}$. Apply a trigonometric ratio to calculate XW.

$$\tan 70° = \frac{XW}{YW}$$
$$\tan 70° = \frac{XW}{2.5}$$
$$2.5(\tan 70°) = XW$$
$$2.5(2.747477419) \approx XW$$
$$6.868693549 \approx XW$$

The perimeter p of the nonagon is the product of the side length $s = 5$ and the number of sides $n = 9$.

$$p = sn = 5(9) = 45 \text{ in}$$

Substitute $a = 6.868693549$ and $p = 45$ into the formula for the area of a regular polygon.

$$\text{area} = \frac{1}{2}ap \approx \frac{1}{2}(6.868693549)(45) \approx 154.5 \text{ in}^2$$

Circles

$A = \pi r^2, C = 2\pi r$

Note: Problems 17.32–17.33 refer to a circle with radius 7.

17.32 Calculate the area of the circle.

The area A of a circle is equal to the product of π and the square of the radius r.

$$A = \pi r^2 = \pi(7)^2 = 49\pi \longleftarrow$$

An answer with a π in it is more accurate than a decimal answer, so don't substitute a decimal approximation (like 3.14159) into your answer for π when you're done.

Note: Problems 17.32–17.33 refer to a circle with radius 7.

17.33 Calculate the circumference of the circle.

The circumference of a circle is the equivalent of the perimeter of a polygon—it measures the distance around the circle. The circumference C of a circle is equal to the product of the radius and 2π.

$$C = 2\pi r = 2\pi(7) = (7 \cdot 2)\pi = 14\pi$$

17.34 Calculate the area of a circle with a circumference of 12.

Substitute $C = 12$ into the formula for the circumference of a circle to identify the corresponding radius.

$$C = 2\pi r$$
$$12 = 2\pi r$$
$$\frac{12}{2\pi} = r$$
$$\frac{6}{\pi} = r$$

Calculate the area A of a circle with radius $r = \dfrac{6}{\pi}$.

This is approximately 11.459.

$$A = \pi r^2 = \pi \left(\frac{6}{\pi} \right)^2 = \pi \cdot \frac{36}{\pi^2} = \frac{36\pi}{\pi^2} = \frac{36}{\pi}$$

The area of a circle with circumference 12 is $\dfrac{36}{\pi}$.

Note: Problems 17.35–17.36 refer to the diagram below, in which $\bigcirc C$ and $\bigcirc D$ are internally tangent at X and the radius of $\bigcirc C$ is 14.

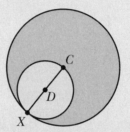

17.35 Calculate the area of the shaded region.

The diameter of $\bigcirc D$ is 14, so the radius of $\bigcirc D$ must be half as long: $r_D = DC = 7$.

Notice that $\overline{CX}$ is a radius of $\bigcirc C$ and a diameter of $\bigcirc D$. The problem states that the radius of $\bigcirc C$ is 14, so $r_C = 14$ and $r_D = 7$ are the radii of the circles.

The shaded region is formed by excluding the region defined by $\bigcirc D$ from the region defined by circle C. Thus, the area of the shaded region (A_S) is equal to A_D (the area of $\bigcirc D$) subtracted from A_C (the area of $\bigcirc C$).

$$A_S = A_C - A_D$$
$$= \pi \left(r_C \right)^2 - \pi \left(r_D \right)^2$$
$$= \pi (14)^2 - \pi (7)^2$$
$$= 196\pi - 49\pi$$
$$= 147\pi$$

Note: Problems 17.35–17.36 refer to the diagram in Problem 17.35, in which the circles are internally tangent at X and the radius of $\bigcirc C$ is 14.

The "hole" where $\bigcirc D$ is excluded is an edge of the shaded region.

17.36 Calculate the perimeter of the shaded region.

The perimeter P_S of the shaded region is equal to the sum of the circumferences of the circles.

$$P_S = 2\pi r_C + 2\pi r_D$$

According to Problem 17.35, $r_C = 14$ and $r_D = 7$.

$$P_S = 2\pi(14) + 2\pi(7) = 28\pi + 14\pi = 42\pi$$

Sectors and Arc Length

$\dfrac{(\text{arc length})^\circ}{360}$ times area or perimeter of the circle

Note: Problems 17.37–17.38 refer to the diagram below, in which $\odot F$ has radius 9.

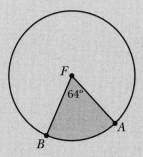

17.37 Calculate the length of $\overset{\frown}{AB}$.

The problem instructs you to calculate the *length* of the arc, not the *measure* of the arc. To calculate the length L of an arc, multiply the circumference of the circle by the ratio of the arc measure (x) to the measure of the entire circle ($360°$).

$$L = \frac{x}{360}(2\pi r)$$

An entire circle measures 360°. This formula asks, "How much of the circumference 2πr corresponds to x° out of 360°?" If x = 180°, for example, then you get half the circumference.

The measure of an arc is defined as the measure of the central angle that intercepts the arc. Therefore, $m\overset{\frown}{AB} = m\angle AFB = 64°$. Substitute $x = 64$ and $r = 9$ into the formula to calculate the length of $\overset{\frown}{AB}$.

$$L = \frac{64}{360}(2\pi \cdot 9) = \frac{\cancel{8} \cdot 8}{\cancel{8} \cdot 45}(18\pi) = \frac{8(18\pi)}{45} = \frac{144\pi}{45} = \frac{16 \cdot \cancel{9} \cdot \pi}{5 \cdot \cancel{9}} = \frac{16\pi}{5}$$

Note: Problems 17.37–17.38 refer to the diagram in Problem 17.37, in which $\odot F$ has radius 9.

17.38 Calculate the area of the shaded sector.

A sector is a portion of a circle bounded by two radii and the arc that connects them. It's shaped like a piece of pie.

As Problem 17.37 indicates, the length of an arc is equal to the circumference of a circle multiplied by $\dfrac{x}{360}$, where x is the measure of the arc in degrees. Similarly, the area of a sector S is equal to the *area* of a circle multiplied by $\dfrac{x}{360}$.

$$S = \frac{x}{360}\left(\pi r^2\right)$$

Substitute $x = m\overset{\frown}{AB} = 64$ and $r = 9$ into the formula to calculate the area of the shaded sector.

$$S = \frac{64}{360}\left(\pi r^2\right) = \frac{\cancel{8} \cdot 8}{\cancel{8} \cdot 45}\left(\pi \cdot 9^2\right) = \frac{8(81\pi)}{45} = \frac{648\pi}{45} = \frac{72 \cdot \cancel{9} \cdot \pi}{5 \cdot \cancel{9}} = \frac{72\pi}{5}$$

17.39 Calculate the area of the shaded region below, given the radius of ⊙L is 3.

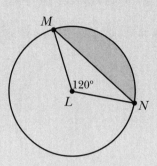

Notice that △LMN is isosceles, because two of its sides ($\overline{LM}$ and $\overline{LN}$) are radii of ⊙L. The base angles of an isosceles triangle are congruent, so $m\angle LMN = m\angle LNM = 30°$.

The angle bisector of the vertex angle of an isosceles triangle is also the perpendicular bisector of the base. In the diagram below, $\overline{LO}$ bisects $\angle MLN$ into two 60° angles and bisects $\overline{MN}$ into two congruent segments.

The angles in the triangle must add up to 180°, and one of the angles is 120°. The two congruent angles must add up to the leftover 60°, and $60 \div 2 = 30$.

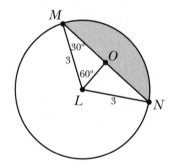

The properties of a 30°-60°-90° triangle state that $LO = \dfrac{1}{2}(ML) = \dfrac{3}{2}$ and $MO = \sqrt{3}(LO) = \dfrac{3}{2}\sqrt{3}$. According to the segment addition postulate, $MN = 2(MO) = 2\left(\dfrac{3}{2}\sqrt{3}\right) = 3\sqrt{3}$. Calculate the area A of △MLN.

$$A = \frac{1}{2}bh = \frac{1}{2}(MN)(LO) = \frac{1}{2}\left(3\sqrt{3}\right)\left(\frac{3}{2}\right) = \frac{9}{4}\sqrt{3}$$

Calculate the area S of the sector bounded by $\overline{LM}$, $\overline{LN}$, and $\overparen{MN}$.

$$S = \frac{120}{360}\left(\pi \cdot 3^2\right) = \frac{1}{3}(9\pi) = 3\pi$$

The area of the shaded region is the area S of the sector minus the area A of the triangle.

$$\text{area of shaded region} = S - A$$
$$= 3\pi - \frac{9}{4}\sqrt{3}$$

Note: Problems 17.40–17.41 refer to the diagram below, in which rectangle QRST is inscribed in $\bigcirc C$.

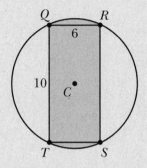

17.40 Calculate the perimeter of the shaded region, and report the answer accurate to the thousandths place.

The shaded region is bounded by two sides of rectangle $QRST$ ($\overline{QT}$ and $\overline{RS}$) and two arcs ($\overparen{QR}$ and $\overparen{TS}$). Opposite sides of a parallelogram are congruent, so $RS = QT = 10$. Congruent chords of a circle intercept congruent arcs. Thus, $\overparen{QR} \cong \overparen{TS}$ because $\overline{QR} \cong \overline{TS}$.

Notice that ΔQST is a right triangle. Use the Pythagorean Theorem to calculate QS.

$$\left(QT\right)^2 + \left(TS\right)^2 = \left(QS\right)^2$$
$$10^2 + 6^2 = \left(QS\right)^2$$
$$100 + 36 = \left(QS\right)^2$$
$$\sqrt{136} = \sqrt{\left(QS\right)^2}$$
$$\sqrt{4 \cdot 34} = QS$$
$$2\sqrt{34} = QS$$

The radius r of $\bigcirc C$ is half the length of diameter $\overline{QS}$, so $r = \frac{1}{2}\left(2\sqrt{34}\right) = \sqrt{34}$. Apply a trigonometric ratio to calculate $m\angle TQS$.

$$\tan\left(m\angle TQS\right) = \frac{TS}{QT}$$

$$\tan\left(m\angle TQS\right) = \frac{6}{10}$$

$$\tan\left(m\angle TQS\right) = \frac{3}{5}$$

$$m\angle TQS = \tan^{-1}\left(\frac{3}{5}\right)$$

$$m\angle TQS \approx 30.96375653°$$

To calculate the lengths of the arcs that bound the shaded region, you must first calculate their measures. Recall that the measures of inscribed angles are half the measures of the arcs they intercept.

$$m\angle TQS = \frac{1}{2}m\widehat{TS}$$

$$30.96375653° \approx \frac{1}{2}m\widehat{TS}$$

$$2\left(30.96375653°\right) \approx m\widehat{TS}$$

$$61.92751306° \approx m\widehat{TS}$$

Apply the arc length formula to calculate the length L of arcs $\widehat{QR}$ and $\widehat{TS}$.

$$L = \frac{x}{360}\left(2\pi r\right)$$

$$\approx \left(\frac{61.92751306}{360}\right)\left(2\pi \cdot \sqrt{34}\right)$$

$$\approx \left(0.1720208696\right)\left(36.636951273\right)$$

$$\approx 6.302320217$$

The perimeter P of the shaded region is the sum of the lengths of the sides and arcs that define the boundaries of the region—two sides with length 10 and two arcs with approximate length 6.302320217.

$$P \approx 2(10) + 2(6.302320217) \approx 32.605$$

Note: Problems 17.40–17.41 refer to the diagram in Problem 17.40, in which rectangle QRST is inscribed in ⊙C.

17.41 Calculate the area of the shaded region, and report the answer accurate to the thousandths place.

The shaded region consists of two sectors and two triangles, as illustrated in the following diagram.

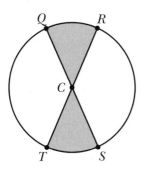
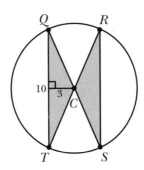

The above sectors are congruent because $\overset{\frown}{QR} \cong \overset{\frown}{ST}$. Substitute $r = \sqrt{34}$ and the measure of the congruent arcs $x \approx 61.92751306°$ (both calculated in Problem 17.40) into the formula for the area S of a sector.

$$S \approx \frac{61.92751306}{360}\left(\pi \cdot \left(\sqrt{34}\right)^2\right) \approx (0.1720208696)(34\pi) \approx 18.374263007$$

Calculate the area A of $\triangle QCT$ and notice that $\triangle QCT \cong \triangle RCS$. ◄

> By the SSS postulate, the triangles are made of four congruent radii and two congruent opposite sides.

$$A = \frac{1}{2}bh = \frac{1}{2}(10)(3) = 5(3) = 15$$

The area A of the entire shaded region is the sum of the areas of two sectors and the areas of two triangles.

$$A = 2(15) + 2(18.374263007) \approx 66.749$$

> The width of rectangle QRST is 6, so the height of $\triangle QCT$ is half of that: $6 \div 2 = 3$.

Similar Figures

$P = x{:}y,\ A = x^2{:}y^2$

17.42 Describe the relationship between the areas and perimeters of similar figures.

The ratio of the perimeters of two similar figures is equal to the ratio of two corresponding sides. In other words, if the scale factor of two similar figures is $x{:}y$, then the ratio of the perimeters is also $x{:}y$. The ratio of the areas, however, is the square of the scale factor, $x^2{:}y^2$.

> This makes sense. If the sides of $\triangle ABC$ are twice as big as the corresponding sides of $\triangle XYZ$, then the sum of the sides of $\triangle ABC$ will be twice as big as the sum of the sides of $\triangle XYZ$ as well.

17.43 Given two similar triangles with a scale factor of 1:3, such that the perimeter of the smaller triangle is 22 and the area of the larger triangle is 140, calculate the perimeter of the larger triangle and the area of the smaller triangle.

Let A_1 and $P_1 = 22$ represent the area and perimeter of the smaller triangle; let $A_2 = 140$ and P_2 represent the area and perimeter of the larger triangle. The scale factor from the smaller to the larger triangle is 1:3, and their perimeters are in the same proportion.

$$\frac{P_1}{P_2} = \frac{1}{3}$$

$$\frac{22}{P_2} = \frac{1}{3}$$

Cross-multiply and solve for P_2.

$$P_2 = 22 \cdot 3 = 66$$

The ratio of the areas of the figures is the square of the scale factor.

$$\frac{A_1}{A_2} = \frac{1^2}{3^2}$$

$$\frac{A_1}{140} = \frac{1}{9}$$

$$9A_1 = 140 \cdot 1$$

$$A_1 = \frac{140}{9}$$

The area of the smaller triangle is $\dfrac{140}{9}$ and the perimeter of the larger triangle is 66.

Note: Problems 17.44–17.46 refer to the diagram below, in which ABCDE and VWXYZ are similar, regular pentagons. The scale factor from ABCDE to VWXYZ is 3:5.

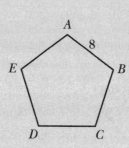
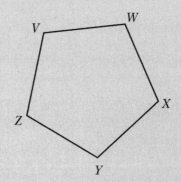

17.44 Calculate the perimeter of *VWXYZ*.

Regular pentagons have five congruent sides, so each side of *ABCDE* has length 8. The perimeter P_1 of *ABCDE* is equal to the product of its side length $s_1 = 8$ and the number of sides $n = 5$.

$$P_1 = s_1(n) = 8(5) = 40$$

The scale factor of *ABCDE* to *VWXYZ* is 3:5, so the ratio of the perimeters $P_1{:}P_2$ is 3:5 as well.

$$\frac{P_1}{P_2} = \frac{3}{5}$$

$$\frac{40}{P_2} = \frac{3}{5}$$

The little ones refer to figure ABCDE—it has perimeter P_1 and side length s_1. Figure VWXYZ has perimeter P_2.

Cross-multiply to solve for P_2.

$$3P_2 = 40 \cdot 5$$
$$3P_2 = 200$$
$$P_2 = \frac{200}{3}$$

The perimeter of $VWXYZ$ is $\dfrac{200}{3}$.

Note: Problems 17.44–17.46 refer to the diagram in Problem 17.44, in which ABCDE and VWXYZ are similar, regular pentagons.

17.45 Calculate the apothems of both figures, accurate to the thousandths place.

Each interior angle of an n-sided regular polygon measures $\left[\dfrac{180(n-2)}{n}\right]^{\circ}$.

Thus, each interior angle of both regular polygons measures $\dfrac{180(3)}{5} = \dfrac{540}{5} = 108^\circ$. Recall that the radii of a regular polygon bisect the interior angles. In the diagram below, $ABCDE$ has center M and $m\angle MCD = m\angle MDC = \dfrac{1}{2}(108^\circ) = 54^\circ$.

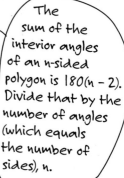

The sum of the interior angles of an n-sided polygon is 180(n – 2). Divide that by the number of angles (which equals the number of sides), n.

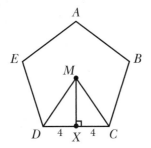

Apply a trigonometric ratio to calculate the length of apothem $\overline{MX}$.

$$\tan \angle MCX = \frac{MX}{XC}$$
$$\tan 54^\circ = \frac{MX}{4}$$
$$4(\tan 54^\circ) = MX$$
$$5.505527682 \approx MX$$

The apothems of similar figures are in the same proportion as corresponding sides of similar figures. The scale factor from $ABCDE$ to $VWXYZ$ is 3:5, so the ratio of the apothems is 3:5 as well. Let x represent the length of the apothem of $VWXYZ$.

$$\frac{5.505527682}{x} \approx \frac{3}{5}$$

$$3x \approx (5.505527682)5$$

$$x \approx \frac{27.52763841}{3}$$

$$x \approx 9.176$$

The apothems of the similar figures are approximately 5.506 and 9.176.

Note: Problems 17.44–17.46 refer to the diagram in Problem 17.44, in which ABCDE and VWXYZ are similar, regular pentagons.

17.46 Calculate the areas of the figures, accurate to the thousandths place.

According to Problem 17.44, the perimeter of *ABCDE* is $p = 40$; according to Problem 17.45, the apothem of *ABCDE* is $a \approx 5.505527682$. Apply the formula for the area of a regular polygon.

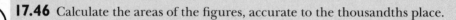

$$\text{area of } ABCDE = \frac{1}{2}ap \approx \frac{1}{2}(5.505527682)(40) \approx 110.11055364$$

The scale factor from *ABCDE* to *VWXYZ* is 3:5, so the ratio of the areas of the similar figures is $3^2:5^2 = 9:25$.

$$\frac{\text{area of } ABCDE}{\text{area of } VWXYZ} = \frac{9}{25}$$

$$\frac{110.11055364}{\text{area of } VWXYZ} \approx \frac{9}{25}$$

$$9(\text{area of } VWXYZ) \approx 25(110.11055364)$$

$$\text{area of } VWXYZ \approx \frac{2{,}752.763841}{9}$$

$$\text{area of } VWXYZ \approx 305.863$$

Therefore, the area of *VWXYZ* is approximately 305.863, and the area of *ABCDE* is approximately 110.111.

You also could have calculated the area of VWXYZ by multiplying its perimeter (from Problem 17.44) by its apothem (from Problem 17.45) and dividing by 2.

Chapter 18
VOLUME AND SURFACE AREA

Measure the inside and outside of three-dimensional solids

This chapter extends the notions of area and perimeter, introduced in Chapter 17, to three-dimensional figures. Instead of calculating the region bounded by a two-dimensional polygon, you will calculate the volume bounded by a prism. Instead of calculating the circumference of a circle, you will determine the surface area of a sphere.

A three-dimensional solid is nothing more than a stack of two-dimensional solids glued together. This chapter revisits some of the formulas from Chapter 17 and explains how to apply them when polygons have not only length and width, but also height.

Prisms

Extend 2-D polygons into 3-D solids

Note: Problems 18.1–18.2 refer to the following diagram, in which $\overline{BA} \perp \overline{AC}$ but $\overline{BE}$ is not an altitude of the prism.

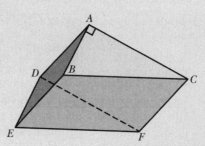

18.1 Identify the bases and lateral faces of the prism.

A prism consists of two congruent bases located in parallel planes that are connected by parallelograms called lateral faces. The bases of this prism are right triangles *ABC* and *DEF*. The lateral faces are □*ABED*, □*BCFE*, and □*ACFD*.

> Lateral means "side," so the lateral faces are sides of the prism, not the top or the bottom. The top and bottom are the bases, congruent polygons that are not necessarily parallelograms.

Note: Problems 18.1–18.2 refer to the prism illustrated in Problem 18.1, in which $\overline{BA} \perp \overline{AC}$ but $\overline{BE}$ is not an altitude of the prism.

18.2 Classify the prism.

The problem states that lateral edges of the prism (the segments at which the lateral faces intersect) are not altitudes. Therefore, the lateral edges are not perpendicular to the bases and the prism is oblique. The bases of the prism are triangles, so it is best classified as an oblique triangular prism.

> If the lateral edges were altitudes, then the lateral faces would be rectangles and you'd have a right triangular prism.

Note: Problems 18.3–18.8 refer to the prism illustrated in the following, in which $\overline{LM} \parallel \overline{ON}$ and $\overline{PQ} \parallel \overline{SR}$.

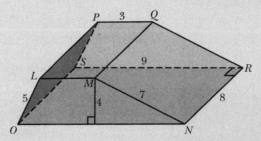

18.3 Classify the prism.

The bases of the prism are quadrilaterals that contain exactly one pair of parallel sides. Thus, *LMNO* and *PQRS* are trapezoids. Because $\overline{NR} \perp \overline{SR}$, the lateral edges of the prism are perpendicular to the base and the lateral faces of the prism are rectangles. Therefore, the solid is a right trapezoidal prism.

> A prism's lateral faces are always parallelograms, but as soon as you know that one of the lateral edges is perpendicular to a base, you can conclude that all the lateral edges are perpendicular to both bases.

Note: Problems 18.3–18.8 refer to the prism illustrated in Problem 18.3, in which $\overline{LM} \parallel \overline{ON}$ and $\overline{PQ} \parallel \overline{SR}$.

18.4 Identify an altitude of the prism.

An altitude of a prism is a segment that represents the distance between the parallel planes containing the bases. Thus, an altitude must be perpendicular to both bases of a prism.

Notice that $\overline{ON} \parallel \overline{SR}$ because they lie on parallel planes, as do all segments that comprise the bases of a prism. Those two lines are intersected by transversal and lateral edge $\overline{NR}$. The diagram states that $\overline{SR} \perp \overline{NR}$. Recall that a transversal perpendicular to one of two parallel lines must be perpendicular to the other line as well, so $\overline{ON} \perp \overline{NR}$.

Because $\overline{NR}$ is perpendicular to both bases, it is an altitude of the prism, and you can conclude that all of the other lateral edges ($\overline{MQ}$, $\overline{LP}$, and $\overline{OS}$) are altitudes as well. ←

> The lateral edges of a right prism are always altitudes of the prism. That is not true for oblique prisms—the lateral edges of those solids are not perpendicular to the bases.

Note: Problems 18.3–18.8 refer to the prism illustrated in Problem 18.3, in which $\overline{LM} \parallel \overline{ON}$ and $\overline{PQ} \parallel \overline{SR}$.

18.5 Calculate the lateral surface area A_L of the prism by adding the areas of the lateral faces.

The prism has four rectangular lateral faces: *ONRS*, *OLPS*, *LMQP*, and *MNRQ*. Calculate the area of each rectangle by multiplying its length by its width. Notice that $NR = MQ = LP = OS$ because the altitudes of a right prism are congruent.

$$\text{area of } ONRS = 9 \cdot 8 = 72$$
$$\text{area of } OLPS = 5 \cdot 8 = 40$$
$$\text{area of } LMQP = 3 \cdot 8 = 24$$
$$\text{area of } MNRQ = 7 \cdot 8 = 56$$

> The lateral surface area does NOT include the bases of the prism. If you add them in you get A_T, the TOTAL surface area—see Problem 18.7.

The lateral area A_L of the prism is the sum of the areas of the sides.

$$A_L = 72 + 40 + 24 + 56 = 192$$

Note: Problems 18.3–18.8 refer to the prism illustrated in Problem 18.3, in which $\overline{LM} \parallel \overline{ON}$ and $\overline{PQ} \parallel \overline{SR}$.

18.6 Verify the lateral surface area calculated in Problem 18.5 by applying the lateral area formula for a right prism.

The formula for the lateral surface area of a right prism is $A_L = ph$, where p is the perimeter of a base and h is the height of the prism. Calculate p by adding the side lengths of base *LMNO*.

$$p = LM + MN + NO + OL = 3 + 7 + 9 + 5 = 24$$

> The bases of a prism are congruent figures, so NO = RS = 9.

The height of a prism is defined as the length of the altitude, so substitute $h = 8$ and $p = 24$ into the lateral area formula.

$$A_L = ph = (24)(8) = 192$$

Note: Problems 18.3–18.8 refer to the prism illustrated in Problem 18.3, in which $\overline{LM} \parallel \overline{ON}$ and $\overline{PQ} \parallel \overline{SR}$.

18.7 Calculate the total surface area A_T of the prism.

The total surface area of the prism is equal to the sum of the lateral surface area A_L and the areas of the bases. The bases of this prism are trapezoids with bases $LM = 3$ and $ON = 9$; the height of the trapezoid is $h = 4$.

$$\text{area of base} = \frac{1}{2}h(b_1 + b_2) = \frac{1}{2}(4)(3+9) = \frac{4}{2}(12) = 2(12) = 24$$

Note that the areas of $LMNO$ and $PQRS$ are equal because the bases of a prism are congruent. Thus, the total surface area of the bases is $2(24) = 48$. Add this value to the lateral surface area calculated in Problem 18.6.

$$A_T = A_L + 48 = 192 + 48 = 240$$

Note: Problems 18.3–18.8 refer to the prism illustrated in Problem 18.3, in which $\overline{LM} \parallel \overline{ON}$ and $\overline{PQ} \parallel \overline{SR}$.

18.8 Calculate the volume of the prism.

The volume V of a right prism is equal to the area of its base B multiplied by its height h: $V = Bh$. According to Problem 18.7, $B = 24$. Recall that this is a right prism, so the lateral edges are altitudes of the solid and their lengths represent the height. Thus, $h = NR = 8$.

$$V = Bh = (24)(8) = 192$$

The volume of a solid measures how much "stuff" can fit inside. To calculate the volume, you're measuring how much stuff makes up one slice of the solid (the area of the base) and then multiplying that by how many slices there are (the height).

Note: Problems 18.9–18.10 refer to the right square prism illustrated below, which contains a rectangular hole that passes through the entire length of the prism.

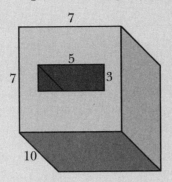

18.9 Calculate the volume of the solid.

The base of the prism is a square with side length $s = 7$, but from that area you must subtract the rectangular hole with length $l = 5$ and width $w = 3$.

$$\text{area of base} = \text{area of square} - \text{area of rectangle}$$
$$= 7 \cdot 7 - 3 \cdot 5$$
$$= 49 - 15$$
$$= 34$$

Substitute $B = 34$ and $h = 10$ into the formula for the volume of a right prism.

$$V = Bh = (34)(10) = 340$$

Remember that you can apply the volume and area formulas $V = Bh$ and $A_L = ph$ only when dealing with right prisms.

Note: Problems 18.9–18.10 refer to the right square prism illustrated in Problem 18.9, which contains a rectangular hole that passes through the entire length of the prism.

18.10 Calculate the total surface area of the solid.

The prism has four rectangular lateral faces with length $l = 10$ and width $w = 7$. Calculate A_L, the lateral surface area.

$$A_L = 4(l \cdot w) = 4(10 \cdot 7) = 4(70) = 280$$

The hole in the solid produces four additional rectangular surfaces, two 3×10 rectangles and two 5×10 rectangles with a total area of $2(30) + 2(50) = 160$.

Problem 18.9 states that the bases have area 34, so the total surface area of both bases is $2(34) = 68$.

The total surface area A_T of the solid is the sum of the lateral surface area, the areas of the faces formed by the rectangular hole, and the areas of the bases.

$$A_T = 280 + 160 + 68 = 508$$

Pyramids

Egypt meets slant height

> The fancy name for a pyramid is "tetrahedron."

Note: Problems 18.11–18.16 refer to a regular square pyramid with base WXYZ, vertex V, and altitude $\overline{VC}$. Assume that WX = 2 and VX = 6.

18.11 Construct a diagram of the pyramid.

A regular pyramid has a base that is a regular polygon—in this case, square *WXYZ*. The lateral edges of a pyramid extend from the vertices of the bases to the vertex *V* of the pyramid, and those edges are congruent in a regular pyramid. The altitude of a regular pyramid is the segment extending from *V* that is perpendicular to the base at its center *C*.

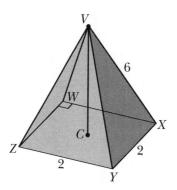

Note: Problems 18.11–18.16 refer to a regular square pyramid with base WXYZ, vertex V, and altitude $\overline{VC}$. Assume that WX = 2 and VX = 6.

18.12 Identify and classify the lateral faces of the pyramid.

The lateral faces of this pyramid are Δ*VZY*, Δ*VYX*, Δ*VXW*, and Δ*VWZ*. Each lateral face is bounded by one side of the square base and two of the lateral edges. The lateral edges of a regular pyramid are congruent, so *VY* = *VX* = *VW* = *VZ* = 6. Because the triangular lateral faces each contain exactly two congruent sides, they are isosceles.

Note: Problems 18.11–18.16 refer to a regular square pyramid with base WXYZ, vertex V, and altitude $\overline{VC}$. Assume that WX = 2 and VX = 6.

18.13 Calculate the slant height of the pyramid.

The slant height is defined as the height of the lateral faces of the pyramid. Consider the following diagram, which illustrates one lateral face of the pyramid, Δ*VYX*. Sides $\overline{VY}$ and $\overline{VX}$ are lateral edges with length 6, and side $\overline{XY}$ is one side of the square base with length 2.

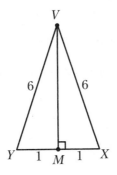

The altitude drawn to the base of a triangle bisects the vertex angle and the base only when the triangle is isosceles.

The height of the triangle is $\overline{VM}$, the segment that bisects $\angle YVX$ and $\overline{YX}$. Apply the Pythagorean Theorem to calculate VM.

$$(YM)^2 + (VM)^2 = (YV)^2$$
$$1^2 + (VM)^2 = 6^2$$
$$(VM)^2 = 36 - 1$$
$$VM = \sqrt{35}$$

The slant height of the pyramid is $\sqrt{35}$.

Note: Problems 18.11–18.16 refer to a regular square pyramid with base WXYZ, vertex V, and altitude $\overline{VC}$. Assume that WX = 2 and VX = 6.

18.14 Calculate the lateral surface area of the pyramid by adding the surface areas of the lateral faces.

Each of the lateral faces of the pyramid is an isosceles triangle with base 2 and height $\sqrt{35}$. Thus, each triangle has area $\frac{1}{2}bh = \frac{1}{2}(2)\left(\sqrt{35}\right) = \sqrt{35}$. The pyramid has four lateral faces, so the total lateral surface area A_L is four times the area of one lateral face: $A_L = 4\sqrt{35}$.

According to Problem 18.13.

Note: Problems 18.11–18.16 refer to a regular square pyramid with base WXYZ, vertex V, and altitude $\overline{VC}$. Assume that WX = 2 and VX = 6.

18.15 Verify the lateral surface area calculated in Problem 18.14 using the lateral surface area formula for a regular pyramid.

The lateral surface area A_L of a regular pyramid is equal to $\frac{1}{2}ps$, where p represents the perimeter of the base and s represents the slant height. This pyramid has a square base with side length 2, so the perimeter of the base is $p = 4(2) = 8$. According to Problem 18.13, the slant height of the pyramid is $s = \sqrt{35}$.

$$A_L = \frac{1}{2}ps = \frac{1}{2}(8)\left(\sqrt{35}\right) = \frac{8}{2}\sqrt{35} = 4\sqrt{35}$$

Whether you use the formula $A_L = \frac{1}{2}ps$ or add the areas of the lateral faces of a pyramid, the result is the same.

Note: Problems 18.11–18.16 refer to a regular square pyramid with base WXYZ, vertex V, and altitude $\overline{VC}$. Assume that WX = 2 and VX = 6.

18.16 Calculate the volume of the pyramid.

> The height of a pyramid is the length of its altitude.

Apply the formula for the volume of a pyramid: $V = \dfrac{1}{3}Bh$, such that B is the area of the base and h is the height of the pyramid. The base of this pyramid is a square with side length 2, so $B = 2 \cdot 2 = 4$. Notice that $\triangle WZX$ is a right triangle with two legs of length 2. Apply the Pythagorean Theorem to calculate ZX, the length of a diagonal of the base.

$$(ZW)^2 + (WX)^2 = (ZX)^2$$
$$2^2 + 2^2 = (ZX)^2$$
$$8 = (ZX)^2$$
$$\sqrt{4 \cdot 2} = \sqrt{(ZX)^2}$$
$$2\sqrt{2} = ZX$$

To calculate the height of the pyramid, consider the following diagram of right triangle VCX.

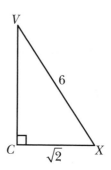

Notice that $\overline{VX}$ is a lateral edge of the pyramid, so $VX = 6$. $\overline{CX}$ is a radius of square base $WXYZ$, half the length of diagonal $\overline{ZX}$.

$$CX = \frac{1}{2}(ZX) = \frac{1}{2}\left(2\sqrt{2}\right) = \sqrt{2}$$

Apply the Pythagorean Theorem to calculate VC, the height of the pyramid.

$$(VC)^2 + (CX)^2 = (VX)^2$$
$$(VC)^2 + \left(\sqrt{2}\right)^2 = 6^2$$
$$(VC)^2 + 2 = 36$$
$$(VC)^2 = 36 - 2$$
$$VC = \sqrt{34}$$

Apply the formula for the volume of a regular pyramid.

$$V = \frac{1}{3}Bh = \frac{1}{3}(4)\left(\sqrt{34}\right) = \frac{4}{3}\sqrt{34}$$

Note: Problems 18.17–18.18 refer to a regular hexagonal pyramid with base ABCDEF and vertex V. Assume that the radius of the base is 5 and the height of the pyramid is 12.

18.17 Calculate the volume of the pyramid.

Consider the following diagram, which illustrates the pyramid and its base, regular hexagon *ABCDEF* with center *X*. Each interior angle of a regular hexagon (with $n = 6$ sides) measures $\frac{180(6-2)}{6} = \frac{180(4)}{6} = 120°$. Recall that the radii of a regular polygon bisect the interior angles, so $m\angle XED = m\angle XDE = 120° \div 2 = 60°$.

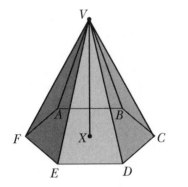 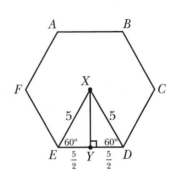

Triangle *XED* is equiangular (all its angles measure 60°) and therefore equilateral, so $XE = XD = ED = 5$. The sides of a regular polygon are congruent, so the perimeter *p* of *ABCDEF* is 6(5) = 30. Apothem $\overline{XY}$ bisects side $\overline{ED}$, forming 30°-60°-90° triangle *XYD*. Thus, $XY = \left(\sqrt{3}\right)(YD) = \frac{5}{2}\sqrt{3}.$

The radius and side length of a polygon aren't usually equal, but they are in the case of regular hexagons.

Calculate the area of the base of the pyramid. Recall that the area of a regular polygon is half the product of its apothem and perimeter.

$$\frac{1}{2}ap = \frac{1}{2}\left(\frac{5}{2}\sqrt{3}\right)(30) = \frac{5 \cdot 30}{2 \cdot 2}\sqrt{3} = \frac{150}{4}\sqrt{3} = \frac{75}{2}\sqrt{3}$$

The volume of a regular pyramid is one-third the product of its height and the area of its base.

$$V = \frac{1}{3}Bh = \frac{1}{3}\left(\frac{75}{2}\sqrt{3}\right)(12) = \frac{75 \cdot 12}{3 \cdot 2}\sqrt{3} = \frac{75 \cdot \cancel{6} \cdot 2}{\cancel{6}}\sqrt{3} = 150\sqrt{3}$$

Note: Problems 18.17–18.18 refer to a regular hexagonal pyramid with base ABCDEF and vertex V. Assume that the radius of the base is 5 and the height of the pyramid is 12.

18.18 Calculate the total surface area of the pyramid.

Consider the following diagram, a right triangle with vertices X (the center of the hexagonal base), D (one vertex of the base), and V (the vertex of the pyramid). The radius of the base is 5, so $XD = 5$; the height of the pyramid is $h = VX = 12$. Side $\overline{VD}$ is a lateral edge of the pyramid.

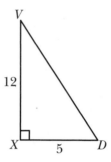

Apply the Pythagorean Theorem to calculate VD.

$$(VX)^2 + (XD)^2 = (VD)^2$$
$$12^2 + 5^2 = (VD)^2$$
$$144 + 25 = (VD)^2$$
$$169 = (VD)^2$$
$$13 = VD$$

Consider $\triangle VED$, a lateral face of the pyramid, in the diagram below. The lateral edges of a regular pyramid are congruent, so $VE = VD = 13$. Side $\overline{ED}$ is a side of the regular hexagonal base $ABCDEF$, so $ED = 5$.

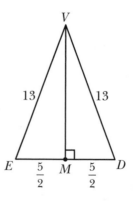

Apply the Pythagorean Theorem to calculate VM, the height of the isosceles triangle and the slant height of the pyramid.

$$(VM)^2 + (MD)^2 = (VD)^2$$

$$(VM)^2 + \left(\frac{5}{2}\right)^2 = 13^2$$

$$(VM)^2 = 169 - \frac{25}{4}$$

$$(VM)^2 = \frac{676 - 25}{4}$$

$$VM = \frac{\sqrt{651}}{\sqrt{4}}$$

$$VM = \frac{\sqrt{651}}{2}$$

Apply the formula for the lateral surface area of a pyramid.

$$A_L = \frac{1}{2}ps = \frac{1}{2}(30)\left(\frac{\sqrt{651}}{2}\right) = \frac{30}{4}\sqrt{651} = \frac{15}{2}\sqrt{651}$$

According to Problem 18.17, the area of the base of the pyramid is $B = \frac{75}{2}\sqrt{3}$.

The total surface area A_T of the pyramid is the sum of the area of the base and the lateral surface area.

$$A_T = B + A_L$$
$$= \frac{75}{2}\sqrt{3} + \frac{15}{2}\sqrt{651}$$

Cylinders and Cones

Solids with circular cross-sections

18.19 Generate the formulas for the lateral surface area and volume of a right cylinder based on the corresponding formulas for a right prism.

A right cylinder, like a right prism, is formed by two congruent bases. Whereas the bases of a prism are congruent polygons, the bases of a cylinder are congruent circles with radius r. ←

> The only difference between right cylinders and right prisms is that the bases of a cylinder are round instead of polygonal.

The lateral surface area of a right prism is equal to the perimeter of its base p multiplied by its height h. A cylinder has a circular base with circumference $2\pi r$, so the lateral surface area of a cylinder is $A_L = ph = 2\pi rh$.

The volume of a right prism is the product of its height h and the area of its base B. The circular base of a cylinder has area πr^2, so the volume of a cylinder is $V = Bh = \pi r^2 h$.

Note: Problems 18.20–18.23 refer to a right cylinder with radius 3 and height 4.

18.20 Draw a diagram illustrating the cylinder.

The bases of the cylinder below, $\bigcirc A$ and $\bigcirc B$, have radius 3. The height of the cylinder is the segment connecting the centers of the circles $\left(\overline{AB}\right)$; it is perpendicular to both circles at their centers.

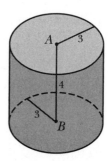

Note: Problems 18.20–18.23 refer to a right cylinder with radius 3 and height 4.

18.21 Calculate the volume of the cylinder.

According to Problem 18.19, the volume V of a right cylinder is equal to the area of the base multiplied by the height h of the cylinder.

$$V = \pi r^2 h = \pi (3)^2 (4) = \pi (9)(4) = 36\pi$$

Note: Problems 18.20–18.23 refer to a right cylinder with radius 3 and height 4.

18.22 Calculate the lateral surface area of the cylinder.

According to Problem 18.19, the lateral surface area A_L of a cylinder is equal to the circumference of its base multiplied by the height of the cylinder.

$$A_L = 2\pi r h = 2\pi (3)(4) = 24\pi$$

Note: Problems 18.20–18.23 refer to a right cylinder with radius 3 and height 4.

18.23 Calculate the total surface area of the cylinder.

The total surface area A_T of a cylinder is the sum of the lateral surface area A_L and the areas of the bases. Note that the bases are congruent circles, so it is sufficient to calculate the area B of one base and multiply it by 2. According to Problem 18.22, $A_L = 24\pi$.

$$A_T = A_L + 2B$$
$$= 24\pi + 2\left(\pi r^2\right)$$
$$= 24\pi + 2 \cdot \pi \cdot 3^2$$
$$= 24\pi + 2 \cdot 9\pi$$
$$= 24\pi + 18\pi$$
$$= 42\pi$$

18.24 Generate the formulas for the lateral surface area and volume of the right cone illustrated below, based on the corresponding formulas for a right pyramid.

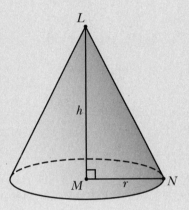

A cone is essentially a pyramid with a circular base—much like a cylinder has the same properties as a prism with a circular (instead of polygonal) base. Thus, the lateral surface area and volume formulas for pyramids may be applied to cones as well.

The volume of a pyramid is equal to one-third the product of its height and the area of its base: $V = \dfrac{1}{3}Bh$. The base of the cone in the illustration is a circle with radius r and area πr^2. Use this value of B to generate the formula for the volume of a cone.

$$V = \frac{1}{3}Bh$$
$$V = \frac{1}{3}\pi r^2 h$$

The lateral surface area of a pyramid is half the product of its slant height and the perimeter of the base: $A_L = \dfrac{1}{2}ps$. Consider right triangle LMN in the diagram, the triangle connecting vertex L, the center of the circular base M, and point N on the circular base.

The hypotenuse of the right triangle is $\overline{LN}$, the slant height of the cone. Apply the Pythagorean Theorem to express the slant height in terms of the radius and height of the cone.

$$(LM)^2 + (MN)^2 = (LN)^2$$
$$h^2 + r^2 = (LN)^2$$
$$\sqrt{r^2 + h^2} = LN$$

Thus, the slant height of the cone is $s = \sqrt{r^2 + h^2}$. The base of the cone is circular, so its perimeter is the circumference of a circle with radius r: $p = 2\pi r$. Substitute s and p into the lateral surface area formula for a pyramid to generate the lateral surface area formula for a cone.

Some books write this formula as $A_L = \pi rs$, where s represents the slant height. Both formulas are useful, depending on what information you're given in a problem.

$$A_L = \frac{1}{2}ps$$
$$A_L = \frac{1}{2}(2\pi r)\left(\sqrt{r^2 + h^2}\right)$$
$$A_L = \frac{2}{2}\pi r\left(\sqrt{r^2 + h^2}\right)$$
$$A_L = \pi r\left(\sqrt{r^2 + h^2}\right)$$

18.25 Calculate the volume of a right cone with height 10, given that the diameter of its base is 6.

A radius is half the length of a diameter, so the radius of the base is $r = 6 \div 2 = 3$. Substitute $r = 3$ and $h = 10$ into the formula for the volume of a cone.

$$V = \frac{1}{3}\pi r^2 h$$
$$= \frac{1}{3}\pi(3)^2(10)$$
$$= \frac{\cancel{3} \cdot 3 \cdot 10 \cdot \pi}{\cancel{3}}$$
$$= 30\pi$$

Note: Problems 18.26–18.27 refer to a right cone with height 14 and volume $\dfrac{4\pi}{21}$.

18.26 Calculate the radius of the base of the cone.

Substitute $V = \dfrac{4\pi}{21}$ and $h = 14$ into the formula for the volume of a cone to calculate the radius.

$$V = \frac{1}{3}\pi r^2 h$$
$$\frac{4\pi}{21} = \frac{1}{3}\pi r^2(14)$$
$$\frac{4\pi}{21} = \frac{14\pi r^2}{3}$$

Cross-multiply and solve for r.

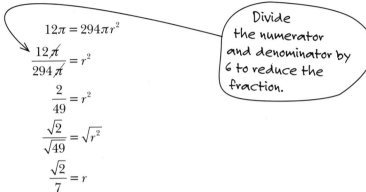

$$12\pi = 294\pi r^2$$

$$\frac{12\cancel{\pi}}{294\cancel{\pi}} = r^2$$

Divide the numerator and denominator by 6 to reduce the fraction.

$$\frac{2}{49} = r^2$$

$$\frac{\sqrt{2}}{\sqrt{49}} = \sqrt{r^2}$$

$$\frac{\sqrt{2}}{7} = r$$

18.27 Calculate the total surface area of the right cone with base $\bigcirc C$, illustrated below.

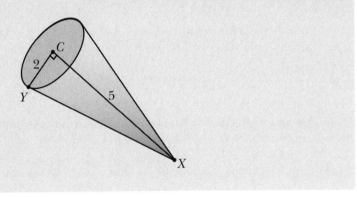

Apply the formula for the lateral surface area A_L of a right cone.

$$A_L = \pi r \sqrt{r^2 + h^2} = \pi(2)\sqrt{2^2 + 5^2} = 2\pi\sqrt{4 + 25} = 2\pi\sqrt{29}$$

The base of the cone has area $B = \pi r^2 = \pi(2)^2 = 4\pi$. The total surface area of the cone A_T is the sum of the lateral surface area and the area of the base.

$$A_T = A_L + B = 2\pi\sqrt{29} + 4\pi$$

Spheres

You'll have a ball!

Note: Problems 18.28–18.29 refer to a sphere with radius 2.

18.28 Calculate the volume of the sphere.

The volume of a sphere is four-thirds the product of π and the cube of the radius: $V = \frac{4}{3}\pi r^3$. Substitute $r = 2$ into the formula.

$$V = \frac{4}{3}\pi r^3 = \frac{4}{3}\pi(2)^3 = \frac{4}{3}\pi(8) = \frac{32}{3}\pi$$

Pyramids, prisms, cylinders, and cones have two kinds of surface area: lateral (which doesn't include the bases) and total (which does). However, spheres don't have bases or sides, so one formula is all you need.

Note: Problems 18.28–18.29 refer to a sphere with radius 2.

18.29 Calculate the surface area of the sphere.

The formula for the surface area A of a sphere is $A = 4\pi r^2$, where r represents the radius. Substitute $r = 2$ into the formula.

$$A = 4\pi r^2 = 4\pi(2)^2 = 4\pi(4) = 16\pi$$

Note: Problems 18.30–18.31 refer to a snowman constructed of three tangent spheres such that the radius of each sphere is half the radius of the sphere upon which it rests.

18.30 Express the total volume of the snowman in terms of r, the radius of the largest sphere.

"Snow sphere" sounds so much less festive than "snowball."

If the largest sphere has radius r, then the sphere stacked on it has radius $\frac{r}{2}$. The smallest sphere on top has a radius half as large as the medium-size sphere: $\frac{1}{2}\left(\frac{r}{2}\right) = \frac{r}{4}$. Calculate the volumes of the three spheres individually.

volume of large sphere: $\frac{4}{3}\pi r^3$

volume of medium sphere: $\frac{4}{3}\pi\left(\frac{r}{2}\right)^3 = \frac{4}{3}\pi\left(\frac{r^3}{8}\right) = \frac{4}{24}\pi r^3 = \frac{1}{6}\pi r^3$

volume of small sphere: $\frac{4}{3}\pi\left(\frac{r}{4}\right)^3 = \frac{4}{3}\pi\left(\frac{r^3}{64}\right) = \frac{4}{192}\pi r^3 = \frac{1}{48}\pi r^3$

The total volume V of the snowman is the sum of the volumes of the three spheres.

$$V = \frac{4}{3}\pi r^3 + \frac{1}{6}\pi r^3 + \frac{1}{48}\pi r^3$$

Simplify the expression using the least common denominator, 48.

$$V = \frac{4}{3} \cdot \frac{16}{16} \cdot \pi r^3 + \frac{1}{6} \cdot \frac{8}{8} \pi r^3 + \frac{1}{48} \pi r^3$$

$$= \frac{64}{48} \cdot \pi r^3 + \frac{8}{48} \pi r^3 + \frac{1}{48} \pi r^3$$

$$= \frac{73}{48} \pi r^3$$

Note: Problems 18.30–18.31 refer to a snowman constructed of three tangent spheres such that the radius of each sphere is half the radius of the sphere upon which it rests.

18.31 Express the total surface area of the snowman in terms of r, the radius of the largest sphere.

According to Problem 18.30, the radii of the spheres are r, $\frac{r}{2}$, and $\frac{r}{4}$. Calculate the surface areas of the spheres independently.

surface area of large sphere: $\quad 4\pi r^2$

surface area of medium sphere: $4\pi\left(\frac{r}{2}\right)^2 = 4\pi\left(\frac{r^2}{4}\right) = \frac{4}{4}\pi r^2 = \pi r^2$

surface area of small sphere: $\quad 4\pi\left(\frac{r}{4}\right)^2 = 4\pi\left(\frac{r^2}{16}\right) = \frac{4}{16}\pi r^2 = \frac{1}{4}\pi r^2$

The total surface area A of the snowman is the sum of the surface areas of the spheres.

$$A = 4\pi r^2 + \pi r^2 + \frac{1}{4}\pi r^2$$

$$= 5\pi r^2 + \frac{1}{4}\pi r^2$$

$$= \frac{20}{4}\pi r^2 + \frac{1}{4}\pi r^2$$

$$= \frac{21}{4}\pi r^2$$

Add $4\pi r^2$ and πr^2 to get $5\pi r^2$ before you convert to common denominators and add the fractions.

18.32 Calculate the volume of the sphere that has a surface area of 72π cm^2.

Surface area is reported in square units, whereas volume is reported in cubed units. The surface area of the sphere is measured in cm^2, so its volume is measured in cm^3. To calculate the volume, first calculate the radius of the sphere by substituting $A = 72\pi$ into the surface area formula.

$$A = 4\pi r^2$$
$$72\pi = 4\pi r^2$$
$$\frac{72\cancel{\pi}}{4\cancel{\pi}} = r^2$$
$$18 = r^2$$
$$\sqrt{9 \cdot 2} = \sqrt{r^2}$$
$$3\sqrt{2} = r$$

$$\left(\sqrt{2}\right)^3 = \sqrt{2} \cdot \sqrt{2} \cdot \sqrt{2}$$
$$= \sqrt{2 \cdot 2 \cdot 2}$$
$$= \sqrt{4 \cdot 2}$$
$$= \sqrt{4} \cdot \sqrt{2}$$
$$= 2\sqrt{2}$$

Substitute $r = 3\sqrt{2}$ into the formula for the volume of a sphere.

$$V = \frac{4}{3}\pi r^3$$
$$= \frac{4}{3}\pi\left(3\sqrt{2}\right)^3$$
$$= \frac{4}{3}\pi(3)^3\left(\sqrt{2}\right)^3$$
$$= \frac{4 \cdot \pi \cdot 27 \cdot 2\sqrt{2}}{3}$$
$$= \frac{216\pi\sqrt{2}}{3}$$
$$= 72\pi\sqrt{2} \text{ cm}^3$$

Note: Problems 18.33–18.36 refer to the illustration below, a right cylindrical can that has the same radius as the three spherical tennis balls it contains, $\frac{3}{2}$ in. Assume that adjacent tennis balls are externally tangent and that the top and bottom tennis balls are tangent to the top and bottom of the container, respectively.

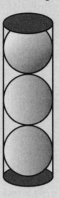

18.33 Calculate the total surface area of the three tennis balls.

All three tennis balls have the same radius, so the spheres are congruent. Calculate the surface area A of one ball with radius $r = \frac{3}{2}$.

$$A = 4\pi r^2 = 4\pi \left(\frac{3}{2}\right)^2 = 4\pi \left(\frac{9}{4}\right) = \frac{36\pi}{4} = 9\pi$$

Multiply the surface area of the ball by 3 to calculate the total surface area A_T of all three congruent spheres.

$$A_T = 3(9\pi) = 27\pi \text{ in}^2$$

Note: Problems 18.33–18.36 refer to the illustration in Problem 18.33, a right cylindrical can that has the same radius as the three spherical tennis balls it contains, $\frac{3}{2}$ in.

18.34 Calculate the volume of the right cylindrical container.

Each ball has a radius of $\frac{3}{2}$ and thus a diameter of $2\left(\frac{3}{2}\right) = 3$. If the balls are externally tangent to each other and to the edges of the container, then the height of the container is equal to 3 times the diameter of a tennis ball: $h = 3(3) = 9$. The problem states that the container and the balls have the same radius, so substitute $h = 9$ and $r = \frac{3}{2}$ into the volume formula for a right cylinder.

$$V = \pi r^2 h = \pi \left(\frac{3}{2}\right)^2 (9) = \pi \left(\frac{9}{4}\right)(9) = \frac{81\pi}{4} \text{ in}^3$$

> A diameter is made up of two radii.

> The container is exactly as tall as a stack of three tennis balls. Each ball has a height of 3 inches, so the total height of the stack is 9 inches.

Note: Problems 18.33–18.36 refer to the illustration in Problem 18.33, a right cylindrical can that has the same radius as the three spherical tennis balls it contains, $\dfrac{3}{2}$ in.

18.35 Calculate the area of the label that surrounds the tube of tennis balls, assuming that the ends of the label extend from the top to the bottom of the cylinder, and assuming that the label does not overlap itself.

The area of the label is exactly equal to the lateral surface area A_L of the cylinder. Substitute $r = \dfrac{3}{2}$ and $h = 9$ (calculated in Problem 18.34) into the appropriate formula.

See Problem 18.22.

$$A_L = 2\pi r h = 2\pi \left(\frac{3}{2}\right)(9) = \frac{54\pi}{2} = 27\pi \ \text{in}^2$$

Note: Problems 18.33–18.36 refer to the illustration in Problem 18.33, a right cylindrical can that has the same radius as the three spherical tennis balls it contains, $\dfrac{3}{2}$ in.

18.36 Calculate the volume of the air in the container.

According to Problem 18.34, the volume V_C of the container is $\dfrac{81\pi}{4}$ in^3.

Calculate the volume V_B of a tennis ball, a sphere with radius $r = \dfrac{3}{2}$.

$$V_B = \frac{4}{3}\pi r^3 = \frac{4}{3}\pi \left(\frac{3}{2}\right)^3 = \frac{4}{3}\pi \left(\frac{27}{8}\right) = \frac{108\pi}{24} = \frac{9\pi}{2} \ \text{in}^3$$

The volume V_A of air in the container is equal to the volume of the container V_C minus the volume occupied by three tennis balls, each with volume V_B.

$$
\begin{aligned}
V_A &= V_C - 3V_B \\
&= \frac{81\pi}{4} - 3\left(\frac{9\pi}{2}\right) \\
&= \frac{81\pi}{4} - \frac{27\pi}{2} \\
&= \frac{81\pi}{4} - \frac{27\pi}{2}\cdot\frac{2}{2} \\
&= \frac{81\pi - 54\pi}{4} \\
&= \frac{27\pi}{4} \ \text{in}^3
\end{aligned}
$$

Similarity of Solids

Area = $x^2:y^2$, volume = $x^3:y^3$

18.37 Complete the following statement.

If two solids are similar, then their bases are _____ and corresponding lengths are _____.

If two solids are similar, then their bases are <u>similar</u> and corresponding lengths are <u>in proportion</u>. ←

> Similar solids are the same shape in different sizes.

18.38 All spheres are similar, so choose two spheres of any radius to identify the relationship between the volume of similar figures.

The problem does not indicate radius values, so you may assign any real numbers to the radii. For instance, let $r_1 = 1$ and $r_2 = 4$ represent the radii of two spheres. Calculate the corresponding volumes V_1 and V_2. Note that the scale factor from the small sphere to the large sphere is the ratio of their radii, 1:4.

> All spheres have the exact shape—they just come in different sizes. Small bowling balls are no less spherical than large bowling balls.

$$V_1 = \frac{4}{3}\pi(r_1)^3 \qquad V_2 = \frac{4}{3}\pi(r_2)^3$$

$$V_1 = \frac{4}{3}\pi(1)^3 \qquad V_2 = \frac{4}{3}\pi(4)^3$$

$$V_1 = \frac{4\pi}{3} \qquad V_2 = \frac{4\pi \cdot 4^3}{3} \text{←}$$

Calculate the ratio of the volumes of the spheres.

$$\frac{V_1}{V_2} = \frac{4\pi}{3} \div \frac{4\pi \cdot 4^3}{3} = \frac{4\pi}{3} \cdot \frac{3}{4\pi \cdot 4^3} = \frac{1}{4^3}$$

The ratio of the radii of the spheres is 1:4, and the ratio of the volumes is $1^3:4^3$. If two similar solids have a scale factor of $x:y$, then their volumes are in the proportion $x^3:y^3$.

> There's a good reason the book doesn't multiply 4 and 4^3 to get 256. In the next step, you divide A_1 and A_2, and leaving things unmultiplied now makes things easier later.

> Dividing by a fraction is the same as multiplying by its reciprocal, so change ÷ to · and flip the second fraction.

Note: Problems 18.39–18.40 refer to a pair of similar right cones such that the height of one cone is 4 times the height of the other.

18.39 Express the lateral surface area of the small cone in terms of r and h, the radius and height of the large cone.

If similar solids have scale factor a:b, the ratio of their surface areas is $a^2:b^2$ and the ratio of their volumes is $a^3:b^3$.

Corresponding lengths of similar solids are in proportion, so the scale factor from the large cone to the small cone is 4:1. Thus, the ratio of the lateral surface areas of the cones is $4^2:1^2 = 16:1$; the lateral surface area of the large cone is 16 times as large as the lateral surface area of the small cone.

A right cone with height h and radius r has lateral surface area $A_L = \pi r \sqrt{r^2 + h^2}$.

Therefore, the lateral surface area of the small cone is $\dfrac{1}{16}\left(\pi r \sqrt{r^2 + h^2}\right)$.

This is the lateral surface area of the large cone because it has height h and a base with radius r.

Note: Problems 18.39–18.40 refer to a pair of similar right cones such that the height of one cone is four times the height of the other.

18.40 Express the volume of the small cone in terms of r and h, the radius and height of the large cone.

The large cone has height h and a base with radius r. Thus, the volume of the large cone is $V = \dfrac{1}{3}\pi r^2 h$. The scale factor from the large cone to the small cone is 4:1, so the volumes of the cones are in the ratio $4^3:1^3 = 64:1$. The volume of the small cone is $\dfrac{1}{64}\left(\dfrac{1}{3}\pi r^2 h\right) = \dfrac{1}{192}\pi r^2 h$.

You're basically dividing the volume of the big cone by 64 to get the volume of the small cone.

Note: Problems 18.41–18.42 refer to a pair of similar right prisms with isosceles right triangle bases. The hypotenuses of the bases of the prisms are 1 and $\sqrt{3}$, and the height of the small prism is 6.

18.41 Calculate the volume of the large prism.

If the hypotenuse of the base of the small prism is 1 and the corresponding length of the large prism is $\sqrt{3}$, then the scale factor is $1:\sqrt{3}$. Therefore, the volumes of the prisms are in the ratio $1^3 : \left(\sqrt{3}\right)^3 = 1 : 3\sqrt{3}$.

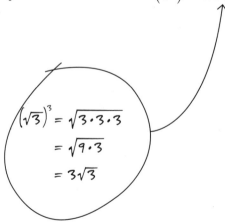

$$\left(\sqrt{3}\right)^3 = \sqrt{3 \cdot 3 \cdot 3}$$
$$= \sqrt{9 \cdot 3}$$
$$= 3\sqrt{3}$$

The volume V of a right prism is equal to Bh, where B represents the area of the base and h is the height of the prism. Consider the small prism, which has height $h = 6$ and an isosceles right triangle base with hypotenuse length 1. Let x represent the lengths of the congruent legs of the base. Apply the Pythagorean Theorem to calculate x.

$$x^2 + x^2 = 1^2$$
$$2x^2 = 1$$
$$x^2 = \frac{1}{2}$$
$$\sqrt{x^2} = \frac{\sqrt{1}}{\sqrt{2}}$$
$$x = \frac{1}{\sqrt{2}} \longleftarrow$$
$$x = \frac{\sqrt{2}}{2}$$

Rationalize this fraction by multiplying the numerator and denominator by $\sqrt{2}$.

The legs of an isosceles triangle are perpendicular—they form the right angle opposite the hypotenuse. Thus, the legs of the triangle represent the base and height of the right triangle. Calculate B, the area of the base.

$$B = \frac{1}{2}bh = \frac{1}{2}\left(\frac{\sqrt{2}}{2}\right)\left(\frac{\sqrt{2}}{2}\right) = \frac{\sqrt{4}}{2 \cdot 2 \cdot 2} = \frac{2}{2 \cdot 2 \cdot 2} = \frac{1}{4}$$

Calculate the volume of the small prism.

$$V = Bh = \frac{1}{4}(6) = \frac{6}{4} = \frac{3}{2}$$

Recall that the volumes of the prisms are in the ratio $1 : 3\sqrt{3}$. Let y represent the volume of the large prism.

$$\frac{\text{volume of small prism}}{\text{volume of large prism}} = \frac{1}{3\sqrt{3}}$$
$$\frac{3/2}{y} = \frac{1}{3\sqrt{3}}$$
$$\frac{3}{2} \cdot 3\sqrt{3} = 1 \cdot y$$
$$\frac{9\sqrt{3}}{2} = y$$

The volume of the large prism is $\dfrac{9\sqrt{3}}{2}$.

Note: Problems 18.41–18.42 refer to a pair of similar right prisms with isosceles right triangle bases. The hypotenuses of the bases of the prisms are 1 and $\sqrt{3}$, and the height of the small prism is 6.

18.42 Calculate the lateral surface area of the large prism.

Consider the small prism, which has an isosceles right triangular base with sides that measure $\dfrac{\sqrt{2}}{2}$, $\dfrac{\sqrt{2}}{2}$, and 1. Calculate the perimeter p of the base.

$$p = \frac{\sqrt{2}}{2} + \frac{\sqrt{2}}{2} + 1 = \frac{2\sqrt{2}}{2} + 1 = \sqrt{2} + 1$$

The lateral surface area A_L of a right prism is equal to the product of its height h and the perimeter p of its base.

$$A_L = ph = \left(\sqrt{2} + 1\right)6 = 6\left(\sqrt{2}\right) + 6(1) = 6\sqrt{2} + 6$$

The scale factor from the small prism to the large prism is $1 : \sqrt{3}$. Thus, the surface areas of the prisms are in the ratio $1^2 : \left(\sqrt{3}\right)^2 = 1 : 3$. Let z represent the lateral surface area of the large prism.

$$\frac{\text{lateral surface area of small prism}}{\text{lateral surface area of large prism}} = \frac{1}{3}$$

$$\frac{6\sqrt{2} + 6}{z} = \frac{1}{3}$$

$$3\left(6\sqrt{2} + 6\right) = z \cdot 1$$

$$18\sqrt{2} + 18 = z$$

The lateral surface area of the large prism is $18\sqrt{2} + 18$, three times the lateral surface area of the small prism.

Chapter 19
COORDINATE GEOMETRY

Points with (x,y) coordinates

Mathematician René Descartes was instrumental in the development of modern geometry, providing a practical arithmetic foundation for the purely abstract and logical concepts rigorously defined by Euclid. The innovation was a coordinate plane—a plane containing two axes (commonly called the *x*- and *y*-axis) measured in fixed units. The Cartesian plane (named for its creator) provides a concrete and visual means by which to validate many of the theorems in the preceding chapters.

The coordinate plane has a horizontal x-axis and a vertical y-axis that intersect at a point called the origin. Any location on the plane can be described using a point (a,b), where a represents the distance from the x-axis and b represents the distance from the y-axis.

In this chapter, you learn how to measure the distance between points and calculate the coordinates of a segment's midpoint, among other things. You then use basic arithmetic to prove some of the theorems that you previously proved logically. There's something satisfying about this. It's nice to know that something is true not only in theory, but also in practice, as verified by cold, hard numbers.

The Coordinate Plane
Review plotting points

All the x- and y-coordinate values are what they appear to be. For example, you can assume that the x-coordinate of C is 4, instead of something weird and deceptive like 4.01.

Note: Problems 19.1–19.6 refer to the coordinate plane below. The coordinates of all points are integers.

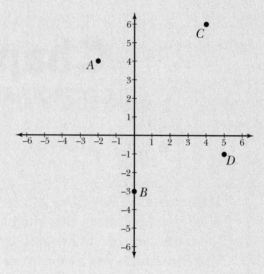

19.1 Identify the coordinates of point C.

Every point in the coordinate plane corresponds to exactly one point (x,y), where x is the distance from the vertical axis and y is the distance from the horizontal axis. Practically speaking, x measures how far right or left a point is from the y-axis (a positive number indicates that the point is right of the axis, whereas a negative number indicates that the point is left of the axis), and y measures how far a point is above or below the x-axis (y is positive when the point is above the axis, and y is negative when the point is below the axis).

x measures right or left distance (right = +, left = –); y measures up and down distance (up = +, down = –).

Point C is 4 units right of the vertical axis, so $x = 4$; C is 6 units above the horizontal axis, so $y = 6$. Therefore, the coordinates of C are $(x,y) = (4,6)$.

Note: Problems 19.1–19.6 refer to the coordinate plane illustrated in Problem 19.1. The coordinates of all points are integers.

19.2 Identify the coordinates of point D.

D is 5 units right of the vertical axis, so $x = 5$; D is 1 unit below the horizontal axis, so $y = -1$. The coordinates of D are $(x,y) = (5,-1)$.

Note: Problems 19.1–19.6 refer to the coordinate plane illustrated in Problem 19.1. The coordinates of all points are integers.

19.3 Identify the coordinates of point B.

Points on the y-axis have an x-coordinate of 0, and points on the x-axis have a y-coordinate of 0.

B lies on the vertical axis, so $x = 0$; B is 3 units below the horizontal axis, so $y = -3$. The coordinates of B are $(x,y) = (0,-3)$.

Note: Problems 19.1–19.6 refer to the coordinate plane illustrated in Problem 19.1. The coordinates of all points are integers.

19.4 Identify the equations of the horizontal and vertical lines that pass through *A*.

The coordinates of *A* are $(x,y) = (-2,4)$. Thus, the equation of the vertical line is $x = -2$, and the equation of the horizontal line is $y = 4$. All vertical lines have equation $x = m$ (where *m* represents the horizontal distance between the vertical line and the *y*-axis), and all horizontal lines have equation $y = n$ (where *n* represents the vertical distance between the horizontal line and the *x*-axis).

Note: Problems 19.1–19.6 refer to the coordinate plane illustrated in Problem 19.1. The coordinates of all points are integers.

19.5 Identify the quadrant of the coordinate plane that does not contain any labeled points.

The *x*- and *y*-axes divide the coordinate plane into four regions called quadrants that are typically numbered using Roman numerals. Quadrant I is the upper-right quadrant (in which $x > 0$ and $y > 0$), quadrant II is the upper-left quadrant (in which $x < 0$ and $y > 0$), quadrant III is the lower-left quadrant (in which $x < 0$ and $y < 0$), and quadrant IV is the lower-right quadrant (in which $x > 0$ and $y < 0$). Quadrant III contains none of the labeled points.

C is in quadrant I, A is in quadrant II, D is in quadrant IV, and B is on the y-axis, so it's not actually in a quadrant.

Note: Problems 19.1–19.6 refer to the coordinate plane illustrated in Problem 19.1. The coordinates of all points are integers.

19.6 Identify a point *E* in quadrant I that lies on the same vertical line as *D*.

The coordinates of *D* are $(x,y) = (5,-1)$. All vertical lines have equation $x = m$, where *m* is the *x*-value of all the points on that vertical line. Point *D* has *x*-value 5, so it lies on the vertical line $x = 5$.

Points in quadrant I have positive *x*- and *y*-values. Therefore, any point $(5,y)$ lies on the line $x = 5$ and is located in the first quadrant as long as $y > 0$. For example, point *E* may have coordinates $(5,11)$ because it has an *x*-coordinate of 5 and its *y*-coordinate is positive.

A lot of correct answers exist for this one: (5,1), (5,2), (5,9.6), and so on. The y-coordinate doesn't have to be an integer. As long as the y-coordinate is positive, the point will be above the x-axis.

Distance Formula
How far from point A to point B?

19.7 Calculate the distance between points $A = (-6,3)$ and $B = (9,3)$.

Points *A* and *B* have the same *y*-coordinate (3), so both lie on the same horizontal line, $y = 3$. According to the ruler postulate, the distance *d* between two points on the same horizontal line is the absolute value of the difference of their *x*-coordinates. (The distance between two points on the same vertical line is the absolute value of the difference of their *y*-coordinates.)

Problem 19.6 states that points with the same x-coordinate are on the same vertical line.

See Problem 4.8.

$$d = |9 - (-6)| = |9 + 6| = |15| = 15$$

The distance between two unique points is always a positive value, which is why absolute values are used.

19.8 Use the following diagram to generate the distance formula, which measures the distance between points $A = (x_1, y_1)$ and $B = (x_2, y_2)$ in the coordinate plane.

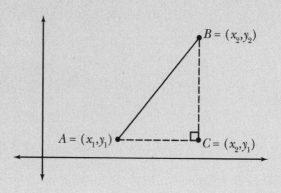

Notice that a dotted horizontal line is drawn through A and a vertical dotted line is drawn through B. The dotted lines intersect at point C, which has the same horizontal distance from the y-axis as B (x_2) and the same vertical distance from the x-axis as A (y_1). The dotted lines are perpendicular, so $\triangle ABC$ is a right triangle.

The distance between A and B is AB, the length of the hypotenuse of the right triangle. You can easily calculate the lengths of the legs of $\triangle ABC$. According to Problem 19.7, the length of a horizontal segment is the absolute value of the difference of the x-coordinates of the endpoints, and the length of a vertical segment is the absolute value of the difference of the y-coordinates of the endpoints.

$$AC = |x_2 - x_1|$$
$$BC = |y_2 - y_1|$$

Let d represent AB, the distance between A and B, and apply the Pythagorean Theorem.

$$(AC)^2 + (BC)^2 = (AB)^2$$
$$(x_2 - x_1)^2 + (y_2 - y_1)^2 = d^2$$
$$\sqrt{(x_2 - x_1)^2 + (y_2 - y_1)^2} = \sqrt{d^2}$$
$$\sqrt{(x_2 - x_1)^2 + (y_2 - y_1)^2} = d$$

The absolute values are gone! They were there to make sure you got positive values, but you don't have to worry about that now because $(x_2 - x_1)$ and $(y_2 - y_1)$ are both squared. Squaring a real number never results in a negative number.

Taking the square root of both sides of an equation usually requires you to place "±" on one side of the equation because the squares of positive and negative numbers are both positive. However, distance is always a positive value, so negative distance values are omitted.

19.9 Calculate the distance between (3,7) and (–2,5).

Substitute $x_1 = 3$, $y_1 = 7$, $x_2 = -2$, and $y_2 = 5$ into the distance formula.

$$d = \sqrt{\left(x_2 - x_1\right)^2 + \left(y_2 - y_1\right)^2}$$
$$= \sqrt{(-2-3)^2 + (5-7)^2}$$
$$= \sqrt{(-5)^2 + (-2)^2}$$
$$= \sqrt{25+4}$$
$$= \sqrt{29}$$

19.10 Calculate the distance between (6,–4) and (–13,–10).

Substitute $x_1 = 6$, $y_1 = -4$, $x_2 = -13$, and $y_2 = -10$ into the distance formula.

$$d = \sqrt{\left(x_2 - x_1\right)^2 + \left(y_2 - y_1\right)^2}$$
$$= \sqrt{(-13-6)^2 + (-10-(-4))^2}$$
$$= \sqrt{(-13-6)^2 + (-10+4)^2}$$
$$= \sqrt{(-19)^2 + (-6)^2}$$
$$= \sqrt{361+36}$$
$$= \sqrt{397}$$

Note: Problems 19.11–19.12 refer to a triangle with vertices L = (–3,5), M = (1,–4), and N = (10,0).

19.11 Prove that the triangle is isosceles.

The sides of the triangle are $\overline{LM}$, $\overline{MN}$, and $\overline{LN}$. Calculate the lengths of the sides individually using the distance formula.

$$LM = \sqrt{\left(1-(-3)\right)^2 + (-4-5)^2} \qquad MN = \sqrt{(10-1)^2 + (0-(-4))^2} \qquad LN = \sqrt{(10-(-3))^2 + (0-5)^2}$$
$$= \sqrt{(1+3)^2 + (-9)^2} \qquad\qquad = \sqrt{(10-1)^2 + (0+4)^2} \qquad\qquad = \sqrt{(10+3)^2 + (0-5)^2}$$
$$= \sqrt{4^2 + (-9)^2} \qquad\qquad\quad = \sqrt{9^2 + 4^2} \qquad\qquad\qquad = \sqrt{13^2 + (-5)^2}$$
$$= \sqrt{16+81} \qquad\qquad\qquad = \sqrt{81+16} \qquad\qquad\qquad = \sqrt{169+25}$$
$$= \sqrt{97} \qquad\qquad\qquad\quad = \sqrt{97} \qquad\qquad\qquad\quad = \sqrt{194}$$

Two sides of the triangle have the same length $\left(LM = MN = \sqrt{97}\right)$, so $\triangle LMN$ is isosceles.

Note: Problems 19.11–19.13 refer to the triangle with vertices L = (–3,5), M = (1,–4), and N = (10,0).

19.12 Prove that $\triangle LMN$ is a right triangle.

Basi-cally, it's a right triangle if the Pythagorean Theorem works. If the square of the big side is larger, the triangle is obtuse; if the sum of the squares of the smaller sides is larger, the triangle is acute.

$\triangle LMN$ is a right triangle only if the square of the longest side is equal to the sum of the squares of the shorter sides.

$$(LM)^2 + (MN)^2 \stackrel{?}{=} (LN)^2$$

Substitute the lengths of the sides, calculated in Problem 19.11, into the equation.

$$\left(\sqrt{97}\right)^2 + \left(\sqrt{97}\right)^2 \stackrel{?}{=} \left(\sqrt{194}\right)^2$$
$$97 + 97 \stackrel{?}{=} 194$$
$$194 = 194 \textbf{ True}$$

$\triangle LMN$ is a right triangle because the lengths of its sides satisfy the Pythagorean Theorem.

Note: Problems 19.11–19.13 refer to the triangle with vertices L = (–3,5), M = (1,–4), and N = (10,0).

19.13 For what value of a does the point $Q = (3,a)$ lie on the perpendicular bisector of $\overline{LN}$?

Points on the perpendicular bisector of a segment are equidistant from the endpoints of that segment. If point Q lies on the perpendicular bisector of $\overline{LN}$, then $QL = QN$. Apply the distance formula to calculate the distances.

Squaring both sides of an equation some-times introduces a false answer, so you should check the answers when you're done by substitut-ing back into the equation with the radicals in it.

$$QL = \sqrt{(3-(-3))^2 + (a-5)^2} \qquad QN = \sqrt{(3-10)^2 + (a-0)^2}$$
$$= \sqrt{(3+3)^2 + (a-5)^2} \qquad\qquad = \sqrt{(-7)^2 + a^2}$$
$$= \sqrt{6^2 + (a-5)^2} \qquad\qquad = \sqrt{49 + a^2}$$
$$= \sqrt{36 + (a-5)^2}$$

Set the distances equal.

$$\sqrt{36 + (a-5)^2} = \sqrt{49 + a^2}$$

Square both sides of the equation to eliminate the radicals.

$$\left(\sqrt{36 + (a-5)^2}\right)^2 = \left(\sqrt{49 + a^2}\right)^2$$
$$36 + (a-5)^2 = 49 + a^2$$
$$36 + a^2 - 10a + 25 = 49 + a^2$$
$$a^2 - 10a + 61 = 49 + a^2$$

Subtract a^2 from both sides of the equation and solve for a.

$$-10a + 61 = 49$$
$$-10a = 49 - 61$$
$$-10a = -12$$
$$a = \frac{-12}{-10}$$
$$a = \frac{6}{5}$$

Substitute $a = \frac{6}{5}$ into the radical equation to verify that it is a valid solution.

$$\sqrt{36 + \left(\frac{6}{5} - 5\right)^2} = \sqrt{49 + \left(\frac{6}{5}\right)^2}$$
$$\sqrt{36 + \left(\frac{6}{5} - \frac{25}{5}\right)^2} = \sqrt{49 + \left(\frac{6}{5}\right)^2}$$
$$\sqrt{36 + \left(-\frac{19}{5}\right)^2} = \sqrt{49 + \left(\frac{6}{5}\right)^2}$$
$$\sqrt{36 + \frac{361}{25}} = \sqrt{49 + \frac{36}{25}}$$
$$\sqrt{\frac{36 \cdot 25}{25} + \frac{361}{25}} = \sqrt{\frac{49 \cdot 25}{25} + \frac{36}{25}}$$
$$\sqrt{\frac{900}{25} + \frac{361}{25}} = \sqrt{\frac{1,225}{25} + \frac{36}{25}}$$
$$\sqrt{\frac{1,261}{25}} = \sqrt{\frac{1,261}{25}} \quad \textbf{True}$$

The equation is true, so $a = \frac{6}{5}$ is a valid solution and the correct answer for this problem.

Note: Problems 19.14–19.16 refer to a quadrilateral with vertices W = (–8,9), X = (–1,10), Y = (4,5), and Z = (–3,4).

19.14 Prove that *WXYZ* is a rhombus.

A rhombus is a quadrilateral with four congruent sides. Use the distance formula to calculate the lengths of sides $\overline{WX}$, $\overline{XY}$, $\overline{YZ}$, and $\overline{ZW}$.

The sides of a polygon are the segments connecting consecutive vertices, so two letters next to each other in the name of the polygon are endpoints of a side, as are the first and last letters in the name.

$$WX = \sqrt{\left(-1-(-8)\right)^2 + (10-9)^2}$$
$$= \sqrt{(-1+8)^2 + 1^2}$$
$$= \sqrt{7^2 + 1^2}$$
$$= \sqrt{49+1}$$
$$= \sqrt{50}$$
$$= 5\sqrt{2}$$

$$XY = \sqrt{\left(4-(-1)\right)^2 + (5-10)^2}$$
$$= \sqrt{(4+1)^2 + (-5)^2}$$
$$= \sqrt{5^2 + (-5)^2}$$
$$= \sqrt{25+25}$$
$$= \sqrt{50}$$
$$= 5\sqrt{2}$$

$$YZ = \sqrt{(-3-4)^2 + (4-5)^2}$$
$$= \sqrt{(-7)^2 + (-1)^2}$$
$$= \sqrt{49+1}$$
$$= \sqrt{50}$$
$$= 5\sqrt{2}$$

$$ZW = \sqrt{\left(-3-(-8)\right)^2 + (4-9)^2}$$
$$= \sqrt{(-3+8)^2 + (-5)^2}$$
$$= \sqrt{(5)^2 + (-5)^2}$$
$$= \sqrt{25+25}$$
$$= \sqrt{50}$$
$$= 5\sqrt{2}$$

All four sides of the quadrilateral have the same length, $5\sqrt{2}$, so the sides are congruent and $WXYZ$ is a rhombus.

Note: Problems 19.13–19.16 refer to the quadrilateral with vertices W = (−8,9), X = (−1,10), Y = (4,5), and Z = (−3,4).

19.15 Calculate the lengths of the diagonals of $WXYZ$.

> If you feel confused and/or ambitious, graph points W, X, Y, and Z on the coordinate plane to help you figure out which segments are the diagonals.

The diagonals of a polygon are line segments connecting nonconsecutive vertices. The diagonals of $WXYZ$ are $\overline{WY}$ and $\overline{XZ}$. Apply the distance formula to calculate their lengths.

$$WY = \sqrt{\left(4-(-8)\right)^2 + (5-9)^2}$$
$$= \sqrt{(4+8)^2 + (-4)^2}$$
$$= \sqrt{12^2 + (-4)^2}$$
$$= \sqrt{144+16}$$
$$= \sqrt{160}$$
$$= 4\sqrt{10}$$

$$XZ = \sqrt{\left(-3-(-1)\right)^2 + (4-10)^2}$$
$$= \sqrt{(-3+1)^2 + (-6)^2}$$
$$= \sqrt{(-2)^2 + (-6)^2}$$
$$= \sqrt{4+36}$$
$$= \sqrt{40}$$
$$= 2\sqrt{10}$$

Note: Problems 19.13–19.16 refer to the quadrilateral with vertices W = (−8,9), X = (−1,10), Y = (4,5), and Z = (−3,4).

19.16 Determine whether $WXYZ$ is a square and justify your answer.

> See Problem 9.17.

A square is both a rhombus and a rectangle; according to Problem 19.14, $WXYZ$ is a rhombus. However, the diagonals of a rectangle must be congruent, and Problem 19.15 proves that the diagonals of $WXYZ$ have different lengths. Therefore, $WXYZ$ is not a square because it is not a rectangle.

Midpoint Formula

Locate a midpoint, given two endpoints

19.17 Describe the relationship between the coordinates of the endpoints of a segment and the coordinates of the midpoint of the segment.

The x-coordinate of a segment's midpoint is the average of the x-coordinates of the segment's endpoints. Similarly, the average of the y-coordinates of a segment's endpoints is the y-coordinate of the midpoint. Thus, the midpoint M of a segment with endpoints (x_1, y_1) and (x_2, y_2) is $M = \left(\dfrac{x_1 + x_2}{2}, \dfrac{y_1 + y_2}{2} \right)$.

> To calculate an average, add up the numbers and divide by how many numbers there are. A segment has two endpoints, so add the two values of x and divide by 2; then do the same for y.

19.18 Calculate the midpoint of the segment with endpoints $(5,11)$ and $(3,-1)$.

Substitute $x_1 = 5$, $y_1 = 11$, $x_2 = 3$, and $y_2 = -1$ into the midpoint formula.

$$M = \left(\frac{x_1 + x_2}{2}, \frac{y_1 + y_2}{2} \right)$$
$$= \left(\frac{5+3}{2}, \frac{11+(-1)}{2} \right)$$
$$= \left(\frac{8}{2}, \frac{10}{2} \right)$$
$$= (4,5)$$

The midpoint of the segment has coordinates $(x,y) = (4,5)$.

> The formula from Problem 19.17.

19.19 Calculate the midpoint of the segment with endpoints $\left(-4, \dfrac{3}{2} \right)$ and $\left(-2, -\dfrac{1}{3} \right)$.

Substitute $x_1 = -4$, $y_1 = \dfrac{3}{2}$, $x_2 = -2$, and $y_2 = -\dfrac{1}{3}$ into the midpoint formula.

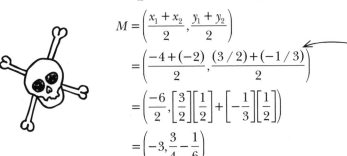

$$M = \left(\frac{x_1 + x_2}{2}, \frac{y_1 + y_2}{2} \right)$$
$$= \left(\frac{-4+(-2)}{2}, \frac{(3/2)+(-1/3)}{2} \right)$$
$$= \left(\frac{-6}{2}, \left[\frac{3}{2} \right]\left[\frac{1}{2} \right] + \left[-\frac{1}{3} \right]\left[\frac{1}{2} \right] \right)$$
$$= \left(-3, \frac{3}{4} - \frac{1}{6} \right)$$

> In the distance formula, you SUBTRACT the x- and y-coordinates. In the midpoint formula, you ADD them.

> Get rid of the y-coordinate denominator by multiplying the numerator and denominator by $\dfrac{1}{2}$. That makes the denominator 1, so you have to deal with only the numerator.

Simplify the y-coordinate using the least common denominator, 12.

$$= \left(-3, \left[\frac{3}{4} \right]\left[\frac{3}{3} \right] - \left[\frac{1}{6} \right]\left[\frac{2}{2} \right] \right)$$
$$= \left(-3, \frac{9}{12} - \frac{2}{12} \right)$$
$$= \left(-3, \frac{7}{12} \right)$$

The midpoint of the segment is $\left(-3, \dfrac{7}{12} \right)$.

Note: Problems 19.20–19.21 refer to the triangle with vertices A = (2,9), B = (–2,–7), and C = (8,1).

19.20 Calculate the coordinates of points L, M, and N, the midpoints of sides $\overline{AB}$, $\overline{BC}$, and $\overline{AC}$, respectively.

> So L is the midpoint of $\overline{AB}$, M is the midpoint of $\overline{BC}$, and N is the midpoint of $\overline{AC}$.

Apply the midpoint formula to each segment individually.

$$L = \left(\frac{x_1 + x_2}{2}, \frac{y_1 + y_2}{2}\right) \qquad M = \left(\frac{x_1 + x_2}{2}, \frac{y_1 + y_2}{2}\right) \qquad N = \left(\frac{x_1 + x_2}{2}, \frac{y_1 + y_2}{2}\right)$$

$$= \left(\frac{2 + (-2)}{2}, \frac{9 + (-7)}{2}\right) \qquad = \left(\frac{-2 + 8}{2}, \frac{-7 + 1}{2}\right) \qquad = \left(\frac{2 + 8}{2}, \frac{9 + 1}{2}\right)$$

$$= \left(\frac{0}{2}, \frac{2}{2}\right) \qquad\qquad = \left(\frac{6}{2}, \frac{-6}{2}\right) \qquad\qquad = \left(\frac{10}{2}, \frac{10}{2}\right)$$

$$= (0,1) \qquad\qquad\qquad = (3,-3) \qquad\qquad\qquad = (5,5)$$

Note: Problems 19.20–19.21 refer to the triangle with vertices A = (2,9), B = (–2,–7), and C = (8,1).

19.21 Calculate the length of the median with endpoint M.

M is the midpoint of $\overline{BC}$, which is opposite $\angle A$. Therefore, the median with endpoint M extends from the opposite vertex, A. Apply the distance formula to calculate the length of median $\overline{AM}$.

$$AM = \sqrt{(3-2)^2 + (-3-9)^2}$$
$$= \sqrt{1^2 + (-12)^2}$$
$$= \sqrt{1 + 144}$$
$$= \sqrt{145}$$

Note: Problems 19.22–19.25 refer to the right triangle with vertices X = (–7,15), Y = (–3,1), and Z = (4,3).

19.22 Identify the hypotenuse of $\triangle XYZ$.

The hypotenuse is the longest side of a right triangle. Apply the distance formula to calculate XY, YZ, and XZ—the lengths of the sides of $\triangle XYZ$—to identify the longest side.

> It's easier to figure out which side is longer if you don't simplify the square roots. Just compare the numbers inside the radicals: 265 > 212 > 53.

$$XY = \sqrt{(-3-(-7))^2 + (1-15)^2} \qquad YZ = \sqrt{(4-(-3))^2 + (3-1)^2} \qquad XZ = \sqrt{(4-(-7))^2 + (3-15)^2}$$

$$= \sqrt{(-3+7)^2 + (-14)^2} \qquad\quad = \sqrt{(4+3)^2 + 2^2} \qquad\qquad = \sqrt{(4+7)^2 + (-12)^2}$$

$$= \sqrt{4^2 + (-14)^2} \qquad\qquad\quad = \sqrt{7^2 + 2^2} \qquad\qquad\qquad = \sqrt{11^2 + (-12)^2}$$

$$= \sqrt{16 + 196} \qquad\qquad\qquad = \sqrt{49 + 4} \qquad\qquad\qquad\quad = \sqrt{121 + 144}$$

$$= \sqrt{212} \qquad\qquad\qquad\qquad = \sqrt{53} \qquad\qquad\qquad\qquad = \sqrt{265}$$

Because $XZ > XY > YZ$, $\overline{XZ}$ is the hypotenuse of $\triangle XYZ$.

Note: Problems 19.22–19.25 refer to the right triangle with vertices X = (–7,15), Y = (–3,1), and Z = (4,3).

19.23 Identify the midpoint M of the hypotenuse.

According to Problem 19.22, $\overline{XZ}$ is the hypotenuse, so substitute $x_1 = -7$, $y_1 = 15$, $x_2 = 4$, and $y_2 = 3$ into the midpoint formula.

$$M = \left(\frac{x_1 + x_2}{2}, \frac{y_1 + y_2}{2} \right)$$
$$= \left(\frac{-7+4}{2}, \frac{15+3}{2} \right)$$
$$= \left(-\frac{3}{2}, \frac{18}{2} \right)$$
$$= \left(-\frac{3}{2}, 9 \right)$$

Note: Problems 19.22–19.25 refer to the right triangle with vertices X = (–7,15), Y = (–3,1), and Z = (4,3).

19.24 Demonstrate that the midpoint of the hypotenuse of $\triangle XYZ$ is equidistant from the vertices. ◀

> Prob-lem 9.31 proves that this theorem is true for all right tri-angles.

Use the distance formula to verify that $XM = YM = ZM$, given M as calculated in Problem 19.23.

$$XM = \sqrt{\left(-\frac{3}{2} - (-7) \right)^2 + (9-15)^2}$$
$$= \sqrt{\left(-\frac{3}{2} + \frac{14}{2} \right)^2 + (-6)^2}$$
$$= \sqrt{\left(\frac{11}{2} \right)^2 + 36}$$
$$= \sqrt{\frac{121}{4} + \frac{144}{4}}$$
$$= \sqrt{\frac{265}{4}}$$
$$= \frac{\sqrt{265}}{2}$$

$$YM = \sqrt{\left(-\frac{3}{2} - (-3) \right)^2 + (9-1)^2}$$
$$= \sqrt{\left(-\frac{3}{2} + \frac{6}{2} \right)^2 + 8^2}$$
$$= \sqrt{\left(\frac{3}{2} \right)^2 + 64}$$
$$= \sqrt{\frac{9}{4} + \frac{256}{4}}$$
$$= \sqrt{\frac{265}{4}}$$
$$= \frac{\sqrt{265}}{2}$$

$$ZM = \sqrt{\left(-\frac{3}{2} - 4 \right)^2 + (9-3)^2}$$
$$= \sqrt{\left(-\frac{3}{2} - \frac{8}{2} \right)^2 + 6^2}$$
$$= \sqrt{\left(-\frac{11}{2} \right)^2 + 36}$$
$$= \sqrt{\frac{121}{4} + \frac{144}{4}}$$
$$= \sqrt{\frac{265}{4}}$$
$$= \frac{\sqrt{265}}{2}$$

> Tech-nically, you don't have to use the distance formula to calcu-late ZM. Because M is the midpoint of the hypotenuse, you know that $\overline{ZM}$ has to be half as long as $\overline{XZ}$, which you already calculated in Problem 19.22.

Notice that ZM is equal to $\frac{\sqrt{265}}{2}$, half the length of $\overline{XZ}$, as calculated in Problem 19.22. ◀

Note: Problems 19.22–19.25 refer to the right triangle with vertices X = (–7,15), Y = (–3,1), and Z = (4,3).

19.25 Identify the center and radius of ⊙C, the circle that circumscribes △XYZ.

Problem 19.24 states that M, the midpoint of the hypotenuse of △XYZ, is equidistant from the vertices of the triangle. The common distance between M and a vertex is $\dfrac{\sqrt{265}}{2}$. Therefore, a circle centered at $M = \left(-\dfrac{3}{2}, 9\right)$ with radius $\dfrac{\sqrt{265}}{2}$ passes through the vertices of △XYZ and therefore circumscribes the triangle.

19.26 Use the diagram of ▱KLMN below to demonstrate that the diagonals of a parallelogram bisect each other.

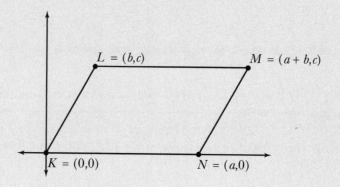

Apply the midpoint formula to calculate midpoint M_1 of diagonal $\overline{KM}$ and midpoint M_2 of diagonal $\overline{LN}$.

$$M_1 = \left(\frac{0+(a+b)}{2}, \frac{0+c}{2}\right) \qquad M_2 = \left(\frac{b+a}{2}, \frac{c+0}{2}\right)$$
$$= \left(\frac{a+b}{2}, \frac{c}{2}\right) \qquad\qquad = \left(\frac{a+b}{2}, \frac{c}{2}\right)$$

> A segment bisector intersects a segment at its midpoint. In this case, each diagonal is intersected by the other diagonal at their midpoints (which overlap). Therefore, each diagonal bisects the other diagonal.

Notice that $M_1 = M_2$, so the diagonals pass through the same point. Two coplanar lines intersect at exactly one point, and in the case of these diagonals, the point of intersection is the midpoint of both diagonals. Thus, each diagonal bisects the other.

Slope of a Line
Change in y divided by change in x

19.27 Define the slope of the line that passes through points (x_1,y_1) and (x_2,y_2).

> Because horizontal lines have a slope of 0.

The slope *m* of a line describes the steepness of the line. Large positive or negative slopes indicate steep lines that rise or fall quickly along the *x*-axis, whereas slopes close to 0 indicate nearly horizontal lines.

The slope m of a line is defined as the difference of the y-coordinates divided by the difference of the x-coordinates, the ratio of the vertical and horizontal changes between the given points.

$$m = \frac{y_2 - y_1}{x_2 - x_1}$$

19.28 Calculate the slope of the line that passes through points (10,–3) and (–4,13).

Substitute $x_1 = 10$, $y_1 = -3$, $x_2 = -4$, and $y_2 = 13$ into the slope formula (presented in Problem 19.27).

$$m = \frac{y_2 - y_1}{x_2 - x_1} = \frac{13 - (-3)}{-4 - 10} = \frac{13 + 3}{-14} = \frac{16}{-14} = -\frac{8}{7}$$

The line passing through points (10,–3) and (–4,13) has slope $-\frac{8}{7}$.

19.29 Calculate the slope of the line that passes through points (–2,–5) and (–7,6).

Substitute $x_1 = -2$, $y_1 = -5$, $x_2 = -7$, and $y_2 = 6$ into the slope formula.

$$m = \frac{y_2 - y_1}{x_2 - x_1} = \frac{6 - (-5)}{-7 - (-2)} = \frac{6 + 5}{-7 + 2} = \frac{11}{-5} = -\frac{11}{5}$$

19.30 Demonstrate that a horizontal line $y = n$ has a slope of 0. Assume that n is a real number.

All points on the horizontal line $y = n$ have y-coordinate n. Thus, the horizontal line passes through points (x_1, n) and (x_2, n). Substitute $y_1 = n$ and $y_2 = n$ into the slope formula and assume that $x_1 \neq x_2$.

$$m = \frac{y_2 - y_1}{x_2 - x_1} = \frac{n - n}{x_2 - x_1} = \frac{0}{x_2 - x_1} = 0$$

> You can plug in actual real number values for x_1 and x_2 if you want. Every real number is a valid x-coordinate for a point on the horizontal line $y = n$ (as long as the y-coordinate is n).

> Zero divided by any nonzero number is equal to 0. You're assuming that x_1 and x_2 are different numbers, so $x_2 - x_1$ will not equal 0.

19.31 Assume that line k passes through points $A = (5,3)$ and $B = (c,-2)$. For what value of c does line k have slope $-\frac{20}{7}$?

Substitute $x_1 = 5$, $y_1 = 3$, $x_2 = c$, $y_2 = -2$, and $m = -\frac{20}{7}$ into the slope formula.

$$m = \frac{y_2 - y_1}{x_2 - x_1}$$

$$-\frac{20}{7} = \frac{-2 - 3}{c - 5}$$

$$-\frac{20}{7} = \frac{-5}{c - 5}$$

$$\frac{20}{7} = \frac{5}{c - 5}$$

> Multiply both sides by –1 to make both fractions positive.

Apply the means-extremes property to solve the proportion.

$$20(c-5) = 7(5)$$
$$20c - 100 = 35$$
$$20c = 35 + 100$$
$$c = \frac{135}{20}$$
$$c = \frac{27}{4}$$

19.32 Assume that line l passes through points (0,4) and (–8,2). Identify the slopes of lines p and n if $l \parallel p$ and $l \perp n$.

> They're opposites (so one is positive and one is negative) and they're reciprocals (so reverse the numerator and denominator).

Apply the slope formula to calculate the slope of l.

$$m = \frac{y_2 - y_1}{x_2 - x_1} = \frac{2 - 4}{-8 - 0} = \frac{-2}{-8} = \frac{1}{4}$$

Parallel lines have the same slope, so the slope of line p is $\frac{1}{4}$. The slopes of perpendicular lines are opposite reciprocals, so the slope of line n is $-\frac{4}{1} = -4$.

19.33 Verify that the triangle with vertices $X = (-7,15)$, $Y = (-3,1)$, and $Z = (4,3)$ is a right triangle using the slopes of its sides.

> This is the triangle from Problems 19.22–19.25.

Calculate the slopes of sides $\overline{XY}$, $\overline{YZ}$, and $\overline{XZ}$.

$$\text{slope of } \overline{XY} = \frac{1 - 15}{-3 - (-7)} \qquad \text{slope of } \overline{YZ} = \frac{3 - 1}{4 - (-3)} \qquad \text{slope of } \overline{XZ} = \frac{3 - 15}{4 - (-7)}$$

$$= \frac{-14}{-3 + 7} \qquad\qquad\qquad = \frac{2}{4 + 3} \qquad\qquad\qquad = \frac{-12}{4 + 7}$$

$$= \frac{-14}{4} \qquad\qquad\qquad\qquad = \frac{2}{7} \qquad\qquad\qquad\qquad = -\frac{12}{11}$$

$$= -\frac{7}{2}$$

The slopes of $\overline{XY}$ and $\overline{YZ}$ are opposite reciprocals, so $\overline{XY} \perp \overline{YZ}$ and the angle formed by those sides is a right angle. Thus, $\triangle XYZ$ is a right triangle.

Note: Problems 19.34–19.35 refer to the quadrilateral with vertices W = (–8,9), X = (–1,10), Y = (4,5), and Z = (–3,4).

19.34 Verify that *WXYZ* is a parallelogram using the slopes of its sides.

Calculate the slopes of $\overline{WX}$, $\overline{XY}$, $\overline{YZ}$, and $\overline{WZ}$.

$$\text{slope of } \overline{WX} = \frac{10-9}{-1-(-8)} \qquad \text{slope of } \overline{XY} = \frac{5-10}{4-(-1)}$$

$$= \frac{1}{-1+8} \qquad\qquad = \frac{-5}{4+1}$$

$$= \frac{1}{7} \qquad\qquad\qquad = \frac{-5}{5}$$

$$\qquad\qquad\qquad\qquad = -1$$

$$\text{slope of } \overline{YZ} = \frac{4-5}{-3-4} \qquad \text{slope of } \overline{WZ} = \frac{4-9}{-3-(-8)}$$

$$= \frac{-1}{-7} \qquad\qquad\qquad = \frac{-5}{-3+8}$$

$$= \frac{1}{7} \qquad\qquad\qquad\quad = \frac{-5}{5}$$

$$\qquad\qquad\qquad\qquad = -1$$

Opposite sides of the quadrilateral have the same slope: $\overline{WX} \parallel \overline{YZ}$ and $\overline{XY} \parallel \overline{WZ}$. A quadrilateral with parallel opposite sides is a parallelogram.

Note: Problems 19.34–19.35 refer to the quadrilateral with vertices W = (–8,9), X = (–1,10), Y = (4,5), and Z = (–3,4).

19.35 Verify that *WXYZ* is a rhombus using the slope formula.

Calculate the slopes of diagonals $\overline{WY}$ and $\overline{XZ}$.

$$\text{slope of } \overline{WY} = \frac{5-9}{4-(-8)} \qquad \text{slope of } \overline{XZ} = \frac{4-10}{-3-(-1)}$$

$$= \frac{-4}{4+8} \qquad\qquad\qquad = \frac{-6}{-3+1}$$

$$= \frac{-4}{12} \qquad\qquad\qquad\quad = \frac{-6}{-2}$$

$$= -\frac{1}{3} \qquad\qquad\qquad\quad = 3$$

The slopes of the diagonals, $-\frac{1}{3}$ and 3, are opposite reciprocals, so the diagonals are perpendicular. According to Problem 19.34, *WXYZ* is a parallelogram. A parallelogram is a rhombus if and only if its diagonals are perpendicular. Thus, *WXYZ* is a rhombus.

19.36 Verify that *KLMN* in the diagram below is a parallelogram by proving that both sides are parallel.

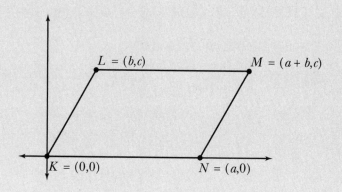

Two sides of *KLMN* are horizontal; the endpoints of $\overline{LM}$ have the same *y*-coordinate, as do the endpoints of $\overline{KN}$. Therefore, the slopes of those sides are equal to 0 and the sides are parallel.

Apply the slope formula to verify that the slopes of $\overline{LK}$ and $\overline{MN}$ are equal, thus ensuring that those opposite sides are also parallel.

$$\text{slope of } \overline{LK} = \frac{0-c}{0-b} \qquad \text{slope of } \overline{MN} = \frac{0-c}{a-(a+b)}$$

$$= \frac{-c}{-b} \qquad\qquad\qquad = \frac{-c}{a-a-b}$$

$$= \frac{c}{b} \qquad\qquad\qquad\quad = \frac{-c}{-b}$$

$$\qquad\qquad\qquad\qquad\qquad = \frac{c}{b}$$

Note: Problems 19.37–19.38 refer to the triangle with vertices A = (2,–1), B = (–4,3), and C = (6,5).

19.37 Calculate the slope of median $\overline{AE}$.

A median is a line segment that extends from a vertex to the midpoint of the opposite side of a triangle. Median $\overline{AE}$ extends from vertex *A* to point *E*, the midpoint of side $\overline{BC}$. Apply the midpoint formula to identify the coordinates of *E*.

$$E = \left(\frac{-4+6}{2}, \frac{3+5}{2}\right) = \left(\frac{2}{2}, \frac{8}{2}\right) = (1,4)$$

Now calculate the slope *m* of $\overline{AE}$ by substituting $x_1 = 2$, $y_1 = -1$, $x_2 = 1$, and $y_2 = 4$ into the slope formula.

$$m = \frac{4-(-1)}{1-2} = \frac{4+1}{-1} = -\frac{5}{1} = -5$$

Note: Problems 19.37–19.38 refer to the triangle with vertices A = (2,–1), B = (–4,3), and C = (6,5).

19.38 Calculate the slope of altitude $\overline{BN}$.

An altitude is a segment that extends from a vertex of the triangle and is perpendicular to the opposite side. Altitude $\overline{BN}$ extends from vertex B to opposite side $\overline{AC}$, intersecting that side at point N. Note that you are not asked to identify the coordinates of N.

If the altitude is perpendicular to the side it intersects, then its slope is the opposite reciprocal of that side. Apply the slope formula to calculate the slope m of $\overline{AC}$.

$$m = \frac{5-(-1)}{6-2} = \frac{6}{4} = \frac{3}{2}$$

The slope of altitude $\overline{BN}$ is $-\dfrac{2}{3}$, the opposite reciprocal of $\dfrac{3}{2}$.

Note: Problems 19.39–19.40 refer to the following diagram, in which $\overleftrightarrow{PQ}$ is tangent to $\bigcirc C$ at P.

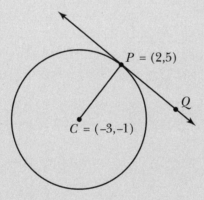

19.39 Calculate the radius of $\bigcirc C$.

Apply the distance formula to calculate CP, the length of the radius.

$$
\begin{aligned}
CP &= \sqrt{\left(2-(-3)\right)^2 + \left(5-(-1)\right)^2} \\
&= \sqrt{(2+3)^2 + (5+1)^2} \\
&= \sqrt{5^2 + 6^2} \\
&= \sqrt{25 + 36} \\
&= \sqrt{61}
\end{aligned}
$$

Note: Problems 19.39–19.40 refer to the diagram in Problem 19.39, in which $\overleftrightarrow{PQ}$ is tangent to $\odot C$ at P.

19.40 Calculate the slope of $\overleftrightarrow{PQ}$.

See Problem 13.21.

Recall that the radius of a circle is perpendicular to a tangent line at the point of tangency. Thus, $\overline{CP} \perp \overleftrightarrow{PQ}$. Calculate the slope m of radius $\overline{CP}$.

$$m = \frac{5-(-1)}{2-(-3)} = \frac{5+1}{2+3} = \frac{6}{5}$$

Because $\overline{CP} \perp \overleftrightarrow{PQ}$, the slope of $\overleftrightarrow{PQ}$ is $-\frac{5}{6}$, the opposite reciprocal of $\frac{6}{5}$.

Note: Problems 19.41–19.46 refer to a quadrilateral with vertices A = (–8,–3), B = (–3,4), C = (3,3), and D = (4,–5).

19.41 Verify that *ABCD* is a trapezoid.

Apply the slope formula to each side of the quadrilateral.

$$\text{slope of } \overline{AB} = \frac{4-(-3)}{-3-(-8)} \qquad \text{slope of } \overline{BC} = \frac{3-4}{3-(-3)}$$

$$= \frac{4+3}{-3+8} \qquad\qquad = \frac{-1}{3+3}$$

$$= \frac{7}{5} \qquad\qquad\qquad = -\frac{1}{6}$$

$$\text{slope of } \overline{CD} = \frac{-5-3}{4-3} \qquad \text{slope of } \overline{AD} = \frac{-5-(-3)}{4-(-8)}$$

$$= \frac{-8}{1} \qquad\qquad\qquad = \frac{-5+3}{4+8}$$

$$= -8 \qquad\qquad\qquad = \frac{-2}{12}$$

$$\qquad\qquad\qquad\qquad\qquad = -\frac{1}{6}$$

Exactly one pair of opposite sides is parallel, $\overline{BC}$ and $\overline{AD}$, so *ABCD* is a trapezoid.

Note: Problems 19.41–19.46 refer to a quadrilateral with vertices A = (–8,–3), B = (–3,4), C = (3,3), and D = (4,–5).

19.42 Identify the endpoints *M* and *N* of the median.

The median of a trapezoid is the segment whose endpoints are the midpoints of the legs. According to Problem 19.41, the bases of *ABCD* are $\overline{BC}$ and $\overline{AD}$, so the legs of the trapezoid are $\overline{AB}$ and $\overline{CD}$. Let *M* represent the midpoint of $\overline{AB}$ and

The legs are the nonparallel sides.

N represent the midpoint of $\overline{CD}$. Apply the midpoint formula to determine the coordinates of M and N.

$$M = \left(\frac{-8-3}{2}, \frac{-3+4}{2} \right) \qquad N = \left(\frac{3+4}{2}, \frac{3-5}{2} \right)$$

$$= \left(-\frac{11}{2}, \frac{1}{2} \right) \qquad\qquad = \left(\frac{7}{2}, \frac{-2}{2} \right)$$

$$= \left(\frac{7}{2}, -1 \right)$$

> **Note: Problems 19.41–19.46 refer to a quadrilateral with vertices A = (–8,–3), B = (–3,4), C = (3,3), and D = (4,–5).**
>
> **19.43** Verify that the median is parallel to the bases of $ABCD$.

According to Problem 19.42, the endpoints of the median are $M = \left(-\frac{11}{2}, \frac{1}{2} \right)$ and $N = \left(\frac{7}{2}, -1 \right)$. Apply the slope formula.

$$\text{slope of } \overline{MN} = \frac{-1-(1/2)}{(7/2)-(-11/2)}$$

$$= \frac{(-2/2)-(1/2)}{(7/2)+(11/2)}$$

$$= \frac{-3/2}{18/2}$$

$$= \frac{-3/2}{9}$$

Write the fraction as a division problem.

$$= -\frac{3}{2} \div 9$$

Write the quotient as a product and reduce the fraction to lowest terms.

$$= -\frac{3}{2} \cdot \frac{1}{9}$$

$$= -\frac{3}{18}$$

$$= -\frac{1}{6}$$

Change division to multiplication, take the reciprocal of the right fraction, and multiply the fractions.

Because $\overline{MN}$ has the same slope as bases $\overline{BC}$ and $\overline{AD}$ (as calculated in Problem 19.41), it is parallel to the bases of the trapezoid.

Note: Problems 19.41–19.46 refer to a quadrilateral with vertices A = (–8,–3), B = (–3,4), C = (3,3), and D = (4,–5).

19.44 Is *ABCD* an isosceles trapezoid? Justify your answer.

Calculate the lengths of the legs of *ABCD*.

$$AB = \sqrt{\left(-3-(-8)\right)^2 + \left(4-(-3)\right)^2}$$
$$= \sqrt{(-3+8)^2 + (4+3)^2}$$
$$= \sqrt{5^2 + 7^2}$$
$$= \sqrt{25+49}$$
$$= \sqrt{74}$$

$$CD = \sqrt{(4-3)^2 + (-5-3)^2}$$
$$= \sqrt{1^2 + (-8)^2}$$
$$= \sqrt{1+64}$$
$$= \sqrt{65}$$

The legs of the trapezoid have different lengths, so *ABCD* is not an isosceles trapezoid.

Note: Problems 19.41–19.46 refer to a quadrilateral with vertices A = (–8,–3), B = (–3,4), C = (3,3), and D = (4,–5).

19.45 Calculate the lengths of the bases of *ABCD*.

Apply the distance formula to calculate the lengths of $\overline{BC}$ and $\overline{AD}$.

$$BC = \sqrt{\left(3-(-3)\right)^2 + (3-4)^2}$$
$$= \sqrt{(3+3)^2 + (3-4)^2}$$
$$= \sqrt{6^2 + (-1)^2}$$
$$= \sqrt{36+1}$$
$$= \sqrt{37}$$

$$AD = \sqrt{\left(4-(-8)\right)^2 + \left(-5-(-3)\right)^2}$$
$$= \sqrt{(4+8)^2 + (-5+3)^2}$$
$$= \sqrt{12^2 + (-2)^2}$$
$$= \sqrt{144+4}$$
$$= \sqrt{148}$$
$$= 2\sqrt{37}$$

Note: Problems 19.41–19.46 refer to a quadrilateral with vertices A = (–8,–3), B = (–3,4), C = (3,3), and D = (4,–5).

19.46 Verify that the median of *ABCD* is half as long as the sum of the lengths of the bases.

Use points M and N (and their coordinates) from Problem 19.42.

Apply the distance formula to calculate the length of the median.

$$MN = \sqrt{\left(\frac{7}{2} - \left(-\frac{11}{2}\right)\right)^2 + \left(-1 - \frac{1}{2}\right)^2}$$

$$= \sqrt{\left(\frac{7}{2} + \frac{11}{2}\right)^2 + \left(-\frac{2}{2} - \frac{1}{2}\right)^2}$$

$$= \sqrt{\left(\frac{18}{2}\right)^2 + \left(-\frac{3}{2}\right)^2}$$

$$= \sqrt{9^2 + \left(-\frac{3}{2}\right)^2}$$

$$= \sqrt{81 + \frac{9}{4}}$$

$$= \sqrt{\frac{324}{4} + \frac{9}{4}}$$

$$= \sqrt{\frac{333}{4}}$$

$$= \frac{\sqrt{9 \cdot 37}}{\sqrt{4}}$$

$$= \frac{3\sqrt{37}}{2}$$

According to Problem 19.45, the lengths of the bases of $ABCD$ are $\sqrt{37}$ and $2\sqrt{37}$. Thus, the sum of the bases is $3\sqrt{37}$ and half the sum is $\frac{3\sqrt{37}}{2}$, the length of median $\overline{MN}$, as calculated above.

Chapter 20

CIRCLES, LINES, AND VECTORS IN THE COORDINATE PLANE

More coordinate geometry

The preceding chapter explored the rudimentary concepts of coordinate geometry, focusing primarily on the characteristics of line segments and lines. This chapter defines lines more rigorously, as infinite sets of coordinate pairs that comply with an equation defined in terms of two unknowns. This chapter also explores the equations of circles and begins an investigation of vectors.

Chapter 19 showed you how to calculate the length, the midpoint, or the slope of a line segment, given its endpoints. In this chapter, you define the lines that contain those segments using the slope-intercept and point-slope formulas for a line. You'll also put lines into standard form.

Speaking of standard form, this chapter explains how to write the equations of circles in standard form (it's easier than you'd think). You'll also learn about vectors, little line segments with one pointy end.

Linear Equations
Point-slope and slope-intercept form

The
y-intercept
of a line is
the y-coordinate
where the line in-
tersects the y-axis.
All y-intercepts have
an x-coordinate of 0,
so the line
$y = mx + b$ passes
through point
(0,b).

20.1 Identify the slope-intercept form of a line and define the variables in the equation.

The slope-intercept form of a line is $y = mx + b$, where m is the slope of the line, b is the y-intercept, and (x,y) is a point on the line.

20.2 Identify the point-slope form of a line and define the variables in the equation.

The point-slope form of a line is $y - y_1 = m(x - x_1)$, where m is the slope of the line and (x_1, y_1) is a point through which the line passes.

x and y are
the coordinates
of some other point
on the line. Here's
the difference: x_1
and y_1 are provided by
the problem, and x and
y are one of the other
infinite pairs of coordi-
nates that satisfy
the equation.

Note: Problems 20.3–20.4 refer to the line with y-intercept 4 and slope –3.

20.3 Write the equation of the line in slope-intercept form.

Substitute $m = -3$ and $b = 4$ into the slope-intercept formula to generate the equation of the line.

$$y = mx + b$$
$$y = -3x + 4$$

Note: Problems 20.3–20.4 refer to the line with y-intercept 4 and slope –3.

20.4 Write the equation of the line in standard form.

The standard form of a line is $Ax + By = C$, such that A, B, and C are integers and $A > 0$. Thus, a line in standard form lists the x-term followed by the y-term left of the equals sign and isolates the constant on the other side of the equation. According to Problem 20.3, the equation of the line in slope-intercept form is $y = -3x + 4$.

A,
B, and C can't
share common fac-
tors, either, so if one
number divides into
each of them evenly,
divide the entire
equation by that
number.

To write the equation in standard form, add $3x$ to both sides of the equation.

$$3x + y = 4$$

The equation is in standard form because the x-term ($3x$) is positive, the x-term precedes the y-term on the left side of the equation, the constant is isolated on the right side of the equation, and the equation contains no fractions or decimals.

The
x-term must be
positive. If it's not,
multiply every term in
the equation by –1 to
change the sign of
each term.

Note: Problems 20.5–20.6 refer to the linear equation with slope $\frac{1}{2}$ that has an x-intercept of 9.

20.5 Use the point-slope formula to write the equation of the line in slope-intercept form.

The *x*-intercept is the point at which a line intersects the *x*-axis. All points on the *x*-axis have a *y*-coordinate of 0, so the line passes through the point (9,0). Substitute $m = \frac{1}{2}$, $x_1 = 9$, and $y_1 = 0$ into the point-slope formula.

$$y - y_1 = m(x - x_1)$$
$$y - 0 = \frac{1}{2}(x - 9)$$
$$y = \frac{1}{2}x - \frac{9}{2}$$

The equation is in slope-intercept form because it is solved for *y*.

> Horizontal lines have the equation y = n, where n is the vertical distance from the x-axis. The x-axis has the equation y = 0 because (unsurprisingly) there is no distance between the x-axis and itself.

> The right side of the equation contains an x-term (whose coefficient is the slope) and a constant (which is the y-intercept).

Note: Problems 20.5–20.6 refer to the linear equation with slope $\frac{1}{2}$ that has an x-intercept of 9.

20.6 Write the equation of the line in standard form.

Problem 20.5 states that the slope-intercept form of the equation is $y = \frac{1}{2}x - \frac{9}{2}$. Isolate the constant $-\frac{9}{2}$ on the right side of the equals sign by subtracting $\frac{1}{2}x$ from both sides of the equation.

$$-\frac{1}{2}x + y = -\frac{9}{2}$$

> Write the x-term before the y-term.

Multiply each term of the equation by 2 to eliminate the fractions.

$$\left(\frac{2}{1}\right)\left(-\frac{1}{2}x\right) + \left(\frac{2}{1}\right)y = \left(\frac{2}{1}\right)\left(-\frac{9}{2}\right)$$
$$-\frac{2}{2}x + \frac{2}{1}y = -\frac{18}{2}$$
$$-x + 2y = -9$$

> If you multiply by –2 instead of 2 in the previous step, you can avoid this step.

The *x*-term of an equation in standard form must be positive, so multiply each of the terms by –1.

$$(-1)(-x) + (-1)(2y) = (-1)(-9)$$
$$x - 2y = 9$$

20.7 Identify the x- and y-intercepts of $4x - 3y = 8$.

The x-intercept of a line has a y-coordinate of 0, so substitute $y = 0$ into the equation and solve for x.

$$4x - 3(0) = 8$$
$$4x - 0 = 8$$
$$4x = 8$$
$$x = \frac{8}{4}$$
$$x = 2$$

The y-intercept of a line has an x-coordinate of 0, so substitute $x = 0$ into the equation and solve for y.

$$4(0) - 3y = 8$$
$$0 - 3y = 8$$
$$-3y = 8$$
$$y = -\frac{8}{3}$$

The x-intercept of the line is 2 and the y-intercept of the line is $-\frac{8}{3}$.

20.8 Write the equation of the line that passes through points $(7,2)$ and $(-5,-1)$ in slope-intercept form.

Begin by calculating the slope m of the line.

$$m = \frac{y_2 - y_1}{x_2 - x_1} = \frac{-1 - 2}{-5 - 7} = \frac{-3}{-12} = \frac{1}{4}$$

The book substitutes $x_1 = 7$ and $y_1 = 2$, but you'll end up with the same final answer if you substitute $x_1 = -5$ and $y_1 = -1$.

Substitute $m = \frac{1}{4}$ and one of the coordinate pairs into the point-slope formula.

$$y - y_1 = m(x - x_1)$$
$$y - 2 = \frac{1}{4}(x - 7)$$
$$y - 2 = \frac{1}{4}x - \frac{7}{4}$$

Solve for y to write the equation in slope-intercept form.

$$y = \frac{1}{4}x - \frac{7}{4} + 2$$
$$y = \frac{1}{4}x - \frac{7}{4} + \frac{8}{4}$$
$$y = \frac{1}{4}x + \frac{1}{4}$$

Note: Problems 20.9–20.10 refer to the rectangle with vertices A = (–4,2), B = (2,5), C = (4,1), and D = (–2,–2).

20.9 Write the equation of $\overline{BC}$ in standard form.

Calculate the slope m of the line that passes through B and C.

$$m = \frac{y_2 - y_1}{x_2 - x_1} = \frac{1 - 5}{4 - 2} = \frac{-4}{2} = -2$$

Or
$x_1 = 2$ and $y_1 = 5$

Substitute $m = -2$, $x_1 = 4$, and $y_1 = 1$ into the point-slope formula.

$$y - y_1 = m(x - x_1)$$
$$y - 1 = -2(x - 4)$$
$$y - 1 = -2x + 8$$

Express the equation in standard form.

$$2x + y = 8 + 1$$
$$2x + y = 9$$

Note: Problems 20.9–20.10 refer to the rectangle with vertices A = (–4,2), B = (2,5), C = (4,1), and D = (–2,–2).

20.10 Write the equation of $\overline{AC}$ in standard form.

Calculate the slope m of the line that passes through A and C.

$$m = \frac{y_2 - y_1}{x_2 - x_1} = \frac{1 - 2}{4 - (-4)} = \frac{-1}{4 + 4} = -\frac{1}{8}$$

Substitute $m = -\frac{1}{8}$, $x_1 = -4$, and $y_1 = 2$ into the point-slope formula.

$$y - y_1 = m(x - x_1)$$
$$y - 2 = -\frac{1}{8}(x - (-4))$$
$$y - 2 = -\frac{1}{8}(x + 4)$$

Multiply both sides of the equation by 8 to eliminate the fraction.

$$8(y - 2) = \left(\frac{8}{1}\right)\left(-\frac{1}{8}\right)(x + 4)$$
$$8(y - 2) = (-1)(x + 4)$$
$$8y - 16 = -x - 4$$

Express the equation in standard form.

$$x + 8y = -4 + 16$$
$$x + 8y = 12$$

> Write the equation of the line that passes through the median's endpoints.

Note: Problems 20.11–20.13 refer to the triangle with vertices X = (–2,1), Y = (3,5), and Z = (9,–4).

20.11 Write the equation of the line containing median $\overline{XQ}$ in standard form.

A median extends from a vertex to the midpoint of the opposite side of a triangle. Because $\overline{YZ}$ is opposite $\angle X$, Q is the midpoint of that segment. Apply the midpoint formula to calculate Q.

$$Q = \left(\frac{x_1 + x_2}{2}, \frac{y_1 + y_2}{2}\right) = \left(\frac{3+9}{2}, \frac{5-4}{2}\right) = \left(\frac{12}{2}, \frac{1}{2}\right) = \left(6, \frac{1}{2}\right)$$

Calculate the slope m of the line passing through X and Q.

> Write the complex fraction as a division problem, convert it to multiplication by flipping the second fraction, and simplify.
> $$-\frac{1}{2} \div 8 = -\frac{1}{2} \cdot \frac{1}{8} = -\frac{1}{16}$$

$$m = \frac{y_2 - y_1}{x_2 - x_1} = \frac{1/2 - 1}{6 - (-2)} = \frac{-1/2}{8} = -\frac{1}{16}$$

Substitute $m = -\dfrac{1}{16}$, $x_1 = -2$, and $y_1 = 1$ into the point-slope formula.

$$y - y_1 = m(x - x_1)$$
$$y - 1 = -\frac{1}{16}(x - (-2))$$
$$y - 1 = -\frac{1}{16}(x + 2)$$

> The first thing you do is multiply both sides by 16 to get rid of the fraction.

Express the equation in standard form.

$$16(y - 1) = \left(\frac{\cancel{16}}{1}\right)\left(-\frac{1}{\cancel{16}}\right)(x + 2)$$
$$16y - 16 = -(x + 2)$$
$$16y - 16 = -x - 2$$
$$x + 16y = -2 + 16$$
$$x + 16y = 14$$

Note: Problems 20.11–20.13 refer to the triangle with vertices X = (–2,1), Y = (3,5), and Z = (9,–4).

20.12 Write the equation of the line containing median $\overline{ZR}$ in standard form.

Implement the same steps you use in Problem 20.11 to identify the line passing through Z (the vertex opposite $\overline{XY}$) and R (the midpoint of $\overline{XY}$). Begin by calculating the coordinates of R.

$$R = \left(\frac{-2+3}{2}, \frac{1+5}{2}\right) = \left(\frac{1}{2}, \frac{6}{2}\right) = \left(\frac{1}{2}, 3\right)$$

> $$\frac{7}{-17/2} = \frac{7}{1} \div \left(-\frac{17}{2}\right) =$$
> $$\frac{7}{1} \cdot \left(-\frac{2}{17}\right) = -\frac{14}{17}$$

Apply the slope formula to calculate the slope m of the line passing through Z and R.

$$m = \frac{3 - (-4)}{(1/2) - 9} = \frac{3 + 4}{(1/2) - (18/2)} = \frac{7}{-17/2} = -\frac{14}{17}$$

Substitute $m = -\dfrac{14}{17}$, $x_1 = 9$, and $y_1 = -4$ into the point-slope formula.

$$y - (-4) = -\frac{14}{17}(x - 9)$$

$$y + 4 = -\frac{14}{17}(x - 9)$$

Write the equation in standard form.

$$17(y + 4) = \left(\frac{\cancel{17}}{1}\right)\left(-\frac{14}{\cancel{17}}\right)(x - 9)$$

$$17y + 68 = -14(x - 9)$$

$$17y + 68 = -14x + 126$$

$$14x + 17y = 126 - 68$$

$$14x + 17y = 58$$

Note: Problems 20.11–20.13 refer to the triangle with vertices X = (–2,1), Y = (3,5), and Z = (9,–4).

20.13 Identify the centroid of $\triangle XYZ$.

The medians of a triangle intersect at a single point C called the centroid. Thus, the linear equations identified in Problems 20.11 and 20.12 form a system of equations, the solution to which is the centroid.

$$\begin{cases} x + 16y = 14 \\ 14x + 17y = 58 \end{cases}$$

Solve the system by eliminating the x-term.

$$\begin{array}{rrrr} -14x & - & 224y & = & -196 \\ 14x & + & 17y & = & 58 \\ \hline & & -207y & = & -138 \end{array}$$

$$y = \frac{-138}{-207}$$

$$y = \frac{-138 \div (-69)}{-207 \div (-69)}$$

$$y = \frac{2}{3}$$

> Multiply the top equation by –14 to get –14x – 224y = –196, and then add like terms to produce an equation that you can solve for y.

Calculate the x-coordinate of the centroid by substituting the y-coordinate into one of the equations of the system.

$$x + 16y = 14$$

$$x + 16\left(\frac{2}{3}\right) = 14$$

$$x + \frac{32}{3} = \frac{42}{3}$$

$$x = \frac{42 - 32}{3}$$

$$x = \frac{10}{3}$$

The centroid of $\triangle XYZ$ is $C = \left(\dfrac{10}{3}, \dfrac{2}{3}\right)$.

Parallel and Perpendicular Lines

Compare the slopes

20.14 Given line l with equation $5x - 3y = 1$ that is parallel to j and perpendicular to k, identify the slopes of j and k.

This is a shortcut that you can use to find the slope, but only when the line is in standard form. You can also solve the equation for y to figure out the slope—it'll be the coefficient of x.

The slope m of a line written in standard form $Ax + By = C$ is $m = -\dfrac{A}{B}$, so the slope of l is $m = -\dfrac{5}{-3} = \dfrac{5}{3}$. Parallel lines have equal slopes, so the slope of j is $\dfrac{5}{3}$. Perpendicular lines have opposite reciprocal slopes, so the slope of k is $-\dfrac{3}{5}$.

20.15 Write the equation of the line parallel to $y = -5x + 8$ that passes through $(-6, 1)$ in slope-intercept form.

It has an x-term, a y-term, and a constant, and it's solved for y.

The linear equation $y = -5x + 8$ is in slope-intercept form, so the coefficient of the x-term is the slope of the line: $m = -5$. Parallel lines have equal slopes, so the slope of the line parallel to $y = -5x + 8$ also has slope -5. Substitute $m = -5$, $x_1 = -6$, and $y_1 = 1$ into the point-slope formula.

$$y - 1 = -5(x - (-6))$$

$$y - 1 = -5(x + 6)$$

$$y - 1 = -5x - 30$$

Solve the equation for y to express it in slope-intercept form.

$$y = -5x - 30 + 1$$

$$y = -5x - 29$$

20.16 Write the equation of the line with x-intercept –1 that is perpendicular to $3x + 4y = 10 - (x - 2y)$ in standard form.

Express the given line in slope-intercept form.

$$3x + 4y = 10 - x - (-2y)$$
$$3x + 4y = 10 - x + 2y$$
$$4y - 2y = -x - 3x + 10$$
$$2y = -4x + 10$$
$$\frac{2}{2}y = -\frac{4}{2}x + \frac{10}{2}$$
$$y = -2x + 5$$

> The slope is right here—the coefficient of x.

The slope of $y = -2x + 5$ is –2. You are instructed to find the perpendicular line, which will have a slope of $\frac{1}{2}$, the opposite reciprocal of –2.

You are given the x-intercept of the line you are asked to identify: –1. Recall that the y-coordinate of an x-intercept is always 0, so a line with x-intercept –1 passes through the point (–1,0). Substitute $m = \frac{1}{2}$, $x_1 = -1$, and $y_1 = 0$ into the point-slope formula.

$$y - 0 = \frac{1}{2}(x - (-1))$$
$$y = \frac{1}{2}(x + 1)$$

Write the equation in standard form.

$$2(y) = \frac{2}{1}\left(\frac{1}{2}\right)(x + 1)$$
$$2y = x + 1$$
$$-1 = x - 2y$$
$$x - 2y = -1$$

> Subtract $2y$ and 1 from both sides of the equation to get $-1 = x - 2y$. Then reverse the sides of the equation.

20.17 Given $W = (-3,-4)$, $X = (2,-1)$, and $\overleftrightarrow{WX} \perp \overleftrightarrow{XY}$, write the equation of $\overleftrightarrow{XY}$ in slope-intercept form.

Calculate the slope m_1 of $\overleftrightarrow{WX}$.

$$m_1 = \frac{-1 - (-4)}{2 - (-3)} = \frac{-1 + 4}{2 + 3} = \frac{3}{5}$$

Let m_2 represent the slope of $\overleftrightarrow{XY}$. Because $\overleftrightarrow{WX} \perp \overleftrightarrow{XY}$, m_2 is the opposite reciprocal of m_1: $m_2 = -\dfrac{5}{3}$. Notice that $\overleftrightarrow{XY}$ passes through X, so substitute $m_2 = -\dfrac{5}{3}$, $x_1 = 2$, and $y_1 = -1$ into the point-slope formula.

$$y - (-1) = -\frac{5}{3}(x - 2)$$

$$y + 1 = -\frac{5}{3}x + \frac{10}{3}$$

$$y = -\frac{5}{3}x + \frac{10}{3} - \frac{3}{3}$$

$$y = -\frac{5}{3}x + \frac{7}{3}$$

20.18 Given $\triangle ABC$ with vertices $A = (-2,0)$, $B = (0,7)$, and $C = (6,-4)$, write the equation of the line containing altitude $\overline{CZ}$ in standard form.

An altitude of a triangle extends from the vertex of a triangle (in this case, C) and is perpendicular to the opposite side (in this case, $\overline{AB}$). Calculate the slope m_1 of side $\overline{AB}$.

$$m_1 = \frac{7 - 0}{0 - (-2)} = \frac{7}{2}$$

> **Z is the point where the altitude and the opposite side intersect.**

The slope m_2 of $\overline{CZ}$ is the opposite reciprocal of m_1: $m_2 = -\dfrac{2}{7}$. Vertex C lies on the altitude, so substitute $m_2 = -\dfrac{2}{7}$, $x_1 = 6$, and $y_1 = -4$ into the point-slope formula.

$$y - (-4) = -\frac{2}{7}(x - 6)$$

$$y + 4 = -\frac{2}{7}(x - 6)$$

> **You can plug points (x_1, y_1) into the point-slope formula only when those points are on the line you're trying to identify, so substituting $(x_1, y_1) = (-2,0)$ or $(0,7)$ will give you the wrong answer.**

Express the equation in standard form.

$$7(y + 4) = \frac{\cancel{7}}{1}\left(-\frac{2}{\cancel{7}}\right)(x - 6)$$

$$7(y + 4) = -2(x - 6)$$

$$7y + 28 = -2x + 12$$

$$2x + 7y = 12 - 28$$

$$2x + 7y = -16$$

20.19 Given the following diagram, in which $\overleftrightarrow{CD}$ is tangent to $\bigcirc L$ and $\bigcirc M$, write the equation of $\overleftrightarrow{CD}$ in slope-intercept form.

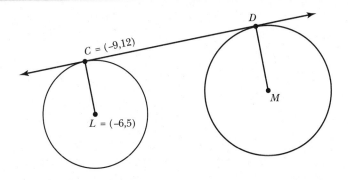

Calculate the slope m_1 of radius $\overline{LC}$.

$$m_1 = \frac{12-5}{-9-(-6)} = \frac{12-5}{-9+6} = \frac{7}{-3} = -\frac{7}{3}$$

A line tangent to a circle is perpendicular to the radius at the point of tangency. Thus, $\overline{LC} \perp \overleftrightarrow{CD}$ at C. The slope m_2 of $\overleftrightarrow{CD}$ is the opposite reciprocal of the slope of $\overline{LC}$: $m_2 = \frac{3}{7}$. Apply the point-slope formula to write the equation of the tangent line. ←

> $\overleftrightarrow{CD}$ passes through point C, so substitute $x_1 = -9$ and $y_1 = 12$ into the point-slope equation.

$$y - 12 = \frac{3}{7}(x - (-9))$$

$$y - 12 = \frac{3}{7}(x + 9)$$

$$y - 12 = \frac{3}{7}x + \frac{27}{7}$$

Solve for y to write the equation in slope-intercept form.

$$y = \frac{3}{7}x + \frac{27}{7} + 12$$

$$y = \frac{3}{7}x + \frac{27}{7} + \frac{84}{7}$$

$$y = \frac{3}{7}x + \frac{111}{7}$$

Note: Problems 20.20–20.21 refer to rhombus ABCD, such that A = (1,12) and C = (9,–2).

20.20 Write the equation of the line containing diagonal $\overline{AC}$ in slope-intercept form.

Calculate the slope m of the line passing through A and C.

$$m = \frac{-2-12}{9-1} = \frac{-14}{8} = -\frac{7}{4}$$

Substitute slope m and the coordinates from either endpoint of the diagonal into the point-slope formula and solve for y.

$$y-(-2)=-\frac{7}{4}(x-9)$$

$$y+2=-\frac{7}{4}x+\frac{63}{4}$$

$$y=-\frac{7}{4}x+\frac{63}{4}-\frac{8}{4}$$

$$y=-\frac{7}{4}x+\frac{55}{4}$$

Note: Problems 20.20–20.21 refer to rhombus ABCD, such that A = (1,12) and C = (9,–2).

20.21 Write the equation of the line containing diagonal $\overline{BD}$ in standard form.

See Problem 9.19.

To write the equation of a line, you must identify its slope and one point through which the line passes. According to Problem 20.20, the line containing diagonal $\overline{AC}$ has equation $y=-\frac{7}{4}x+\frac{55}{4}$; the slope of that line is $-\frac{7}{4}$. Recall that the diagonals of a rhombus are perpendicular and therefore have opposite reciprocal slopes. Thus, the slope of $\overline{BD}$ is $\frac{4}{7}$.

A rhombus is a parallelogram, and the diagonals of parallelograms bisect each other. Hence, the midpoint of $\overline{AC}$ is also the midpoint of $\overline{BD}$. Calculate the common midpoint M by substituting the endpoints of $\overline{AC}$ into the midpoint formula.

$$M=\left(\frac{1+9}{2},\frac{12-2}{2}\right)=\left(\frac{10}{2},\frac{10}{2}\right)=(5,5)$$

Diagonal $\overline{BD}$ has slope $\frac{4}{7}$ and passes through point (5,5). Apply the point-slope formula to generate the equation of the line.

$$y-5=\frac{4}{7}(x-5)$$

Express the equation in standard form.

$$7(y-5)=\frac{\cancel{7}}{1}\left(\frac{4}{\cancel{7}}\right)(x-5)$$

$$7(y-5)=4(x-5)$$

$$7y-35=4x-20$$

$$-35+20=4x-7y$$

$$-15=4x-7y$$

$$4x-7y=-15$$

Note: Problems 20.22–20.24 refer to the triangle with vertices X = (–5,0), Y = (1,–6), and Z = (4,8).

20.22 Write the equation of k, the perpendicular bisector of $\overline{XY}$, in standard form.

Calculate the slope m of $\overline{XY}$.

$$m = \frac{-6-0}{1-(-5)} = \frac{-6-0}{1+5} = \frac{-6}{6} = -1$$

The slope of k is 1, the opposite reciprocal of the slope of $\overline{XY}$. Because k bisects $\overline{XY}$, it passes through the midpoint M of the segment.

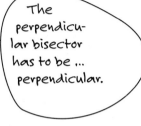

The perpendicular bisector has to be ... perpendicular.

$$M = \left(\frac{-5+1}{2}, \frac{0-6}{2}\right) = \left(\frac{-4}{2}, \frac{-6}{2}\right) = (-2,-3)$$

Substitute $m = 1$, $x_1 = -2$, and $y_1 = -3$ into the point-slope formula to generate the equation of k.

$$y-(-3) = 1(x-(-2))$$
$$y+3 = x+2$$

Express the equation in standard form.

$$3-2 = x-y$$
$$1 = x-y$$
$$x-y = 1$$

Note: Problems 20.22–20.24 refer to the triangle with vertices X = (–5,0), Y = (1,–6), and Z = (4,8).

20.23 Write the equation of j, the perpendicular bisector of $\overline{XZ}$, in standard form.

Calculate the slope m of $\overline{XZ}$.

$$m = \frac{8-0}{4-(-5)} = \frac{8-0}{4+5} = \frac{8}{9}$$

The opposite reciprocal of m is the slope of perpendicular bisector j: $-\frac{9}{8}$. The bisector intersects $\overline{XZ}$ at its midpoint M.

$$M = \left(\frac{-5+4}{2}, \frac{0+8}{2}\right) = \left(\frac{-1}{2}, \frac{8}{2}\right) = \left(-\frac{1}{2}, 4\right)$$

Apply the point-slope formula to generate the equation of line j.

$$m = -\frac{9}{8}, \quad x_1 = -\frac{1}{2}, \text{ and } \quad y_1 = 4$$

$$y-4 = -\frac{9}{8}\left(x-\left(-\frac{1}{2}\right)\right)$$

$$y-4 = -\frac{9}{8}\left(x+\frac{1}{2}\right)$$

$$y-4 = -\frac{9}{8}x - \frac{9}{16}$$

Write the equation in standard form.

$$16(y-4) = \frac{2 \cdot \cancel{8}}{1}\left(-\frac{9}{\cancel{8}}x\right) + \frac{\cancel{16}}{1}\left(-\frac{9}{\cancel{16}}\right)$$

$$16y - 64 = -18x - 9$$

$$18x + 16y = -9 + 64$$

$$18x + 16y = 55$$

Note: Problems 20.22–20.24 refer to the triangle with vertices X = (–5,0), Y = (1,–6), and Z = (4,8).

20.24 Identify the circumcenter C of $\triangle XYZ$.

The circumcenter of a triangle is the point at which the perpendicular bisectors of the sides intersect. Problems 20.22–20.23 generate the equations of two perpendicular bisectors of $\triangle XYZ$ and thus represent a system of equations whose solution is point C, the circumcenter.

$$\begin{cases} x - y = 1 \\ 18x + 16y = 55 \end{cases}$$

Multiply the first equation by 16, add the equations of the system, and solve the resulting equation for x.

$$\begin{array}{rcrcr} 16x & - & 16y & = & 16 \\ 18x & + & 16y & = & 55 \\ \hline 34x & + & 0 & = & 71 \end{array}$$

$$x = \frac{71}{34}$$

Substitute $x = \dfrac{71}{34}$ into the first equation of the system and solve for y.

$$x - y = 1$$

$$\frac{71}{34} - y = 1$$

$$\frac{71}{34} - 1 = y$$

$$\frac{71}{34} - \frac{34}{34} = y$$

$$\frac{37}{34} = y$$

The circumcenter of $\triangle XYZ$ is $C = \left(\dfrac{71}{34}, \dfrac{37}{34}\right)$.

Equation of a Circle

Given its center and radius

20.25 Identify the standard form equation for a circle with radius r and center (h,k).

The equation for the standard form of a circle is $(x-h)^2 + (y-k)^2 = r^2$. Notice that it contains the opposites of the coordinates of the center, $-h$ and $-k$.

So if the center of a circle is $(1,-8)$, then the standard form equation will contain -1 and 8.

20.26 Given that $\odot C$ is centered at the origin and has radius 4, write the equation of $\odot C$ in standard form.

The origin of the coordinate plane is the point $(0,0)$, at which the x-axis and y-axis intersect. Substitute $h = 0$, $k = 0$, and $r = 4$ into the standard form equation presented in Problem 20.25.

$$(x-h)^2 + (y-k)^2 = r^2$$
$$(x-0)^2 + (y-0)^2 = 4^2$$
$$x^2 + y^2 = 16$$

20.27 Write the equation of the circle with center $(7,-6)$ and radius 9 in standard form.

Substitute $h = 7$, $k = -6$, and $r = 9$ into the standard form equation for a circle.

$$(x-h)^2 + (y-k)^2 = r^2$$
$$(x-7)^2 + (y-(-6))^2 = 9^2$$
$$(x-7)^2 + (y+6)^2 = 81$$

20.28 Identify the center and radius of the circle with equation

$$(x+8)^2 + \left(y-\frac{1}{4}\right)^2 = 10.$$

The x-coordinate of the center is the opposite of the constant in the squared quantity containing x: -8. Similarly, the y-coordinate of the center is the opposite of the constant in the squared quantity containing y: $\frac{1}{4}$. Thus, the center of the circle is $\left(-8,\frac{1}{4}\right)$. The radius of the circle is the square root of the constant isolated on the right side of the equation: $r = \sqrt{10}$.

20.29 Given $\bigcirc A$ with equation $(x + 2)^2 + (y - 10)^2 = 12$, write the equation of concentric $\bigcirc B$, which has a radius 4 units larger than $\bigcirc A$.

The center of $\bigcirc A$ is $(h,k) = (-2,10)$. Concentric circles have the same center, so $(-2,10)$ is also the center of $\bigcirc B$. Calculate the radius r_A of $\bigcirc A$.

$$r_A = \sqrt{12} = \sqrt{4 \cdot 3} = 2\sqrt{3}$$

The radius r_B of $\bigcirc B$ is 4 units longer than r_A, the radius of $\bigcirc A$.

$$r_B = r_A + 4 = 2\sqrt{3} + 4$$

Substitute $(h,k) = (-2,10)$ and $r = 2\sqrt{3} + 4$ into the standard form equation of a circle.

$$\left(x - (-2)\right)^2 + \left(y - 10\right)^2 = \left(2\sqrt{3} + 4\right)^2$$
$$(x + 2)^2 + \left(y - 10\right)^2 = \left(2\sqrt{3} + 4\right)^2$$

> You can square the radius if you feel like it:
>
> $(2\sqrt{3} + 4)^2$
>
> $= (2\sqrt{3} + 4)(2\sqrt{3} + 4)$
>
> $= (2\sqrt{3})(2\sqrt{3}) + 4(2\sqrt{3})$
> $\quad + 4(2\sqrt{3}) + (4)(4)$
>
> $= 4(3) + 8\sqrt{3} + 8\sqrt{3} + 16$
>
> $= 12 + 16\sqrt{3} + 16$
>
> $= 28 + 16\sqrt{3}$

Note: Problems 20.30–20.32 refer to $\bigcirc C$, which has diameter $\overline{AB}$ such that A = (3,–11) and B = (–9,15).

20.30 Calculate the center of $\bigcirc C$.

The midpoint of a diameter is the center of a circle. Apply the midpoint formula to $\overline{AB}$.

$$C = \left(\frac{3 - 9}{2}, \frac{-11 + 15}{2}\right) = \left(\frac{-6}{2}, \frac{4}{2}\right) = (-3, 2)$$

Note: Problems 20.30–20.32 refer to $\bigcirc C$, which has diameter $\overline{AB}$ such that A = (3,–11) and B = (–9,15).

20.31 Calculate the radius of $\bigcirc C$.

The radius of a circle is the distance between its center and a point on the circle. According to Problem 20.30, the center of the circle is $C = (-3,2)$. Apply the distance formula to calculate AC.

$$AC = \sqrt{\left(x_2 - x_1\right)^2 + \left(y_2 - y_1\right)^2}$$
$$= \sqrt{(-3 - 3)^2 + (2 - (-11))^2}$$
$$= \sqrt{(-6)^2 + (2 + 11)^2}$$
$$= \sqrt{(-6)^2 + 13^2}$$
$$= \sqrt{36 + 169}$$
$$= \sqrt{205}$$

> You could also calculate CB, because B is another point on the circle. Either way, you'll get the same distance and therefore the same radius.

Note: Problems 20.30–20.32 refer to $\odot C$, which has diameter $\overline{AB}$ such that $A = (3,-11)$ and $B = (-9,15)$.

20.32 Write the equation of $\odot C$ in standard form.

According to Problems 20.30 and 20.31, the center of $\odot C$ is $(h,k) = (-3,2)$ and the radius is $r = \sqrt{205}$. Substitute these values into the standard form equation for a circle.

$$(x-h)^2 + (y-k)^2 = r^2$$
$$(x-(-3))^2 + (y-2)^2 = (\sqrt{205})^2$$
$$(x+3)^2 + (y-2)^2 = 205$$

Note: Problems 20.33–20.34 refer to $\odot X$ that contains point Y, given X = (–1,2) and Y = (7,4).

20.33 Write the equation of $\odot X$ in standard form.

Note that X is the center of the circle and Y is a point on the circle, so $\overline{XY}$ is a radius of $\odot X$. Apply the distance formula to calculate the radius.

$$r = \sqrt{(7-(-1))^2 + (4-2)^2}$$
$$= \sqrt{(7+1)^2 + 2^2}$$
$$= \sqrt{64+4}$$
$$= \sqrt{68}$$

No need to simplify this radical. You're just going to square it in the next step.

Substitute $h = -1$, $k = 2$, and $r = \sqrt{68}$ into the equation for the standard form of a circle.

$$(x-h)^2 + (y-k)^2 = r^2$$
$$(x-(-1))^2 + (y-2)^2 = (\sqrt{68})^2$$
$$(x+1)^2 + (y-2)^2 = 68$$

Note: Problems 20.33–20.34 refer to $\odot X$ that contains point Y, given X = (–1,2) and Y = (7,4).

20.34 Write the equation of j, the line tangent to $\odot X$ at point Y, in standard form.

Calculate the slope m_1 of radius $\overline{XY}$.

$$m_1 = \frac{4-2}{7-(-1)} = \frac{4-2}{7+1} = \frac{2}{8} = \frac{1}{4}$$

The slope m_2 of j is the opposite reciprocal of m_1 because a line tangent to a circle is perpendicular to a radius at the point of tangency: $m_2 = -\frac{4}{1} = -4$. Note

To use the point-slope formula, you need (surprise!) a point and a slope. Line j is tangent to the circle at Y, so Y lies on the circle AND on the tangent line.

that j passes through Y, so substitute $x_1 = 7$, $y_1 = 4$, and $m = -4$ into the point-slope formula to generate the equation of tangent line j.

$$y - y_1 = m(x - x_1)$$
$$y - 4 = -4(x - 7)$$
$$y - 4 = -4x + 28$$

Express the equation in standard form.

$$4x + y = 28 + 4$$
$$4x + y = 32$$

Vectors

Arrows in the coordinate plane

20.35 Describe the difference between a vector and a ray.

The length of a vector is called the magnitude. You can use the distance formula (or the Py-thagorean Theorem) to calculate it.

Both a vector and a ray are defined by two points. The point listed first in the name of a vector or ray indicates the point at which it begins. However, the second point in the name of a vector marks the end of the vector, whereas the second point in the name of a ray merely identifies a point through which the infinitely long ray passes.

Because vectors have a finite length, it is common to move them about the coordinate plane to perform vector operations. Thus, the points that define a vector are used to discern important information about that vector (such as its slope and magnitude), but those points need not be fixed on the coordinate plane. For instance, vectors are often written in component form, which changes their initial points to (0,0), the origin.

Note: Problems 20.36–20.39 refer to $\overrightarrow{AB}$, the vector with initial point A = (–2,4) and terminal point B = (3,1).

20.36 Sketch $\overrightarrow{AB}$.

Plot points A and B in the coordinate plane and draw an arrow that begins at A and ends at B. The arrowhead of the vector is known as the head, and the initial point is known as the tail.

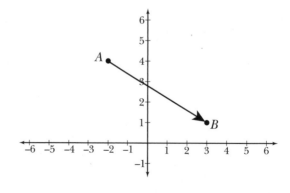

Note: Problems 20.36–20.39 refer to $\overrightarrow{AB}$, the vector with initial point A = (–2,4) and terminal point B = (3,1).

20.37 Explain why $\overrightarrow{AB}$ is not in standard position.

A vector is in standard position when its tail is at the origin. The tail of $\overrightarrow{AB}$ is A = (–2,4).

Note: Problems 20.36–20.39 refer to $\overrightarrow{AB}$, the vector with initial point A = (–2,4) and terminal point B = (3,1).

20.38 Express $\overrightarrow{AB}$ in component form.

To draw $\overrightarrow{AB}$ in standard position (that is, to move its tail to the origin without affecting the slope or magnitude of the vector), calculate the differences of the x- and y-coordinates of the terminal and initial points.

$$\overrightarrow{AB} = \left\langle x_2 - x_1, y_2 - y_1 \right\rangle$$
$$= \left\langle 3 - (-2), 1 - 4 \right\rangle$$
$$= \left\langle 3 + 2, 1 - 4 \right\rangle$$
$$= \left\langle 5, -3 \right\rangle$$

When a vector is in standard position, you know that it starts at (0,0), so you need only one point to define it, the point where it stops. Write that point as <5,–3> instead of (5,–3) to indicate that it's a vector in component form and not just a point.

If the tail of $\overrightarrow{AB}$ is moved to the origin, the head of the vector moves to the point (5,–3). The coordinates of the head of a vector in standard position is the component form of the vector: <5,–3>.

Note: Problems 20.36–20.39 refer to $\overrightarrow{AB}$, the vector with initial point A = (–2,4) and terminal point B = (3,1).

20.39 Calculate $\left\| \overrightarrow{AB} \right\|$, the magnitude of $\overrightarrow{AB}$.

Some books write the magnitude using one set of vertical bars: $\left| \overrightarrow{AB} \right|$.

The magnitude—or length—of a vector is the distance between its initial and terminal points. If a vector **v** is in component form <a,b>, then its magnitude is $\|\mathbf{v}\| = \sqrt{a^2 + b^2}$. According to Problem 20.38, $\overrightarrow{AB} = \left\langle 5, -3 \right\rangle$.

$$\left\| \overrightarrow{AB} \right\| = \sqrt{a^2 + b^2}$$
$$= \sqrt{5^2 + (-3)^2}$$
$$= \sqrt{25 + 9}$$
$$= \sqrt{34}$$

Note: Problems 20.40–20.43 refer to vectors $\overline{CD}$ and $\overline{VW}$, such that C = (–6,–5), D = (0,2), V = (–1,1), and W = (3,–4).

20.40 Express $\overline{CD}$ and $\overline{VW}$ in component form.

Apply the technique described in Problem 20.38—calculate the differences of the coordinates of the terminal and initial points of each vector.

$$\overline{CD} = \left\langle x_2 - x_1, y_2 - y_1 \right\rangle \qquad \overline{VW} = \left\langle x_2 - x_1, y_2 - y_1 \right\rangle$$
$$= \left\langle 0 - (-6), 2 - (-5) \right\rangle \qquad = \left\langle 3 - (-1), -4 - 1 \right\rangle$$
$$= \left\langle 0 + 6, 2 + 5 \right\rangle \qquad = \left\langle 3 + 1, -4 - 1 \right\rangle$$
$$= \left\langle 6, 7 \right\rangle \qquad = \left\langle 4, -5 \right\rangle$$

Note: Problems 20.40–20.43 refer to vectors $\overline{CD}$ and $\overline{VW}$, such that C = (–6,–5), D = (0,2), V = (–1,1), and W = (3,–4).

20.41 Calculate $\overline{CD} + \overline{VW}$.

When two vectors are written in component form, you can calculate the vector sum by adding the corresponding coordinates. According to Problem 20.40, $\overline{CD} = \left\langle 6, 7 \right\rangle$ and $\overline{VW} = \left\langle 4, -5 \right\rangle$.

$$\overline{CD} + \overline{VW} = \left\langle x_1 + x_2, y_1 + y_2 \right\rangle$$
$$= \left\langle 6 + 4, 7 - 5 \right\rangle$$
$$= \left\langle 10, 2 \right\rangle$$

Adding two vectors in component form generates a new vector in component form—in this case, a vector with terminal point (10,2).

Note: Problems 20.40–20.43 refer to vectors $\overline{CD}$ and $\overline{VW}$, such that C = (–6,–5), D = (0,2), V = (–1,1), and W = (3,–4).

20.42 Sketch $\overline{CD} + \overline{VW}$ to verify the sum calculated in Problem 20.41.

According to Problem 20.40, $\overline{CD} = \left\langle 6, 7 \right\rangle$. Thus, when drawn in standard form, the tail of $\overline{CD}$ is at the origin and the head of the vector is at point (6,7), as illustrated below. Problem 20.40 also states that $\overline{VW} = \left\langle 4, -5 \right\rangle$. Draw the tail of $\overline{VW}$ at the head of $\overline{CD}$ and draw the head of $\overline{VW}$ 4 units to the right of and 5 units below its head at point (10,2).

You know that $\vec{VW} = \langle 4,-5 \rangle$. That means you can recreate vector $\vec{VW}$ anywhere in the coordinate plane by choosing a point for the tail and then plotting the point for the head +4 units right and –5 units down from there.

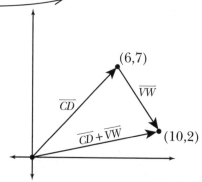

The sum is represented by the vector with initial point (0,0), the initial point of $\overrightarrow{CD}$, and terminal point (10,2), the terminal point of $\overrightarrow{VW}$. Thus, $\overrightarrow{CD} + \overrightarrow{VW} = \langle 10,2 \rangle$.

Note: Problems 20.40–20.43 refer to vectors $\overrightarrow{CD}$ and $\overrightarrow{VW}$, such that C = (–6,–5), D = (0,2), V = (–1,1), and W = (3,–4).

20.43 Calculate $\left\| \overrightarrow{CD} + \overrightarrow{VW} \right\|$.

> See Problem 20.39.

According to Problems 20.41 and 20.42, $\overrightarrow{CD} + \overrightarrow{VW} = \langle 10,2 \rangle$. Recall that the magnitude of a vector $\mathbf{v} = \langle a,b \rangle$ in component form is $\|\mathbf{v}\| = \sqrt{a^2 + b^2}$.

$$\left\| \overrightarrow{CD} + \overrightarrow{VW} \right\| = \sqrt{10^2 + 2^2} = \sqrt{100 + 4} = \sqrt{104} = \sqrt{4 \cdot 26} = 2\sqrt{26}$$

Note: Problems 20.44–20.46 refer to vectors $\mathbf{u} = \langle 3,1 \rangle$ and $\mathbf{v} = \langle -2,-4 \rangle$.

20.44 Calculate $3\mathbf{v}$.

> Instead of using endpoints with a little arrow on top to name a vector, you can use a single letter, like v—just like you can name a line using two points on the line or with a single lowercase letter.

A real number multiplied by a vector is known as a scalar; in this problem, vector $\mathbf{v}$ is multiplied by scalar 3. To calculate $3\mathbf{v}$, multiply the x- and y-values of the vector in component form by 3.

$$3\mathbf{v} = 3\langle -2,-4 \rangle = \langle 3(-2), 3(-4) \rangle = \langle -6,-12 \rangle$$

Note: Problems 20.44–20.46 refer to vectors $\mathbf{u} = \langle 3,1 \rangle$ and $\mathbf{v} = \langle -2,-4 \rangle$.

20.45 Verify that the magnitude of $3\mathbf{v}$ is three times the magnitude of $\mathbf{v}$.

> In other words, verify that multiplying by a scalar N makes the vector N times as long.

Calculate the magnitude of $\mathbf{v}$.

$$\|\mathbf{v}\| = \sqrt{(-2)^2 + (-4)^2} = \sqrt{4 + 16} = \sqrt{20} = \sqrt{4 \cdot 5} = 2\sqrt{5}$$

According to Problem 20.44, $3\mathbf{v} = \langle -6,-12 \rangle$. Calculate $\|3\mathbf{v}\|$.

$$\|3\mathbf{v}\| = \sqrt{(-6)^2 + (-12)^2} = \sqrt{36 + 144} = \sqrt{180} = \sqrt{36 \cdot 5} = 6\sqrt{5}$$

Notice that $3\|\mathbf{v}\| = 3\left(2\sqrt{5}\right) = 6\sqrt{5}$, which is equal to $\|3\mathbf{v}\|$. Thus, $3\|\mathbf{v}\| = \|3\mathbf{v}\|$.

Note: Problems 20.44–20.46 refer to vectors $\mathbf{u} = \langle 3,1 \rangle$ and $\mathbf{v} = \langle -2,-4 \rangle$.

20.46 Calculate $5\mathbf{u} - \mathbf{v}$.

> Subtracting a vector is the same as adding a negative vector. This works with numbers, too. Instead of subtracting 6 from 10, you can add –6 to 10 and get the same result: $10 - 6 = 10 + (-6) = 4$.

Subtracting $\mathbf{v}$ from $5\mathbf{u}$ is equivalent to adding $(-1)\mathbf{v}$ to $5\mathbf{u}$.

$$
\begin{aligned}
5\mathbf{u} - \mathbf{v} &= 5\mathbf{u} + (-1)(\mathbf{v}) \\
&= 5\langle 3,1 \rangle + (-1)\langle -2,-4 \rangle \\
&= \langle 5(3), 5(1) \rangle + \langle (-1)(-2), (-1)(-4) \rangle \\
&= \langle 15,5 \rangle + \langle 2,4 \rangle \\
&= \langle 15+2, 5+4 \rangle \\
&= \langle 17,9 \rangle
\end{aligned}
$$

Chapter 21
TRANSFORMATIONS AND SYMMETRY

Move, rotate, reflect, and dilate figures

A geometric transformation is a formal reassignment of points in the coordinate plane, often with a clear visual consequence. Most commonly, translations reflect, move, rotate, or dilate segments and figures, as this chapter explores in some detail.

You may already have a good idea of how reflections and rotations affect a figure visually, but this chapter explains those concepts mathematically. Before you start this chapter, make sure you work through the area and perimeter problems in Chapter 17 and the coordinate geometry problems in Chapters 19 and 20 to really understand translations.

Transformation Terminology

Image, map, and isometry

A mapping is basically a function that you plug points into (instead of numbers) and get new points out of.

21.1   Complete the following statement.

A mapping in the coordinate plane assigns a point (called the _____) new coordinates that correspond to a new point (called the _____). _____ mappings are known as transformations.

A mapping in the coordinate plane assigns a point (called the <u>pre-image</u>) new coordinates that correspond to a new point (called the <u>image</u>). <u>One-to-one</u> mappings are known as transformations.

In a one-to-one mapping, each transformed point corresponds to only one untransformed point. In other words, given an original point, you can tell exactly what the new mapped point will be and vice versa.

21.2   What characteristic of a transformation classifies it as an isometry?

An isometry is a transformation in which the lengths of mapped segments are preserved. For instance, given pre-images A and B that form segment $\overline{AB}$, if an isometric transformation is applied that produces images A' and B' and the corresponding segment $\overline{A'B'}$, you are assured that $\overline{AB} \cong \overline{A'B'}$.

21.3   Express the following statement in function and mapping notation.

R maps pre-image A to image A'.

In function notation, mapping R is written $R(A) = A'$. In mapping notation, R is written $R{:}A \rightarrow A'$.

Note: Problems 21.4–21.6 refer to mapping P, defined below.

$$P{:}(x,y) \rightarrow (2x-1, y+3)$$

21.4   Calculate the image of $(-5,6)$.

Let A represent point $(-5,6)$. Note that A is a pre-image of the mapping because it has not yet been altered according to the expressions that define mapping P. Substitute $x=-5$ and $y=6$ into $(2x-1,y+3)$ to calculate the coordinates A' of the image to which P maps $A = (-5,6)$.

$$A' = (2x-1, y+3)$$
$$= (2(-5)-1, 6+3)$$
$$= (-10-1, 6+3)$$
$$= (-11,9)$$

P maps $A = (-5,6)$ to $A' = (-11,9)$.

Note: Problems 21.4–21.6 refer to mapping P, defined in Problem 21.4.

21.5 Calculate the pre-image of $\left(8, \dfrac{1}{5}\right)$.

The pre-image of $\left(8, \dfrac{1}{5}\right)$ is the coordinate pair (x,y) that generates the point

$\left(8, \dfrac{1}{5}\right)$ when x is substituted into $2x - 1$ and y is substituted into $y + 3$. ←

$$2x - 1 = 8 \qquad\qquad y + 3 = \dfrac{1}{5}$$
$$2x = 8 + 1 \qquad\qquad y = \dfrac{1}{5} - \dfrac{3}{1} \cdot \dfrac{5}{5}$$
$$2x = 9 \qquad\qquad y = \dfrac{1}{5} - \dfrac{15}{5}$$
$$x = \dfrac{9}{2} \qquad\qquad y = -\dfrac{14}{5}$$

> Set $2x - 1$ equal to the x-coordinate of the image (8) and $y + 3$ equal to the y-coordinate. Solve the equations for x and y.

The pre-image of $\left(8, \dfrac{1}{5}\right)$ is $\left(\dfrac{9}{2}, -\dfrac{14}{5}\right)$.

Note: Problems 21.4–21.6 refer to mapping P, defined in Problem 21.4.

21.6 Is mapping *P* a transformation? Explain your answer.

A transformation is a one-to-one mapping—a mapping in which every pre-image corresponds to exactly one image and vice versa. As Problem 21.4 demonstrates, substituting any pre-image (a,b) into $(2x - 1, y + 3)$ generates exactly one corresponding image point. ←

> When you plug ANY real number into $2x - 1$, you get one answer because $2x - 1$ is a function of x.

Problem 21.5 demonstrates that any image (c,d) has exactly one corresponding pre-image (x,y) such that $2x - 1 = c$ and $y + 3 = d$. Each pre-image corresponds to exactly one image and vice versa, so the mapping is one-to-one and therefore a transformation.

> You end up with two linear equations, which have one solution each no matter what c and d are.

21.7 Given isometry *J*:regular pentagon $ABCDE \rightarrow$ regular pentagon $A'B'C'D'E'$, prove $ABCDE \cong A'B'C'D'E'$.

> Each interior angle of a regular polygon measures 180°(n − 2) divided by n, where n is the number of sides. In this case, both pentagons have 108° interior angles.

Statement	Reason
1. *J*:regular pentagon $ABCDE \rightarrow$ regular pentagon $A'B'C'D'E'$ is an isometry	1. Given
2. $\angle A \cong \angle B \cong \angle C \cong \angle D = \angle E \cong \angle A' \cong \angle B' \cong \angle C' \cong \angle D' \cong \angle E'$	2. The interior angles of all regular pentagons are congruent ←
3. $\overline{AB} \cong \overline{A'B'}$; $\overline{BC} \cong \overline{B'C'}$; $\overline{CD} \cong \overline{C'D'}$; $\overline{DE} \cong \overline{D'E'}$; $\overline{AE} \cong \overline{A'E'}$	3. An isometry preserves the lengths of mapped segments
4. $ABCDE \cong A'B'C'D'E'$	4. Congruent figures have congruent corresponding angles and sides

Reflections

Using a line as a mirror

21.8 Complete the following statement.

Given $R_l:P \to P'$, Rl is a reflection in line l if l is the _____ of every segment _____ and $R_l:$ ____ $\to$ ____ when P lies on l.

> Unless the pre-image is actually ON line l, in which case transformation R leaves the point alone and doesn't assign it a new set of coordinates.

Given $R_l:P \to P'$, Rl is a reflection in line l if l is the <u>perpendicular bisector</u> of every segment $\overline{PP'}$ and $R_l: \underline{P} \to \underline{P}$ when P lies on l. In other words, if you connect every pre-image to the corresponding image, all the resulting segments have the same perpendicular bisector, line l.

Note: Problems 21.9–21.14 refer to a triangle with vertices A = (–3,–1), B = (0,3), and C = (4,2).

21.9 Let $\triangle A'B'C'$ be the reflection of $\triangle ABC$ across the x-axis. Identify the coordinates of the vertices of $\triangle A'B'C'$.

The x-axis is the horizontal line $y = 0$. According to Problem 21.8, $y = 0$ is the perpendicular bisector of the segments connecting corresponding pre-images and images. Thus, $\overline{AA'}$, $\overline{BB'}$, and $\overline{CC'}$ are vertical segments.

> A and A' are endpoints of a vertical segment, so they have the same x-value ($x = -3$). Their y-values are opposites: A is 1 unit below the x-axis ($y = -1$), so A' is 1 unit above the x-axis ($y = 1$).

Consider points A and A'. Because the x-axis is the perpendicular bisector of $\overline{AA'}$, the distance from A to the x-axis must be equal to the distance from the x-axis to A'. Thus, A' is 1 unit above the x-axis: $A' = (-3,1)$.

Similarly, pre-images B and C share x-coordinates with corresponding images B' and C', and have opposite y-coordinates: $B' = (0,-3)$ and $C' = (4,-2)$.

Note: Problems 21.9–21.14 refer to a triangle with vertices A = (–3,–1), B = (0,3), and C = (4,2).

21.10 Let $\triangle A'B'C'$ be the reflection of $\triangle ABC$ across the x-axis, as described in Problem 21.9. Graph both triangles in the coordinate plane.

Plot and connect the vertices of $\triangle ABC$. Graph $\triangle A'B'C'$ by connecting the points identified in Problem 21.9: $A' = (-3,1)$, $B' = (0,-3)$, and $C' = (4,-2)$.

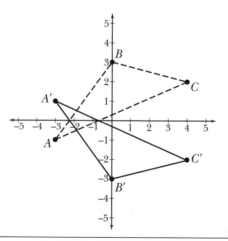

Note: Problems 21.9–21.14 refer to a triangle with vertices A = (–3,–1), B = (0,3), and C = (4,2).

21.11 Let ΔA'B'C' be the reflection of ΔABC across the y-axis. Identify the coordinates of the vertices of ΔA'B'C'.

> In Problems 21.11–21.12, you're reflecting across the y-axis instead of the x-axis.

The y-axis is the perpendicular bisector of all segments connecting the pre-images to the corresponding images. Thus, the vertical line $x = 0$ bisects horizontal segments $\overline{AA'}$, $\overline{BB'}$, and $\overline{CC'}$.

Consider points A and A'. Because the y-axis is the perpendicular bisector of $\overline{AA'}$, the distance from A to the y-axis must be equal to the distance from the y-axis to A'. Thus, A' is 3 units right of the y-axis: A' = (3,–1).

Similarly, pre-images B and C share y-coordinates with corresponding images B' and C', and have opposite x-coordinates: B' = (0,3) and C' = (–4,2).

> B is on the y-axis, the line you're reflecting across. That means the transformation does not affect B. Its coordinates stay the same because the opposite of 0 is 0.

Note: Problems 21.9–21.14 refer to a triangle with vertices A = (–3,–1), B = (0,3), and C = (4,2).

21.12 Let ΔA'B'C' be the reflection of ΔABC across the y-axis, as described in Problem 21.11. Graph both triangles in the coordinate plane.

Plot and connect the vertices of ΔABC. Graph ΔA'B'C' by connecting the points identified in Problem 21.11: A' = (3,–1), B' = (0,3), and C' = (–4,2).

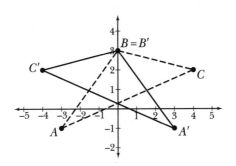

Note: Problems 21.9–21.14 refer to a triangle with vertices A = (–3,–1), B = (0,3), and C = (4,2).

21.13 Let ΔA'B'C' be the reflection of ΔABC across the line y = x. Identify the vertices of ΔA'B'C'.

> The new y-value is the old x-value (and vice versa).

Reflecting coordinates across the line $y = x$ reverses the x- and y-coordinates. In other words, reflecting pre-image (a,b) produces image (b,a). Thus, A' = (–1,–3), B' = (3,0), and C' = (2,4).

$y = x$ is in slope-intercept form, where $m = 1$ and $b = 0$. That means the line has y-intercept 0 (it goes through the origin) and slope 1 (so the line forms a 45° angle with the positive x- and y-axes).

Note: Problems 21.9–21.14 refer to a triangle with vertices A = (–3,–1), B = (0,3), and C = (4,2).

21.14 Let $\triangle A'B'C'$ be the reflection of $\triangle ABC$ across the line $y = x$, as described in Problem 21.13. Graph both triangles in the coordinate plane.

Plot and connect the vertices of $\triangle ABC$. Graph $\triangle A'B'C'$ by connecting the points identified in Problem 21.13: $A' = (-1,-3)$, $B' = (3,0)$, and $C' = (2,4)$. The line of reflection, $y = x$, is the dotted line in the graph below.

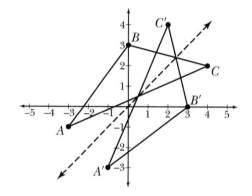

R_k means "the reflection in line k." So you're reflecting A across line k, producing A'.

21.15 Let k be the line $y = -\dfrac{1}{4}x + 1$. Given $R_k:P \to P'$, calculate the coordinates of A', the image of $A = (0,3)$.

The segment connecting pre-image A and image A' is perpendicular to k. Note that the slope of k is $-\dfrac{1}{4}$, so the slope of $\overline{AA'}$ is 4. Substitute $m = 4$ and the known endpoint of $\overline{AA'}$ ($x_1 = 0$ and $y_1 = 3$) into the point-slope formula of a line.

$$y - y_1 = m\left(x - x_1\right)$$
$$y - 3 = 4\left(x - 0\right)$$
$$y = 4x + 3$$

The midpoint M of $\overline{AA'}$ is the point at which k and the line that contains $\overline{AA'}$ (identified above) intersect. Thus, M is the solution to the following system of equations.

Because k bisects $\overline{AA'}$.

$$\begin{cases} y = -\dfrac{1}{4}x + 1 \\ y = 4x + 3 \end{cases}$$

Solve the system using the substitution method.

$$4x + 3 = -\frac{1}{4}x + 1$$

$$4x + \frac{1}{4}x = 1 - 3$$

$$\frac{16}{4}x + \frac{1}{4}x = -2$$

$$\frac{17}{4}x = -2$$

$$x = \left(\frac{4}{17}\right)(-2)$$

$$x = -\frac{8}{17}$$

Plug the second equation ($y = 4x + 3$) into the first equation and then solve the resulting equation for x.

Calculate the corresponding value of y by substituting $x = -\frac{8}{17}$ into one equation of the system.

$$y = -\frac{1}{4}x + 1$$

$$y = -\frac{1}{4}\left(-\frac{8}{17}\right) + 1$$

$$y = \frac{2 \cdot \cancel{4}}{\cancel{4} \cdot 17} + 1$$

$$y = \frac{2}{17} + \frac{17}{17}$$

$$y = \frac{19}{17}$$

The midpoint of $\overline{AA'}$ is $M = \left(-\frac{8}{17}, \frac{19}{17}\right)$. Therefore, M is the same distance from both A and A'. Identify Δx and Δy, the horizontal and vertical distances you travel from M to A, by calculating the differences of the x- and y-coordinates of M and A.

To go from M to A, you go RIGHT a certain distance (called Δx or "delta x") and you go UP a certain distance (called Δy or "delta y"). To calculate these values, subtract the x's from each other and the y's from each other. RIGHT and UP represent positive directions, so x and y should be positive.

$$\Delta x = x_A - x_M = 0 - \left(-\frac{8}{17}\right) = 0 + \frac{8}{17} = \frac{8}{17} \qquad \Delta y = y_A - y_M = 3 - \frac{19}{17} = \frac{51}{17} - \frac{19}{17} = \frac{32}{17}$$

If you travel $\frac{8}{17}$ units right and $\frac{32}{17}$ units up to reach A from M, then you must travel the opposite direction from M to reach image A', $\frac{8}{17}$ units left and $\frac{32}{17}$ units down.

$$A' = \left(x_M - \Delta x, y_M - \Delta y\right)$$

$$A' = \left(-\frac{8}{17} - \frac{8}{17}, \frac{19}{17} - \frac{32}{17}\right)$$

$$A' = \left(-\frac{16}{17}, -\frac{13}{17}\right)$$

> "Trans-lation" and "glides" refer to the same kind of transformation. "Glide reflec-tions" are slightly different, as ex-plained in Problem 21.22.

Translations and Glide Reflections
Slides with and without flips

21.16 Describe the relationship between the translation $G:(x,y) \rightarrow (x + a, y + b)$ and the vector $\mathbf{v} = <a,b>$.

Glide G moves each pre-image P horizontally a units and vertically b units, essentially moving a figure from one part of the coordinate plane to another. A translation does not affect the shape of a figure because all points are moved a congruent distance and in the same direction, so the transformation is an isometry.

> See Problem 20.39 for more information about calculating the length (magnitude) of a vector.

Recall that a vector describes a fixed length and direction. In this problem, $\mathbf{v}$ is a vector that travels a units horizontally and b units vertically—a total distance of $\|\mathbf{v}\| = \sqrt{a^2 + b^2}$. Therefore, if the tail of $\mathbf{v}$ is placed at any pre-image P of glide G, the head of $\mathbf{v}$ will identify the corresponding image P'; both the vector and the translation describe movement of the same distance in the same direction.

Problems 21.17–21.19 refer to the translation $G:(x,y) \rightarrow (x + 4, y - 3)$.

21.17 Find the image of $A = (-3,-9)$.

Substitute $x = -3$ and $y = -9$ into the coordinate pair that defines the translation, $(x + 4, y - 3)$.

$$A' = (-3 + 4, -9 - 3) = (1,-12)$$

Problems 21.17–21.19 refer to the translation $G:(x,y) \rightarrow (x + 4, y - 3)$.

21.18 Find the pre-image of $B' = (7,-1)$.

> You can also set the expressions that define G equal to the coordi-nates of B': $x + 4 = 7$ and $y - 3 = -1$. Then solve for x and y to get $x = 3$ and $y = 2$.

Glide G adds 4 to the x-coordinate and subtracts 3 from the y-coordinate of the pre-image to produce the image. Therefore, to generate the pre-image given the image, subtract 4 from the x-coordinate and add 3 to the y-coordinate.

$$B = (7 - 4, -1 + 3) = (3,2)$$

Problems 21.17–21.19 refer to the translation $G:(x,y) \rightarrow (x + 4, y - 3)$.

21.19 Given A, A', B, and B' as defined in Problems 21.17–21.18 and $R_k:A' \rightarrow B$, write the equation of line k in slope-intercept form.

If reflection R_k maps A' to B by reflecting A' across line k, then line k is the perpendicular bisector of $\overline{A'B}$. Recall that $A' = (1,-12)$ and $B = (3,2)$. Calculate the slope m and the midpoint M of $\overline{A'B}$.

$$m = \frac{2-(-12)}{3-1} \qquad M = \left(\frac{1+3}{2}, \frac{-12+2}{2}\right)$$

$$m = \frac{2+12}{2} \qquad M = \left(\frac{4}{2}, \frac{-10}{2}\right)$$

$$m = \frac{14}{2} \qquad M = (2,-5)$$

$$m = 7$$

The slope of k is $-\dfrac{1}{7}$, the opposite reciprocal of m. Substitute this slope and $M = (2,-5)$, the point at which k intersects $\overline{A'B}$, into the point-slope formula.

$$y-(-5) = -\frac{1}{7}(x-2)$$

$$y+5 = -\frac{1}{7}x + \frac{2}{7}$$

Solve for y to express the equation in slope-intercept form.

$$y = -\frac{1}{7}x + \frac{2}{7} - 5$$

$$y = -\frac{1}{7}x + \frac{2}{7} - \frac{35}{7}$$

$$y = -\frac{1}{7}x - \frac{33}{7}$$

Note: Problems 21.20–21.21 refer to the following transformations.

$$G{:}(x,y) \to (x + a, y + b) \quad G{:}P \to P' \quad G{:}P' \to P''$$

21.20 Identify Z'' given $a = 10$, $b = -5$, and $Z = (-2,11)$.

Substitute $a = 10$ and $b = -5$ into the formula that defines G.

$$G:(x,y) \to (x+a, y+b)$$

$$G:(x,y) \to (x+10, y+(-5))$$

$$G:(x,y) \to (x+10, y-5)$$

Calculate Z'.

$$Z' = (-2 + 10, 11 - 5) = (8,6)$$

> $Z = (-2,11)$, so plug $x = -2$ and $y = 11$ into G.

The problem states that $G{:}P' \to P''$, so apply translation G to Z' to generate Z''.

$$Z'' = (x + 10, y - 5) = (8 + 10, 6 - 5) = (18,1)$$

> Plugging Z into G gives you Z with one prime on it. Plug Z with one prime into G, and you get Z with two primes.

Note: Problems 21.20–21.21 refer to the following transformations.

$$G:(x,y) \rightarrow (x + a, y + b) \quad G:P \rightarrow P' \quad G:P' \rightarrow P''$$

21.21 Identify a and b, given $P = (-8,3)$ and $P' = (14,-4)$.

Translation G adds a to the x-coordinate and adds b to the y-coordinate of P to generate P'.

$$-8 + a = 14 \quad \text{and} \quad 3 + b = -4$$

Solve the equations.

$$a = 14 + 8 \qquad b = -4 - 3$$
$$a = 22 \qquad b = -7$$

21.22 The vertices of $\triangle ABC$ are $A = (0,0)$, $B = (2,1)$, and $C = (-3,4)$. Assume that glide reflection G translates pre-images 2 units down and then reflects them across the y-axis. Graph $\triangle A'B'C'$.

Translate each point down 2 units by subtracting 2 from each y-coordinate.

$$A' = (0, 0 - 2) \qquad B' = (2, 1 - 2) \qquad C' = (-3, 4 - 2)$$
$$A' = (0, -2) \qquad B' = (2, -1) \qquad C' = (-3, 2)$$

To complete the glide reflection, reflect the points across the y-axis by taking the opposite of the x-coordinate.

$$A' = (0, -2) \qquad B' = (-2, -1) \qquad C' = (3, 2)$$

The following diagram includes $\triangle ABC$ and the result of glide reflection G, $\triangle A'B'C'$.

A glide re-flection first glides (moves) a set of points and then reflects (flips) them across a line. The line you reflect across will be parallel to the di-rection you moved the points. In this problem, you move the points down a vertical line and then reflect over a vertical line.

See Problem 21.11.

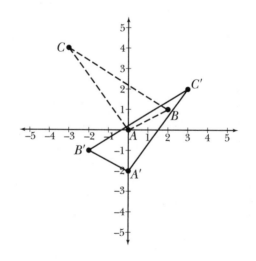

Rotations

Spin a figure around its center

21.23 Describe the transformation $\Re_{Q,45°}$.

The script letter $\Re$ indicates a rotation (whereas an italicized *R* typically indicates a reflection) about point *Q* through 45°. Thus, applying $\Re_{Q,45°}$ to pre-image *A* rotates the point 45° counterclockwise about point *Q*. Note that *A* and the corresponding image *A'* are the same distance from *Q*. ←

> Imagine a circle with center Q that passes through A. The new point A' is located on that circle, 45° or $\frac{45}{360} = \frac{1}{8}$th of the way around the circle counterclockwise.

21.24 Describe the transformation $\Re_{P,-135°}$.

This is a rotation about point *P* through −135°, a rotation of 135° in the clockwise direction.

> *Note: Problems 21.25–21.29 refer to the following diagram, $\bigcirc O$ split into eight congruent sectors by the x-axis, the y-axis, line y = x, and line y = −x.*

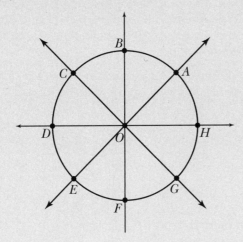

21.25 What is the image of *E* under the rotation $\Re_{O,90°}$?

If *E* is rotated about point *O*, the center of the circle, 90° in the counter-clockwise direction, then $m\angle E'OE = 90°$. Note that the circle is divided into eight congruent sectors, so the central angle of each sector measures 360° ÷ 8 = 45°.

> The sides of angle E'OE contain E and its image E', and the vertex of the angle is O.

Rotating 90° means rotating through two 45° sectors in the counterclockwise direction, from *E* to *G*. Thus, $\Re_{O,90°}(E) = G$.

> *Note: Problems 21.25–21.29 refer to the diagram in Problem 21.25: $\bigcirc O$ split into eight congruent sectors by the x-axis, the y-axis, line y = x, and line y = −x.*

21.26 What is the image of *A* under the rotation $\Re_{O,-45°}$?

Rotating *A* about point *O* through one 45° sector in the clockwise direction results in point *H*; thus, $\Re_{O,-45°}(A) = H$.

> This time, you're given the RESULT, not the starting point.

Note: Problems 21.25–21.29 refer to the diagram in Problem 21.25: $\bigcirc O$ split into eight congruent sectors by the x-axis, the y-axis, line y = x, and line y = –x.

21.27 What is the pre-image of G under the rotation $\Re_{O,-135°}$?

Rotating the pre-image about point O through three 45° sectors in the clockwise direction results in point G. Therefore, rotating three sectors in the counterclockwise direction from image G results in the pre-image, B. Thus, $\Re_{O,-135°}(B) = G$.

> 3(45) = 135

Note: Problems 21.25–21.29 refer to the diagram in Problem 21.25: $\bigcirc O$ split into eight congruent sectors by the x-axis, the y-axis, line y = x, and line y = –x.

21.28 For what value of x is the following statement true? Assume 0 < x < 360.

$$\Re_{O,x°}(A) = \Re_{O,-x°}(A)$$

> 0 and 360 would have been good answers because if you rotate 0°, you stay at A, and if you rotate 360° in either direction, you end up at A, where you started.

You are asked to determine the angle $x°$ through which you can rotate A about point O that produces the same image whether you rotate A in the clockwise or counterclockwise direction. Notice that rotating A 180° in either direction about O results in point E—both rotations travel exactly halfway around the circle and meet at E. Thus, $x = 180°$ because $\Re_{O,180°}(A) = \Re_{O,-180°}(A) = E$.

> This is true for any point A through H. If you rotate 180° in either direction from that point, you end up in the same place.

Note: Problems 21.25–21.29 refer to the diagram in Problem 21.25: $\bigcirc O$ split into eight congruent sectors by the x-axis, the y-axis, line y = x, and line y = –x.

21.29 Identify two values of y, one positive and one negative, for which the following statement is true.

$$\Re_{O,135°}(H) = \Re_{O,y°}(H)$$

Rotating H about point O through 135° in the counterclockwise direction produces image C. Add 360°—a full circular rotation about point O—to reach the same image. Thus, rotating H about point O through $y = 135° + 360° = 495°$ produces image C as well.

Just as *adding* 360° to the given angle produces the same image, *subtracting* 360° *from* the given angle results in the same image. Thus, rotating H about point O through $y = 135° - 360° = -225°$ also results in image C, as illustrated below.

> If you travel 360° around a circle, you end up right where you started. You can also add 360(2) = 720°, 360(3) = 1,080°, and so on to 135° and end up at point C.

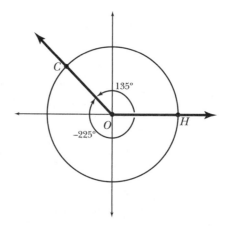

21.30 Given point $P = (2,-1)$ and a triangle with vertices $X = (5,2)$, $Y = (4,0)$, and $Z = (1,3)$, draw the image of the triangle, $\triangle X'Y'Z'$, under the half-turn H_p.

Consider the following graph that contains $\triangle XYZ$ and P, the point about which $\triangle XYZ$ will be rotated. To plot images X', Y', and Z', reverse the horizontal and vertical distances between P and each of the points.

> A half-turn is a rotation of 180°. The direction doesn't matter because rotating 180° about a point produces the same image whether that rotation is clockwise or counterclock-wise.

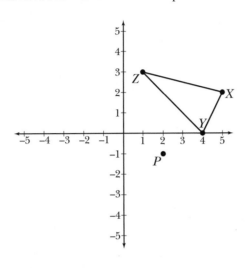

> For example, to reach X from P, you travel 3 units up and 3 units right: $(2 + 3, -1 + 3) = (5,2)$. Thus, to reach X' from P, you should travel in the opposite direction, 3 units down and 3 units left.

> Start with $P = (2,-1)$. If you add 3 to the x- and y-coordinates, you get $X = (5,2)$.

$$X' = (2 - 3, -1 - 3) = (-1,-4)$$

Similarly, you travel 2 units right and 1 unit up to reach Y from P, so subtract 2 from the x-coordinate of P and subtract 1 from the y-coordinate to generate the image of Y.

$$Y' = (2 - 2, -1 - 1) = (0,-2)$$

Finally, you travel 1 unit left and 4 units up to reach Z from P. Thus, Z' is 1 unit right of and 4 units below P.

$$Z' = (2 + 1, -1 - 4) = (3,-5)$$

The image of $\triangle XYZ$ under H_p is illustrated below.

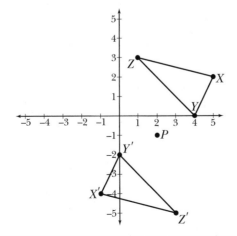

Notice that *P* is the midpoint of the segments that connect pre-images to the corresponding images: $\overline{XX'}$, $\overline{YY'}$, and $\overline{ZZ'}$.

Dilations
Enlarge or ensmall things

21.31 Describe the transformation $D_{P,2}$.

The transformation $D_{P,2}$ is a dilation with center *P* and scale factor $k = 2$. Under this dilation, pre-image *A* is mapped to image *A'* such that *P*, *A*, and *A'* are collinear and $\overrightarrow{PA'}$ is twice as long as $\overrightarrow{PA}$. Note that both vectors have the same endpoint (*P*) and slope.

> This dilation moves the pre-image twice as far from P, along the line that connects center P to the pre-image.

21.32 Describe the transformation $D_{M,-1/3}$.

This dilation has center *M* and scale factor $k = -\dfrac{1}{3}$. Under this dilation, pre-image *B* is mapped to image *B'* such that *P*, *B*, and *B'* are collinear, but $\overrightarrow{PB'}$ is one-third as long as $\overrightarrow{PB}$ and the vectors lie on opposite rays.

> Because the scale factor is negative, the arrows connecting P to B and B' face in opposite directions. They still lie on the same line, but if you travel up and right to go from P to B, you go down and left to go from P to B'.

Note: Problems 21.33–21.35 refer to points X = (–4,2), Y = (0,–6), and P = (2,0).

21.33 Identify the image of *X* under the transformation $D_{P,-1/2}$.

Consider the following diagram, which plots points *P*, *X*, and *Y*. To reach *X* from point *P*, you travel 6 units left and 2 units up.

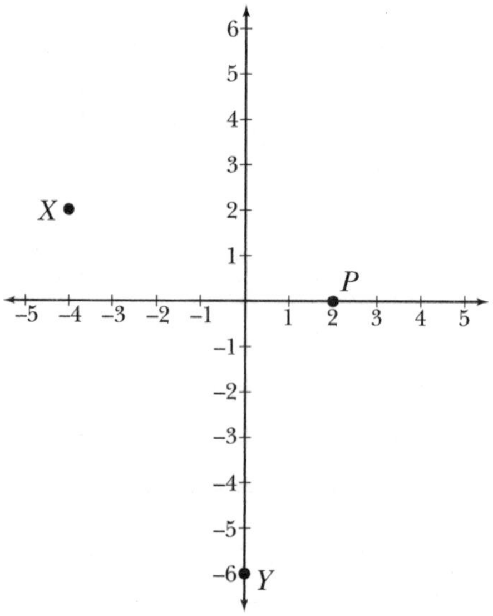

> The path from P to X is 6 left and 2 up, so the path from P to X' is half of each number in the opposite direction: 3 right and 1 down.

The dilation $D_{P,-1/2}$ has center *P* and scale factor $-\dfrac{1}{2}$, so to reach image *X'* from *P*, you travel half as far in the opposite direction: $\dfrac{1}{2}(6) = 3$ units right and $\dfrac{1}{2}(2) = 1$ unit down.

$$X' = (2 + 3, 0 - 1) = (5, -1)$$

> **Start** with $P = (2,0)$. X' is 3 units right of P, so add 3 to the x-coordinate; X' is 1 unit down, so subtract 1 from the y-coordinate.

Note: Problems 21.33–21.35 refer to points X = (–4,2), Y = (0,–6), and P = (2,0).

21.34 Identify the image of Y under the transformation $D_{P,-1/2}$.

To reach Y from P, you travel 2 units left and 6 units down. The scale factor is $-\frac{1}{2}$, so you travel half as far from P in the opposite direction to reach Y': $\frac{1}{2}(2) = 1$ unit right and $\frac{1}{2}(6) = 3$ units up.

$$Y' = (2 + 1, 0 + 3) = (3, 3)$$

Note: Problems 21.33–21.35 refer to points X = (–4,2), Y = (0,–6), and P = (2,0).

21.35   Given X' and Y' as calculated in Problems 21.33–21.34, demonstrate that $\frac{X'Y'}{XY} = \frac{1}{2}$.

> In other words, prove that X' and Y' are not only half as far from P as X and Y, but that they also create a segment that's half as long.

Apply the distance formula to calculate XY and $X'Y'$.

$$XY = \sqrt{(0-(-4))^2 + (-6-2)^2}$$
$$= \sqrt{(0+4)^2 + (-6-2)^2}$$
$$= \sqrt{4^2 + (-8)^2}$$
$$= \sqrt{16+64}$$
$$= \sqrt{80}$$
$$= 4\sqrt{5}$$

$$X'Y' = \sqrt{(3-5)^2 + (3-(-1))^2}$$
$$= \sqrt{(3-5)^2 + (3+1)^2}$$
$$= \sqrt{(-2)^2 + 4^2}$$
$$= \sqrt{4+16}$$
$$= \sqrt{20}$$
$$= 2\sqrt{5}$$

Calculate the quotient $\frac{X'Y'}{XY}$.

$$\frac{X'Y'}{XY} = \frac{2\sqrt{5}}{4\sqrt{5}} = \frac{2}{4}\cdot\frac{\sqrt{5}}{\sqrt{5}} = \frac{1}{2}\cdot 1 = \frac{1}{2}$$

Thus, $\frac{X'Y'}{XY} = \frac{1}{2}$. A segment with endpoints A and B and length AB that is contracted according to the dilation $D_{P,k}$ will map to a segment $\overline{A'B'}$ that has length $|k|\cdot AB$.

> So the endpoints are k times as far away from P and the segments they define will be k times as long. A positive k means an expansion, and a negative k (like here) means a contraction.

Problems 21.36–21.38 refer to the dilation $D_{N,2}$ given N = (–3,2), A = (0,4), B = (1,3), and C = (–2,0), as illustrated below.

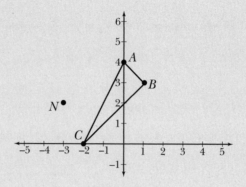

21.36 Draw the image of $\triangle ABC$ under the dilation.

Apply the technique described in Problems 21.33–21.34. Each image should be two times farther from N than the corresponding pre-image because the scale factor of the dilation is 2. Also note that the scale factor is positive, so you travel in the same direction from P to reach both the pre-images and the images.

Consider points N = (–3,2) and A = (0,4). To reach A from center N, you travel 3 units right and 2 units up. Thus, to reach A' from N, you travel 2(3) = 6 units right and 2(2) = 4 units up.

$$A' = (–3 + 6, 2 + 4) = (3,6)$$

Apply the same technique to identify the images of B and C. To reach B from center N, you travel 4 units right and 1 unit up. Thus, to reach B' from N, you travel 2(4) = 8 units right and 2(1) = 2 units up.

$$B' = (–3 + 8, 2 + 2) = (5,4)$$

To reach C from center N, you travel 1 unit right and 2 units down. Thus, to reach C' from N, you travel 2(1) = 2 units right and 2(2) = 4 units down.

$$C' = (–3 + 2, 2 – 4) = (–1,–2)$$

Draw the expanded triangle A'B'C' under the dilation.

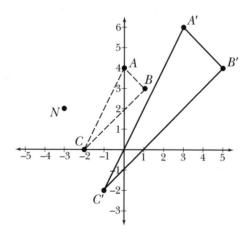

Problems 21.36–21.38 refer to the dilation $D_{N,2}$ given $N = (-3,2)$, $A = (0,4)$, $B = (1,3)$, and $C = (-2,0)$.

21.37 Demonstrate that $\triangle ABC \sim \triangle A'B'C'$.

Calculate the lengths of the sides of the triangles using the distance formula.

$$AB = \sqrt{(1-0)^2 + (3-4)^2} \qquad BC = \sqrt{(-2-1)^2 + (0-3)^2} \qquad AC = \sqrt{(-2-0)^2 + (0-4)^2}$$
$$= \sqrt{1^2 + (-1)^2} \qquad\qquad = \sqrt{(-3)^2 + (-3)^2} \qquad\qquad = \sqrt{(-2)^2 + (-4)^2}$$
$$= \sqrt{1+1} \qquad\qquad\qquad = \sqrt{9+9} \qquad\qquad\qquad = \sqrt{4+16}$$
$$= \sqrt{2} \qquad\qquad\qquad\quad = \sqrt{18} \qquad\qquad\qquad\quad = \sqrt{20}$$
$$\qquad\qquad\qquad\qquad\qquad = 3\sqrt{2} \qquad\qquad\qquad\quad = 2\sqrt{5}$$

$$A'B' = \sqrt{(5-3)^2 + (4-6)^2} \qquad B'C' = \sqrt{(-1-5)^2 + (-2-4)^2} \qquad A'C' = \sqrt{(-1-3)^2 + (-2-6)^2}$$
$$= \sqrt{2^2 + (-2)^2} \qquad\qquad = \sqrt{(-6)^2 + (-6)^2} \qquad\qquad = \sqrt{(-4)^2 + (-8)^2}$$
$$= \sqrt{4+4} \qquad\qquad\qquad = \sqrt{36+36} \qquad\qquad\qquad = \sqrt{16+64}$$
$$= \sqrt{8} \qquad\qquad\qquad\quad = \sqrt{72} \qquad\qquad\qquad\quad = \sqrt{80}$$
$$= 2\sqrt{2} \qquad\qquad\qquad = 6\sqrt{2} \qquad\qquad\qquad = 4\sqrt{5}$$

Corresponding sides of $\triangle ABC$ and $\triangle A'B'C'$ are in proportion.

$$\frac{A'B'}{AB} = \frac{2\sqrt{2}}{\sqrt{2}} = \frac{2}{1} = 2 \qquad \frac{B'C'}{BC} = \frac{6\sqrt{2}}{3\sqrt{2}} = \frac{6}{3} = 2 \qquad \frac{A'C'}{AC} = \frac{4\sqrt{5}}{2\sqrt{5}} = \frac{4}{2} = 2$$

Thus, $\triangle ABC \sim \triangle A'B'C'$ by the SSS similarity postulate. The scale factor from $\triangle A'B'C'$ to $\triangle ABC$ is $\frac{2}{1} = 2$, the same scale factor as dilation $D_{N,2}$.

Problems 21.36–21.38 refer to the dilation $D_{N,2}$ given $N = (-3,2)$, $A = (0,4)$, $B = (1,3)$, and $C = (-2,0)$.

21.38 Calculate the ratio of the areas of $\triangle ABC$ and $\triangle A'B'C'$.

Notice that $\triangle ABC$ and $\triangle A'B'C'$ are right triangles because they satisfy the Pythagorean Theorem. (Problem 21.37 calculates the lengths of the sides.)

$$(AB)^2 + (BC)^2 = (AC)^2 \qquad\qquad (A'B')^2 + (B'C')^2 = (A'C')^2$$
$$\left(\sqrt{2}\right)^2 + \left(3\sqrt{2}\right)^2 = \left(2\sqrt{5}\right)^2 \qquad\qquad \left(2\sqrt{2}\right)^2 + \left(6\sqrt{2}\right)^2 = \left(4\sqrt{5}\right)^2$$
$$2 + 9 \cdot 2 = 4 \cdot 5 \qquad\qquad\qquad\quad 4 \cdot 2 + 36 \cdot 2 = 16 \cdot 5$$
$$2 + 18 = 20 \qquad\qquad\qquad\qquad 8 + 72 = 80$$
$$20 = 20 \qquad\qquad\qquad\qquad\quad 80 = 80$$

The area of a triangle is half the product of its base and height.

$$\text{area of } \triangle ABC = \frac{1}{2}bh \qquad\qquad \text{area of } \triangle A'B'C' = \frac{1}{2}bh$$

$$= \frac{1}{2}(BC)(AB) \qquad\qquad = \frac{1}{2}(B'C')(A'B')$$

$$= \frac{1}{2}\left(3\sqrt{2}\right)\left(\sqrt{2}\right) \qquad\qquad = \frac{1}{2}\left(6\sqrt{2}\right)\left(2\sqrt{2}\right)$$

$$= \frac{1}{2}\left(3\sqrt{4}\right) \qquad\qquad = \frac{1}{2}\left(12\sqrt{4}\right)$$

$$= \frac{1}{2}(3 \cdot 2) \qquad\qquad = \frac{1}{2}(12 \cdot 2)$$

$$= 3 \qquad\qquad\qquad = 12$$

> The base and height of a right triangle are its legs, the perpendicular sides. It doesn't matter which you call the base and which you call the height.

Calculate the ratio of the areas of the triangles.

$$\frac{\text{area of } \triangle A'B'C'}{\text{area of } \triangle ABC} = \frac{12}{3} = 4$$

The ratio of the areas of the triangles is 4, the square of the scale factor of $D_{N,2}$.

> Problem 18.39 stated that the ratio of the areas of similar figures with scale factor a:b is $a^2:b^2$. According to Problem 21.37, the scale factor of these triangles is 2:1, so the ratio of the areas is $2^2:1^2 = 4:1$.

Symmetry

Including point, line, and rotational

21.39 The following parallelogram exhibits line symmetry. Draw all four lines in which the figure is symmetric.

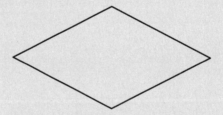

A figure has line symmetry if reflecting it across a line maps the figure onto itself. In other words, the reflection and the original figure have exactly the same shape and size. Lines of symmetry l, j, k, and m are illustrated below, as R_j, R_k, R_l, and R_m map the parallelogram onto itself.

> The names of the points may change, but the figure you end up with should look exactly like the figure you started with.

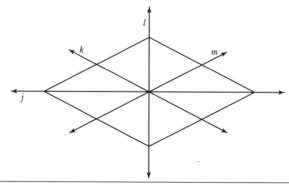

21.40 Complete the following drawing so that it is symmetric in lines p and q.

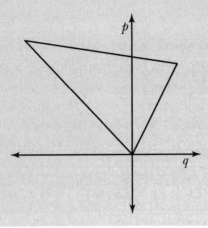

Reflect the figure across line p, and then reflect it across line q to produce the figure below.

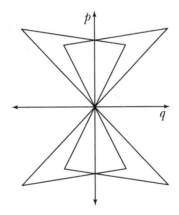

Note: Problems 21.41–21.42 refer to the following figure.

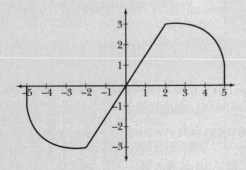

21.41 Identify the positive rotational symmetry exhibited by the figure.

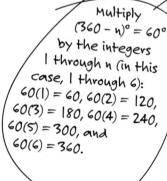

This is called the identity rotation because the identity of each point is unchanged when you rotate 360°.

Rotating the figure 180° about $A = (0,0)$ maps the figure onto itself, so the figure has rotational symmetry about point A through 180°: $\Re_{A,180°}$. Note that all figures have rotational symmetry when rotated 360° about their center because each point is mapped to its original coordinate pair.

Note: Problems 21.41–21.42 refer to the figure in Problem 21.41.

21.42 Is the figure point-symmetric? Why or why not?

A figure has point symmetry if a half-turn maps the figure onto itself. Recall that a half-turn is a 180° rotation about a point. According to Problem 21.41, the figure has rotational symmetry $\Re_{A,180°}$. Thus, the half-turn H_A maps the figure onto itself and the figure is point-symmetric about $A = (0,0)$.

A figure transformed by a half-turn H_A does opposite things on each side of A. Just to the right of A is a line that slopes up and right. Therefore, just to the left of A is a line that slopes down and left.

21.43 Identify the rotational symmetries exhibited by a regular hexagon with center P.

A regular polygon with n sides has n unique rotational symmetries because it can be divided into n isosceles triangles with vertex angle $(360 \div n)°$. Thus, a regular hexagon with $n = 6$ sides and center P has six rotational symmetries. Note that $(360 \div 6)° = 60°$, so the figure is symmetric when rotated through 60°, 120°, 180°, 240°, 280°, and 360°. Thus, the figure is symmetric under the rotations $\Re_{P,60°}$, $\Re_{P,120°}$, $\Re_{P,180°}$, $\Re_{P,240°}$, $\Re_{P,300°}$, and $\Re_{P,360°}$.

Multiply $(360 - n)° = 60°$ by the integers 1 through n (in this case, 1 through 6): $60(1) = 60$, $60(2) = 120$, $60(3) = 180$, $60(4) = 240$, $60(5) = 300$, and $60(6) = 360$.

In the following figure, a regular hexagon is divided into six congruent isosceles triangles, one of which is shaded for reference. The hexagon is rotated 60° at a time in the counterclockwise direction, moving the shaded triangle but producing a hexagon that has the same size and shape, as well as the same orientation, as the original hexagon.

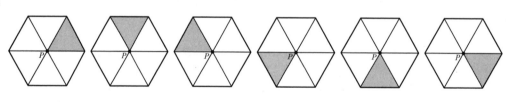

Chapter 22
TRUTH TABLES

Wood furniture that refuses to lie

Truth tables provide a formal way to investigate the validity of compound statements. This chapter begins with an introduction of logical symbology (including a symbolic representation of conjunction and disjunction), explores the construction of a truth table, and graduates to the proof of logical equivalence and the validity of arguments.

Truth tables are basically charts that contain columns of T's and F's (which stand for "true" and "false"). You start with the basics, statements like, "If p is true and q is false, is (p or q) true or false?" You then transition into conditional (if-then) statements, answering the eternal question "If p is false and q is true, is $p \rightarrow q$ true or false?" Finally, you explore arguments and learn what in the world modus ponens and modus tollens mean.

Symbology

What do the arrowheads and sideways L's mean?

22.1 Describe the difference between simple and compound statements, and give an example of each.

> Either it's Wednesday when you read this or it's not. If it is, then the statement is true. If it's Sunday, however, then the statement is false.

A simple statement makes an assertion that is either true or false, a determination that is typically made independently of any other statements. Consider this simple statement: "It is Wednesday." Simple statements are represented by letters like p and q. For instance, if p represents the previous simple statement, then p is false on Mondays.

A compound statement contains two or more simple statements joined by connectives like "and" and "or." Consider this compound statement: "It is Wednesday or it is summer." Two simple statements ("it is Wednesday" and "it is summer") are joined by a connective ("or").

22.2 Identify the connective symbolized by $\wedge$, and describe the circumstances under which the compound statement $r \wedge s$ is true.

> You might think that an AND statement is true as long as the simple statements have the same truth value, when they're both true or both false. However, that's not the way it works. Both statements must be true for an AND statement to be true.

The symbol $\wedge$ represents the connective "and," also known as a conjunction. The compound statement $r \wedge s$, comprised of simple statements r and s, is true only when both r and s are true. Note that $r \wedge s$ is false when r and s are both false.

22.3 Identify the connective symbolized by $\vee$, and describe the circumstances under which the compound statement $p \vee q$ is true.

The symbol $\vee$ represents the connective "or," also known as a disjunction. More specifically, it represents an "inclusive or," which means that $p \vee q$ is true as long as one of the simple statements is true. Thus, if p is false but q is true, then $p \vee q$ is true.

> Think in terms of a restaurant. Entrées are an "exclusive or"—you can have either the chicken OR the fish. However, appetizers and dessert are an "inclusive or." You can order the breadsticks to start or have cake for dessert— or if you're really hungry, you can order both.

Understand that the logical "or" connective does not require that exactly one of the statements be true. In other words, if p is true and q is true, then $p \vee q$ is true as well—not only has one of the statements met the truth requirement, but both have. Thus, $p \vee q$ is false only when p and q are both false.

Note: Problems 22.4–22.7 refer to the following simple statements.

> p: It is raining.
> q: Soccer practice is cancelled.

22.4 Assuming that it is raining and soccer practice is canceled, determine the truth value of the compound statement $p \wedge q$.

If it is raining, then simple statement p is true. Furthermore, practice is cancelled, so statement q is true. Because p and q are both true, $p \wedge q$ is also true.

Note: Problems 22.4–22.7 refer to the simple statements in Problem 22.4.

22.5 Assume that it is raining, but the soccer coach decided to hold practice in the gymnasium and therefore did not cancel practice. Determine the truth value of the compound statement $p \wedge q$.

It is raining, so p is true. However, soccer practice is *not* cancelled, so q is false. Thus, $p \wedge q$ is false. A conjunction is true only when the simple statements it joins are both true. In this case, only p is true, so $p \wedge q$ is false.

Note: Problems 22.4–22.7 refer to the simple statements in Problem 22.4.

22.6 Assume that it is not raining, but the soccer coach has to cancel practice due to a faculty meeting. Determine the truth value of the compound statement $p \vee q$.

It is not raining, so p is false. Due to a scheduling conflict, practice has been cancelled, so q is true. The statement $p \vee q$ is true because at least one of the simple statements it connects is true. ←

> Practice is cancelled, so statement q is true.

Note: Problems 22.4–22.7 refer to the simple statements in Problem 22.4.

22.7 Assume that it is not raining and soccer practice is not cancelled. Determine the truth value of the compound statement $p \vee q$.

It is not raining, so p is false; practice is not cancelled, so q is also false. Recall that a disjunction (like $p \vee q$) is true when at least one of the simple statements it connects is true. In this case, both statements are false, so $p \vee q$ is false.

22.8 Identify the logical symbol $\neg$, and identify the truth values of the statements $\neg r$ and $\neg(\neg r)$, assuming that r is true.

> Some books use ~ to represent negation.

The symbol $\neg$ represents negation, and it reverses the truth value of a statement. For instance, the problem states that r is a true statement, so the negation $\neg r$ must be false. Assuming that $\neg r$ is false, then the negation of that statement, $\neg(\neg r)$, is true.

22.9 Express the following compound statement symbolically.

It is raining and I do not have my umbrella.

Let p represent the simple statement "it is raining," and let q represent the simple statement "I have my umbrella." Join p and $\neg q$ with the "and" connective ($\wedge$).

$$p \wedge (\neg q)$$

> Statement q states that you have an umbrella, but the problem states you don't, so you have to negate q. Technically, you could make "I don't have an umbrella" statement q, but I think the point of the problem is to use the negation symbol.

22.10 Express the following compound statement symbolically.

It is not raining and I have my umbrella, or it is snowing and I do not have a scarf.

Let *p*, *q*, *r*, and *s* represent the following statements.

p: It is raining

q: I have my umbrella

r: It is snowing

s: I have a scarf

The given statement consists of two compound statements. Express each symbolically.

It is not raining and I have my umbrella: $\neg p \wedge q$

It is snowing and I do not have a scarf: $r \wedge \neg s$

Notice that the two compound statements are joined with the connective "or"; thus, you should join the symbolic representations of those statements with ∨.

$$(\neg p \wedge q) \vee (r \wedge \neg s)$$

There's an OR between "I have my umbrella" and "it is snowing."

Basic Truth Tables
Including tautology and contradiction

Note: Problems 22.11–22.13 complete the following truth table for the compound statement $p \vee q$, one column at a time.

p	*q*	$p \vee q$

22.11 Complete the *p* column of the truth table.

Given a compound statement containing two simple statements, in this case *p* and *q*, the first two columns represent the truth values of those statements. To explore all possible combinations of truth values for *p* and *q*, you must construct a truth table in a deliberate, organized manner.

From left to right

A compound statement involving two simple statements consists of four rows (not including the row of labels "*p*, *q*, and $p \vee q$" already provided). In each column, half the rows are labeled "*T*" for true and half are labeled "*F*" for false, but each column accomplishes this in a slightly different manner.

Conventionally, the top half of the rows in the *p* column are labeled as true and the bottom half are labeled as false, as illustrated in the following.

p	*q*	*p* ∨ *q*
T		
T		
F		
F		

Note: Problems 22.11–22.13 complete the truth table in Problem 22.11, one column at a time.

22.12 Complete the *q* column of the truth table.

As in the *p* column, the *q* column will contain two true values and two false values. Whereas the *p* column listed the values in groups of two, the groups in column *q* are half as large. Instead of two *T*'s followed by two *F*'s, alternate one *T* followed by one *F* until the chart is complete.

p	*q*	*p* ∨ *q*
T	*T*	
T	*F*	
F	*T*	
F	*F*	

> If column p is TTFF (groups of two), then the next column is TFTF (groups of one). All truth tables follow this pattern: each column has true/false groups half as large as the column to its left.

Every truth table containing two simple statements (like *p* and *q*) begins with the two columns illustrated above.

Note: Problems 22.11–22.13 complete the truth table in Problem 22.11, one column at a time.

22.13 Complete the *p* ∨ *q* column of the truth table.

Consider each row of the truth table independently, determining whether *p* ∨ *q* is true or false using only the truth values of *p* and *q* in that row. For instance, in the first row, *p* and *q* are both true, so *p* ∨ *q* is true. Similarly, the second and third rows of the truth table produce true values for *p* ∨ *q* because one of the simple statements is true in each case. Only the fourth row makes *p* ∨ *q* false, because *p* and *q* are both false.

> An OR statement is true as long as ONE of the statements is true. In this case, p and q are both true.

p	*q*	*p* ∨ *q*
T	*T*	*T*
T	*F*	*T*
F	*T*	*T*
F	*F*	*F*

Note: Problems 22.14–22.16 complete the following truth table for the compound statement ¬p ∧ q.

p	q	$\neg p$	$\neg p \wedge q$

22.14 Explain why this truth table contains four columns instead of three columns like the truth table in Problems 22.11–22.13.

> *This makes things easier. When you're trying to figure out whether ¬p ∧ q is true, you actually have a ¬p column that you can look at and compare with q.*

All truth tables begin with one column for each of the simple statements present. Compound statement $\neg p \wedge q$ contains two simple statements (p and q), which comprise the first two columns of the truth table. Notice, however, that $\neg p \wedge q$ compares q to the negation of p. Thus, a third column is added that contains the values of $\neg p$.

Though the rightmost column should represent the statement you are investigating (in this case, $\neg p \wedge q$), you may add as many columns in the middle of the truth table as you want. Typically, any simple or compound statement that appears in the rightmost column should be represented by a corresponding column within the truth table.

Note: Problems 22.14–22.16 complete the truth table in Problem 22.14.

22.15 Complete the p, q, and $\neg p$ columns of the truth table.

The truth tables for all compound statements comprised of two simple statements (in this case p and q) begin with the same two columns, as explained in Problems 22.11–22.12.

> *¬p is not a third, unique simple statement. It's just the negation of p.*

p	q	$\neg p$	$\neg p \wedge q$
T	T		
T	F		
F	T		
F	F		

The $\neg p$ column of the truth table contains the opposite truth value of p in each row. The first two rows in the p column are true and the last two rows are false. Therefore, the first two rows in the $\neg p$ column are false and the last two rows are true.

p	q	$\neg p$	$\neg p \wedge q$
T	T	F	
T	F	F	
F	T	T	
F	F	T	

Note: Problems 22.14–22.16 complete the truth table in Problem 22.14.

22.16 Complete the ¬$p \wedge q$ column in the truth table.

The "and" (∧) connective produces a true statement only when both of the simple statements are true. In the first two rows, ¬p is false, so ¬$p \wedge q$ must be false. In the third row, ¬p and q are both true, so ¬$p \wedge q$ is true. Finally, ¬$p \wedge q$ is false in the fourth row because q is false.

> In the second row, q is also false, so that row is as false as false gets.

p	q	¬p	¬$p \wedge q$
T	T	F	F
T	F	F	F
F	T	T	T
F	F	T	F

According to the truth table, ¬$p \wedge q$ is true only when p is false and q is true.

Note: Problems 22.17–22.19 prove one of de Morgan's Laws: $\neg(p \wedge q) \equiv (\neg p) \vee (\neg q)$.

22.17 Construct a truth table for $\neg(p \wedge q)$, the compound statement left of the equivalence symbol.

Compound statement $\neg(p \wedge q)$ contains two simple statements, p and q, each of which is represented by a column in the truth table. You should also include a column for $p \wedge q$ as well as the statement itself, $\neg(p \wedge q)$. Complete the first two columns of the table using the conventional pattern described in Problems 22.11–22.12.

> The statements on either side of the equivalence symbol ≡ have the same truth values, given corresponding values for p and q. In other words, if a false p and true q make the left statement true, they should make the right statement true as well.

p	q	$p \wedge q$	$\neg(p \wedge q)$
T	T		
T	F		
F	T		
F	F		

The third column is true only if both p and q are true. Thus, only the first row of the third column is true.

> Before you can determine whether ¬(p ∧ q) is true, it's helpful to know whether p ∧ q is true.

p	q	$p \wedge q$	$\neg(p \wedge q)$
T	T	T	
T	F	F	
F	T	F	
F	F	F	

The final column of the truth table is the negation of the third column.

p	q	$p \wedge q$	$\neg(p \wedge q)$
T	T	T	F
T	F	F	T
F	T	F	T
F	F	F	T

Note: Problems 22.17–22.19 prove one of de Morgan's Laws: $\neg(p \wedge q) \equiv (\neg p) \vee (\neg q)$.

22.18 Construct the truth table for $(\neg p) \vee (\neg q)$, the compound statement right of the equivalence symbol.

> The first columns should be the simple statements, and the last column should be the statement you're building the truth table for.

Aside from the standard columns for p, q, and the compound statement $(\neg p) \vee (\neg q)$ itself, this truth table should also include columns for $\neg p$ and $\neg q$ because they appear in the compound statement.

p	q	$\neg p$	$\neg q$	$(\neg p) \vee (\neg q)$
T	T			
T	F			
F	T			
F	F			

The third and fourth columns of the truth table are the negations of columns p and q, respectively.

p	q	$\neg p$	$\neg q$	$(\neg p) \vee (\neg q)$
T	T	F	F	
T	F	F	T	
F	T	T	F	
F	F	T	T	

The final column is true in each row that either $\neg p$ or $\neg q$ is true. Thus, the final column is true in each row except the first, in which both $\neg p$ and $\neg q$ are false.

p	q	$\neg p$	$\neg q$	$(\neg p) \vee (\neg q)$
T	T	F	F	F
T	F	F	T	T
F	T	T	F	T
F	F	T	T	T

Note: *Problems 22.17–22.19 prove one of de Morgan's Laws:* $\neg(p \wedge q) \equiv (\neg p) \vee (\neg q)$.

22.19 Explain why $\neg(p \wedge q) \equiv (\neg p) \vee (\neg q)$.

This symbol means "is logically equivalent to."

The statement $\neg(p \wedge q)$ is logically equivalent to $(\neg p) \vee (\neg q)$ because they have the same truth values, given corresponding values for p and q. Consider the following truth table, which contains p, q, and the final columns from the truth tables in Problems 22.17 and 22.18.

p	q	$\neg(p \wedge q)$	$(\neg p) \vee (\neg q)$
T	T	F	F
T	F	T	T
F	T	T	T
F	F	T	T

Columns of a truth table that have equivalent values in every row represent logically equivalent statements.

Note: *Problem 22.20 proves the second of de Morgan's Laws:* $\neg(p \vee q) \equiv (\neg p) \wedge (\neg q)$.

22.20 Prove that the statements are logically equivalent using a single truth table.

The truth table begins with columns for the simple statements p and q, includes columns for the statements you are trying to prove equivalent, and also includes columns for each component of those statements, as illustrated below.

p	q	$p \vee q$	$\neg(p \vee q)$	$\neg p$	$\neg q$	$(\neg p) \wedge (\neg q)$
T	T					
T	F					
F	T					
F	F					

The third column is true when at least one of the first two columns is true. Thus, it is false only in the fourth row, when both p and q are false.

p	q	$p \vee q$	$\neg(p \vee q)$	$\neg p$	$\neg q$	$(\neg p) \wedge (\neg q)$
T	T	T				
T	F	T				
F	T	T				
F	F	F				

The fourth column is a negation of the third column.

p	q	$p \vee q$	$\neg(p \vee q)$	$\neg p$	$\neg q$	$(\neg p) \wedge (\neg q)$
T	T	T	F			
T	F	T	F			
F	T	T	F			
F	F	F	T			

The fifth and sixth columns are negations of the first two columns, respectively.

p	q	$p \vee q$	$\neg(p \vee q)$	$\neg p$	$\neg q$	$(\neg p) \wedge (\neg q)$
T	T	T	F	F	F	
T	F	T	F	F	T	
F	T	T	F	T	F	
F	F	F	T	T	T	

The final column is true only when the fifth and sixth columns are true. This occurs only in the fourth row.

p	q	$p \vee q$	$\neg(p \vee q)$	$\neg p$	$\neg q$	$(\neg p) \wedge (\neg q)$
T	T	T	F	F	F	F
T	F	T	F	F	T	F
F	T	T	F	T	F	F
F	F	F	T	T	T	T

You can conclude that $\neg(p \vee q) \equiv (\neg p) \wedge (\neg q)$ because the shaded columns, which represent the equivalent statements, have the same truth values in every row.

22.21 Prove that $(p \wedge q) \wedge (\neg p)$ is a contradiction.

> A contradiction is a compound statement that is false in all cases. The right column contains nothing but F's.

Construct a truth table that has columns representing the compound statement $(p \wedge q) \wedge (\neg p)$, the simple statements p and q contained therein, and the components of the compound statement $p \wedge q$ and $\neg p$.

p	q	$p \wedge q$	$\neg p$	$(p \wedge q) \wedge (\neg p)$
T	T			
T	F			
F	T			
F	F			

The third column is true only when p and q are both true.

p	q	$p \wedge q$	$\neg p$	$(p \wedge q) \wedge (\neg p)$
T	T	T		
T	F	F		
F	T	F		
F	F	F		

The fourth column is the negation of the first column.

p	q	$p \wedge q$	$\neg p$	$(p \wedge q) \wedge (\neg p)$
T	T	T	F	
T	F	F	F	
F	T	F	T	
F	F	F	T	

The final column is true only when the third and fourth columns are both true.

p	q	$p \wedge q$	$\neg p$	$(p \wedge q) \wedge (\neg p)$
T	T	T	F	F
T	F	F	F	F
F	T	F	T	F
F	F	F	T	F

Because the truth values for the compound statement are false in all cases (according to the rightmost column of the truth table), $(p \wedge q) \wedge (\neg p)$ is a contradiction.

Note: Problems 22.22–22.24 complete the following truth table for the compound statement $\left[(p \vee \neg r) \vee \neg (p \wedge \neg q)\right]$.

p	q	r						$(p \vee \neg r) \vee \left[\neg (p \wedge \neg q)\right]$

22.22 Identify the missing labels for columns 4–8.

The compound statement $(p \vee \neg r) \vee \left[\neg (p \wedge \neg q)\right]$ is a disjunction of two compound statements, one disjunction and one conjunction. The parenthetical compound statements contain negations of simple statements that should be represented in the truth table: $\neg r$ and $\neg q$. Include columns for the parenthetical statements as well: $p \vee \neg r$ and $p \wedge \neg q$. Notice that the given statement contains the negation of the conjunction $p \wedge \neg q$, so include a column representing $\neg (p \wedge \neg q)$ as well.

An OR joins two compound statements. The left group has another OR connective, and the right group has an AND connective. It's a compound statement made up of compound statements.

p	q	r	$\neg r$	$\neg q$	$p \vee \neg r$	$p \wedge \neg q$	$\neg(p \wedge \neg q)$	$(p \vee \neg r) \vee \neg(p \wedge \neg q)$

Note: Problems 22.22–22.24 complete the truth table in Problem 22.22.

22.23 Complete the first three columns of the truth table, which contain the truth values for p, q, and r.

When constructing a truth table containing two simple statements, Problem 22.11 indicated that the top half of the p column should contain true values and the bottom half should contain false values. The truth table in this problem will contain eight rows of truth values, so the top four rows of p should be true and the bottom four rows should be false.

The groups in the q column should be half as large as the groups in the p column (as noted in Problem 22.12). Instead of having four consecutive true values followed by four consecutive false values, list groups of two. Finally, the r column should contain groups that are again half as large.

A truth table containing two simple statements has $2^2 = 4$ rows. A truth table containing three simple statements has $2^3 = 8$ rows.

So when there are three simple statements (like p, q, and r), the first column is TTTTFFFF, the second is TTFFTTFF, and the third is TFTFTFTF.

p	q	r	$\neg r$	$\neg q$	$p \vee \neg r$	$p \wedge \neg q$	$\neg(p \wedge \neg q)$	$(p \vee \neg r) \vee \neg(p \wedge \neg q)$
T	T	T						
T	T	F						
T	F	T						
T	F	F						
F	T	T						
F	T	F						
F	F	T						
F	F	F						

A tautology is the opposite of a contradiction; it's always true, and the rightmost column of its truth table contains nothing but T's.

Note: Problems 22.22–22.24 complete the truth table in Problem 22.22.

22.24 Demonstrate that $(p \vee \neg r) \vee \neg(p \wedge \neg q)$ is a tautology by completing the remaining columns of the truth table.

The fourth and fifth columns are negations of the third and second columns, respectively.

p	q	r	$\neg r$	$\neg q$	$p \vee \neg r$	$p \wedge \neg q$	$\neg(p \wedge \neg q)$	$(p \vee \neg r) \vee \neg(p \wedge \neg q)$
T	T	T	F	F				
T	T	F	T	F				
T	F	T	F	T				
T	F	F	T	T				
F	T	T	F	F				
F	T	F	T	F				
F	F	T	F	T				
F	F	F	T	T				

The $p \vee \neg r$ column is true when the p column is true, when the $\neg r$ column is true, or when both are true. Because this truth table contains so many rows and columns, shading is used to identify the columns of interest in each step.

p	q	r	$\neg r$	$\neg q$	$p \vee \neg r$	$p \wedge \neg q$	$\neg(p \wedge \neg q)$	$(p \vee \neg r) \vee \neg(p \wedge \neg q)$
T	T	T	F	F	T			
T	T	F	T	F	T			
T	F	T	F	T	T			
T	F	F	T	T	T			
F	T	T	F	F	F			
F	T	F	T	F	T			
F	F	T	F	T	F			
F	F	F	T	T	T			

The $p \wedge \neg q$ column is true only when the p and $\neg q$ columns are both true.

p	q	r	$\neg r$	$\neg q$	$p \vee \neg r$	$p \wedge \neg q$	$\neg(p \wedge \neg q)$	$(p \vee \neg r) \vee \neg(p \wedge \neg q)$
T	T	T	F	F	T	F		
T	T	F	T	F	T	F		
T	F	T	F	T	T	T		
T	F	F	T	T	T	T		
F	T	T	F	F	F	F		
F	T	F	T	F	T	F		
F	F	T	F	T	F	F		
F	F	F	T	T	T	F		

The $\neg(p \land \neg q)$ column is the negation of the column immediately preceding it.

p	q	r	$\neg r$	$\neg q$	$p \lor \neg r$	$p \land \neg q$	$\neg(p \land \neg q)$	$(p \lor \neg r) \lor \neg(p \land \neg q)$
T	T	T	F	F	T	F	T	
T	T	F	T	F	T	F	T	
T	F	T	F	T	T	T	F	
T	F	F	T	T	T	T	F	
F	T	T	F	F	F	F	T	
F	T	F	T	F	T	F	T	
F	F	T	F	T	F	F	T	
F	F	F	T	T	T	F	T	

The final column is true if either (or both) of the values in the sixth and eighth columns are true.

p	q	r	$\neg r$	$\neg q$	$p \lor \neg r$	$p \land \neg q$	$\neg(p \land \neg q)$	$(p \lor \neg r) \lor \neg(p \land \neg q)$
T	T	T	F	F	T	F	T	T
T	T	F	T	F	T	F	T	T
T	F	T	F	T	T	T	F	T
T	F	F	T	T	T	T	F	T
F	T	T	F	F	F	F	T	T
F	T	F	T	F	T	F	T	T
F	F	T	F	T	F	F	T	T
F	F	F	T	T	T	F	T	T

The statement $(p \lor \neg r) \lor \neg(p \land \neg q)$ is a tautology because it is true regardless of whether its component simple statements are true or false.

Conditional Statements

If-then becomes $p \to q$

22.25 Express the following statement symbolically.

If I have enough gas, then I will drive.

Let p represent the simple statement "I have enough gas," and let q represent the simple statement "I will drive." The conditional statement "If I have enough gas, then I will drive" is expressed symbolically as $p \to q$.

p is called the hypothesis or antecedent, and q is called the conclusion or consequent.

22.26 Express the following statement symbolically.

I will pay you back only if John pays me what he owes me.

Let *p* represent the simple statement "I will pay you back," and let *q* represent the simple statement "John pays me what he owes me." The phrase "only if" in a conditional statement precedes the hypothesis, so *q* is the hypothesis and *p* is the conclusion. The symbolic representation of this conditional statement is $q \rightarrow p$.

> The word "if" usually comes right before the hypothesis, even when "if" appears in the middle of the statement.

22.27 Express the following statement symbolically.

Running around the bases leisurely implies that you just hit a home run.

Let *r* represent the simple statement "you are running around the bases leisurely," and let *s* represent the simple statement "you just hit a home run." The word "implies" is similar to a "then" statement because it indicates the conclusion. In other words, "*p* implies *q*" is equivalent to "if *p*, then *q*." The symbolic representation of this statement is $r \rightarrow s$.

22.28 Express the following statement symbolically.

Your exam grade is 100% if and only if you get every question correct.

Let *p* represent the simple statement "your exam grade is 100%," and let *q* represent the simple statement "you got every question correct." The presence of the phrase "if and only if" indicates a biconditional statement, which is represented by the symbol ↔. The symbolic representation of the statement is $p \leftrightarrow q$.

> Biconditional statements work both ways, so p must imply q and q must imply p: $p \leftrightarrow q \equiv (p \rightarrow q) \wedge (q \rightarrow p)$.

22.29 Construct a truth table for the conditional statement $p \rightarrow q$.

The truth table contains three columns, one for each simple statement (*p* and *q*) and one for the conditional statement $p \rightarrow q$.

p	q	$p \rightarrow q$
T	T	
T	F	
F	T	
F	F	

A conditional statement is false only when the hypothesis is true and the conclusion is false.

> In other words, T → F is false.

p	q	$p \rightarrow q$
T	T	T
T	F	F
F	T	T
F	F	T

22.30 Based on the truth table constructed in Problem 22.29, identify the conditions under which the following statement is false. Explain your answer.

If I am sleepy, then I am going to bed.

Let p represent the hypothesis "I am sleepy," and let q represent the conclusion "I am going to bed." According to Problem 22.29, $p \to q$ is false only when p is true and q is false: "if I am sleepy, then I am *not* going to bed." This is the opposite of the original conclusion given the same hypothesis, so $p \to \neg q$ is false.

Notice that a conditional statement is true when its hypothesis is false. Thus, "if I am not sleepy, then I am going to bed" and "if I am not sleepy, I am not going to bed" are considered true. This is because the original statement described a specific condition ("I am sleepy"), so when that condition is not met, you do not have the necessary information to dismiss the conclusion as false.

> Maybe you went to bed because you had a terrible headache, not because you were sleepy. That would make $\neg p \to q$ true.

22.31 Construct a truth table that proves that a statement and its contrapositive are logically equivalent.

> See Problem 5.4.

The contrapositive of the conditional statement $p \to q$ is $(\neg q) \to (\neg p)$. In the following truth table, notice that the shaded columns—representing the statement and its contrapositive—have the same truth values in each row. Thus, $p \to q \equiv (\neg q) \to (\neg p)$.

p	q	$p \to q$	$\neg q$	$\neg p$	$(\neg q) \to (\neg p)$
T	T	T	F	F	T
T	F	F	T	F	F
F	T	T	F	T	T
F	F	T	T	T	T

22.32 Construct a truth table that proves the converse and inverse of a statement are logically equivalent.

> See Problems 5.3 and 5.5.

The converse of $p \to q$ is $q \to p$, and the inverse of $p \to q$ is $(\neg p) \to (\neg q)$. The shaded columns of the following truth table, which represent the converse and inverse of $p \to q$, have the same truth values in each row. Thus, $q \to p \equiv (\neg p) \to (\neg q)$.

> This is false because the hypothesis q is true and the conclusion p is false.

p	q	$q \to p$	$\neg p$	$\neg q$	$(\neg p) \to (\neg q)$
T	T	T	F	F	T
T	F	T	F	T	T
F	T	F	T	F	F
F	F	T	T	T	T

22.33 Construct a truth table for the biconditional statement $p \leftrightarrow q$.

The biconditional statement $p \leftrightarrow q$ is true when the truth values of p and q match. Thus, $p \leftrightarrow q$ is true when p and q are both true or p and q are both false, as illustrated by the following truth table.

p	q	$p \leftrightarrow q$
T	T	T
T	F	F
F	T	F
F	F	T

22.34 Prove $p \rightarrow (\neg q) \equiv \neg(p \wedge q)$.

The shaded columns of the following truth table have the same truth values in corresponding rows, so $p \rightarrow (\neg q) \equiv \neg(p \wedge q)$.

p	q	$\neg q$	$p \rightarrow (\neg q)$	$p \wedge q$	$\neg(p \wedge q)$
T	T	F	F	T	F
T	F	T	T	F	T
F	T	F	T	F	T
F	F	T	T	F	T

22.35 Construct a truth table for the conditional statement $(p \leftrightarrow q) \rightarrow (p \vee q)$.

p	q	$p \leftrightarrow q$	$p \vee q$	$(p \leftrightarrow q) \rightarrow (p \vee q)$
T	T	T	T	T
T	F	F	T	T
F	T	F	T	T
F	F	T	F	F

22.36 Construct a truth table for the conditional statement $\left[(p \rightarrow r) \vee (q \rightarrow r)\right] \rightarrow \left[(p \rightarrow r) \wedge (q \rightarrow r)\right]$.

p	q	r	$p \rightarrow r$	$q \rightarrow r$	$[(p \rightarrow r) \vee (q \rightarrow r)]$	$[(p \rightarrow r) \wedge (q \rightarrow r)]$	$[(p \rightarrow r) \vee (q \rightarrow r)] \rightarrow [(p \rightarrow r) \wedge (q \rightarrow r)]$
T	T	T	T	T	T	T	T
T	T	F	F	F	F	F	T
T	F	T	T	T	T	T	T
T	F	F	F	T	T	F	F
F	T	T	T	T	T	T	T
F	T	F	T	F	T	F	F
F	F	T	T	T	T	T	T
F	F	F	T	T	T	T	T

> Flip back to Problems 22.22–22.24 to review the structure of a truth table that includes three simple statements (p, q, and r).

Arguments
Prove you're right without shouting

22.37 Under what conditions is a mathematical argument valid?

A mathematical argument consists of one or more statements that you must accept as true, called "premises." This concept should not be foreign to a geometry student because each proof in this book begins with given information that is assumed true.

> So one way to prove an argument invalid is to find a way for the conclusion to be false when all the premises are true.

The second component of an argument is the conclusion statement. An argument is valid only if the conclusion is always true when the premises are true. Often, a truth table is the most efficient means by which to determine the validity of an argument.

Note: Problems 22.38–22.39 refer to the following argument.

$$p \wedge q$$

$$\overline{\therefore p \vee q}$$

> This symbol means "therefore." This argument basically says, "Given p AND q is true, you can therefore conclude that p OR q is true."

22.38 Identify the premise and conclusion of the argument.

The premises of an argument—the statements that are assumed true—appear above the horizontal line. This argument has one premise: $p \wedge q$. The conclusion of an argument is the statement that appears below the horizontal line: $p \vee q$.

Note: Problems 22.38–22.39 refer to the argument in Problem 22.38.

22.39 Explain why this is a valid argument.

> An AND statement is true only when the two things it connects are true—in this case, p and q.

The premise of the argument states that $p \wedge q$ is true. For the premise to be true, p and q must both be true. The conclusion statement $p \vee q$ is also true when p and q are both true. Therefore, the argument is valid.

22.40 Explain why the following argument is invalid.

$$p \vee q$$
$$\underline{q}$$
$$\therefore \neg p$$

This argument consists of two premises, the first of which states that either p or q (or both p and q) are true: $p \vee q$. The second premise simply states that q is true. The conclusion is that p must be a false statement.

Temporarily assume the opposite of the conclusion, that p is true. If p is true and q is true, then the first premise, $p \vee q$, is also true. Therefore, it is possible for the conclusion to be false when each of the premises is true, so this argument is invalid.

> This is the second premise. Even though you're temporarily assuming that the conclusion is false, all the premises still must be true.

Note: Problems 22.41–22.42 refer to the following modus ponens argument.

$$p \rightarrow q$$
$$\underline{p}$$
$$\therefore q$$

> Modus ponens is Latin for "method of affirming." This argument is also called the law of detachment.

22.41 Construct a truth table for the argument that contains the following columns: p, q, and $p \rightarrow q$.

Recall that the conditional statement $p \rightarrow q$ is false only when p is true and q is false.

p	q	$p \rightarrow q$
T	T	T
T	F	F
F	T	T
F	F	T

Note: Problems 22.41–22.42 refer to the modus ponens argument in Problem 22.41.

22.42 Determine whether the argument is valid or invalid, and justify your answer using the truth table generated in Problem 22.41.

You must assume that the premises of an argument are true. The premises of this argument are $p \rightarrow q$ and p. Consider the following truth table, generated in Problem 22.41. Of the four possible combinations of truth values listed, the premises are both true only in the first row. Notice that q is also true in that row. Therefore, when $p \rightarrow q$ and p are true, then q must also be true; the argument is valid.

p	q	$p \rightarrow q$
T	T	T
T	F	F
F	T	T
F	F	T

Modus tollens is Latin for "method of denial." This argument is also called the method of indirect reasoning.

Note: Problems 22.43–22.44 refer to the following modus tollens argument.

$$p \rightarrow q$$
$$\frac{\neg q}{\therefore \neg p}$$

22.43 Construct a truth table for the argument.

The columns of the truth table should include the premises and conclusion, as well as elements of those statements (including $\neg p$ and $\neg q$).

p	q	$p \rightarrow q$	$\neg q$	$\neg p$
T	T	T	F	F
T	F	F	T	F
F	T	T	F	T
F	F	T	T	T

Note: Problems 22.43–22.44 refer to the modus tollens argument in Problem 22.43.

22.44 Determine whether the argument is valid or invalid, and justify your answer using the truth table generated in Problem 22.43.

Consider the following truth table, generated in Problem 22.43; the premises $p \rightarrow q$ and $\neg q$ are both true only in the shaded row.

p	q	$p \rightarrow q$	$\neg q$	$\neg p$
T	T	T	F	F
T	F	F	T	F
F	T	T	F	T
F	F	T	T	T

Because the conclusion $\neg p$ is also true in the shaded row, the argument is valid.

22.45 Use a truth table to determine whether the following argument is valid or invalid. Explain your answer.

$$p \rightarrow q$$
$$\frac{p \wedge q}{\therefore p}$$

Consider the following truth table. The premises $p \rightarrow q$ and $p \wedge q$ are both true only in the shaded row. Because the conclusion p is also true in that row, the argument is valid.

p	q	$p \rightarrow q$	$p \wedge q$
T	T	T	T
T	F	F	F
F	T	T	F
F	F	T	F

22.46 Use a truth table to determine whether the following argument is valid or invalid. Explain your answer.

$$p \Leftrightarrow q$$
$$\underline{\neg (q \rightarrow r)}$$
$$\therefore q$$

The shaded row in the following truth table represents the only instance in which premises $p \Leftrightarrow q$ and $\neg (q \rightarrow r)$ are true. Because the conclusion q is also true in that row, the argument is valid.

p	q	r	$p \Leftrightarrow q$	$q \rightarrow r$	$\neg (q \rightarrow r)$
T	T	T	T	T	F
T	T	F	T	F	T
T	F	T	F	T	F
T	F	F	F	T	F
F	T	T	F	T	F
F	T	F	F	F	T
F	F	T	T	T	F
F	F	F	T	T	F

22.47 Use a truth table to determine whether the following argument is valid or invalid. Explain your answer.

$$p \to r$$
$$p \lor q$$
$$\underline{\neg p}$$
$$\therefore r$$

The shaded rows in the following truth table represent the instances in which premises $p \to r$, $p \lor q$, and $\neg p$ are true. One of the shaded rows presents a true conclusion given true premises, but the other presents a false conclusion, so the argument is invalid.

This time, two rows represent a set of true premises. Take a look at row 6—all the premises are true, but the conclusion r is false. That makes this an invalid argument even though the conclusion is true in row 5.

p	q	r	$p \to r$	$p \lor q$	$\neg p$
T	T	T	T	T	F
T	T	F	F	T	F
T	F	T	T	T	F
T	F	F	F	T	F
F	T	T	T	T	T
F	T	F	T	T	T
F	F	T	T	F	T
F	F	F	T	F	T

Appendix A
Properties of Equality

Associative Property of Addition: $(a + b) + c = a + (b + c)$

Associative Property of Multiplication: $(ab)c = a(bc)$

Commutative Property of Addition: $a + b = b + a$

Commutative Property of Multiplication: $ab = ba$

Additive Identity: $a + 0 = 0 + a = a$

Multiplicative Identity: $a(1) = 1(a) = a$

Additive Inverse: $a + (-a) = -a + a = 0$

Multiplicative Inverse: $a \cdot \dfrac{1}{a} = \dfrac{1}{a} \cdot a = 1$

Addition Property: If $x = y$ and $w = z$, then $x + w = y + z$.

Subtraction Property: If $x = y$ and $w = z$, then $x - w = y - z$.

Multiplication Property: If $x = y$ and $w = z$, then $xw = yz$.

Division Property: If $x = y$ and $w \neq 0$, then $\dfrac{x}{w} = \dfrac{y}{w}$.

Reflexive Property: $x = x$

Distributive Property: $a(b + c) = ab + ac$

Symmetric Property: If $a = b$, then $b = a$.

Transitive Property: If $a = b$ and $b = c$, then $a = c$.

Substitution Property: If $y = 10$ and $x = y + z$, then $x = 10 + z$.

Appendix B
Properties of Inequality

Transitive: If $a > b$ and $b > c$, then $a > c$.

Substitution: If $a < b$ and $b = c$, then $a < c$.

Addition/Subtraction

* If unequals are added to equals, the sum is unequal in the same order.

$$\text{If } a = b \text{ and } x < y, \text{ then } a + x < b + y.$$

* If unequals are added to unequals in the same order, the sums are unequal in the same order.

$$\text{If } a < b \text{ and } x < y, \text{ then } a + x < b + y.$$

* If equals are subtracted from unequals, the difference is unequal in the same order.

$$\text{If } a = b \text{ and } x < y, \text{ then } x - a < y - b.$$

* If unequals are subtracted from equals, the difference is unequal in the opposite order.

$$\text{If } a = b \text{ and } x < y, \text{ then } a - x > b - y.$$

Multiplication/Division

* If unequals are multiplied or divided by a negative number, the product or quotient is unequal in the opposite order.

$$\text{If } a < b \text{ and } c < 0, \text{ then } ca > cd \text{ and } \frac{a}{c} > \frac{b}{c}.$$

Appendix C
Important Geometry Theorems

Chapter 5

* Vertical angles are congruent.

* Complements/supplements of congruent angles are congruent.

Chapter 6

* When parallel lines are intersected by a transversal, corresponding angles are congruent, alternate interior angles are congruent, and same-side interior angles are supplementary.

* Two lines perpendicular to the same line are parallel.

* Points on the perpendicular bisector of a segment are equidistant from the endpoints of that segment.

* Points on an angle bisector are equidistant from the sides of the angle that is bisected.

Chapter 7

* The segment connecting a vertex to a centroid is twice as long as the segment connecting the centroid to the midpoint of the opposite side.

* The sum of the measures of the angles of a triangle is 180°.

* Acute angles of a right triangle are complementary.

* The measure of the exterior angle of a triangle is equal to the sum of the measures of the remote interior angles.

* The sum of the interior angle measures of an n-sided regular polygon is $180°(n - 2)$.

* The sum of the exterior angle measures of a regular polygon is 360°.

Chapter 8

* Major triangle congruence justifications: SSS postulate, SAS postulate, ASA postulate, AAS theorem, and HL theorem.
* Corresponding parts of congruent triangles are congruent (CPCTC).
* Isosceles triangle theorem: Base angles of an isosceles triangle are congruent.
* The angle bisector of the vertex angle of an isosceles triangle is the perpendicular bisector of the base.
* Equilateral triangles are equiangular, and vice versa.

Chapter 9

* Ways to prove a quadrilateral is a parallelogram: both pairs of opposite sides are parallel, both pairs of opposite sides are congruent, both pairs of opposite angles are congruent, the diagonals bisect each other, one pair of opposite sides is both congruent and parallel.
* Diagonals of a rectangle are congruent.
* Diagonals of a rhombus are perpendicular.
* Triangle midpoint theorem: A line that bisects one side of a triangle and is parallel to another side bisects the remaining side of the triangle.
* The midpoint of a right triangle is equidistant from the vertices.
* Both pairs of base angles of an isosceles trapezoid are congruent.
* The diagonals of an isosceles trapezoid are congruent.
* The median of a trapezoid is parallel to the bases and is half as long as the sum of the lengths of the bases.
* The diagonals of a kite are perpendicular, and exactly one of the diagonals of a kite bisects the other.

Chapter 10

* Triangle inequality theorem: The sum of the lengths of two sides of a triangle must be greater than the length of the third side.
* The larger a side of a triangle, the larger the opposite angle (and vice versa).
* Two-triangle inequality theorems: SAS and SSS (see Problems 10.40 and 10.41).

Chapter 11

* If two similar figures have scale factor x:y, their perimeters are in the ratio x:y and their areas are in the ratio $x^2:y^2$.
* Triangle similarity theorems: AA, SSS, SAS.
* Corresponding angles of similar triangles are congruent, and corresponding sides are in proportion (CSSTP).
* Triangle proportionality theorem: If a line is parallel to one side of a triangle and it intersects the other two sides, it splits those sides proportionally.
* Three or more parallel lines divide two transversals proportionally.
* Triangle angle bisector theorem: An angle bisector of a triangle divides the opposite side into segments that are proportional to the adjacent sides.

Chapter 12

* The altitude of a right triangle drawn to the hypotenuse divides the right triangle into right triangles that are similar to each other and the original triangle.
* The altitude of a right triangle drawn to the hypotenuse has a length that is the geometric mean of the segments of the hypotenuse.
* The altitude of a right triangle drawn to the hypotenuse divides the hypotenuse such that each leg of the right triangle is the geometric

mean between the hypotenuse and the hypotenuse segment adjacent to the leg.

* Pythagorean Theorem: If a and b are the legs of a right triangle and c is the hypotenuse, then $a^2 + b^2 = c^2$.
* The hypotenuse of a 45°-45°-90° triangle is $\sqrt{2}$ times as long as the legs.
* The shortest side of a 30°-60°-90° triangle is half as long as the hypotenuse, and the side opposite the 60° angle is $\sqrt{3}$ times as long as the shortest side.

Chapter 13
* If a line is perpendicular to the radius of a circle at a point on the circle, then the line is tangent to the circle at that point.
* Two segments tangent to a circle that extend from a common exterior point are congruent.
* The measure of an inscribed angle is half the measure of the arc it intercepts.
* An inscribed angle that intercepts a semicircle is a right angle.

Chapter 14
* Congruent chords intercept congruent arcs.
* A diameter perpendicular to a chord bisects the arc intercepted by that chord.
* Chords equidistant from the center of a circle are congruent.
* The measure of an angle formed by two chords intersecting inside a circle is half the sum of the measures of the intercepted arcs.
* If two chords intersect inside a circle, the product of the lengths of one chord is equal to the product of the lengths of the other.
* The measure of an angle formed by a chord and a tangent is half the measure of the arc it intercepts.

* An angle formed by tangents or secants that extend from a common exterior point has a measure equal to the difference of the measures of the intercepted arcs.
* If two secants to a circle extend from a common exterior point, the product of the lengths of each secant and its exterior segment are equal.
* If a secant and a tangent extend from a common exterior point, the product of the lengths of the secant and its exterior segment is equal to the square of the length of the tangent.

Chapter 18

* If similar solids have scale factor x:y, then the surface areas of the solids are in the ratio $x^2:y^2$ and the volumes of the solids are in the ratio $x^3:y^3$.

Chapter 21

* If $R_k: P \rightarrow P'$ is a reflection in line k, then k is the perpendicular bisector of every segment $\overline{PP'}$.
* If $R_{A,x°}: P \rightarrow P'$ is a rotation about point A through x°, then $m\angle PAP' = x°$.

Appendix D
Important Geometry Formulas

Area/Perimeter of 2D Figures

* **Rectangle:** area = (length)(width)

  perimeter = $2(l + w)$

* **Square:** area = $(side)^2$

  perimeter = $4(side)$

* **Triangle:** area = $\frac{1}{2}(base)(height)$

* **Heron's formula:**

  area = $\sqrt{s(s-a)(s-b)(s-c)}$ when $s = \frac{1}{2}(a + b + c)$

* **Parallelogram:** area = (base)(height)

* **Trapezoid:** area = $\frac{1}{2}h(b_1 + b_2)$

* **Regular polygon:** $\frac{1}{2}(apothem)(perimeter)$

* **Circle:** area = πr^2

  circumference = $2\pi r$

* **Arc length** = $\dfrac{arc\ measure}{360}(2\pi r)$

* **Area of a sector** = $\dfrac{arc\ measure}{360}(\pi r^2)$

Volume (V)/Lateral Surface Area (LSA) of Solids

* Prisms: Volume = (base area)(height)

LSA = (perimeter of base)(height)

* Pyramid: Volume = $\frac{1}{3}$(base area)(height)

LSA = $\frac{1}{2}$(perimeter of base)(slant height)

* Cylinder: Volume = $\pi r^2 h$

LSA = $2\pi r h$

* Cone: Volume = $\frac{1}{3}\pi r^2 h$

LSA = $\pi r \left(\sqrt{r^2 + h^2}\right)$

* Sphere: Volume = $\frac{4}{3}\pi r^3$

Surface area = $4\pi r^2$

Coordinate Geometry

* Pythagorean Theorem: $a^2 + b^2 = c^2$

* Interior angle of an n-sided regular polygon: $\frac{180(n-2)}{n}$

* Distance formula: $\sqrt{(x_2 - x_1)^2 + (y_2 - y_1)^2}$

* Midpoint formula: $\left(\frac{x_1 + x_2}{2}, \frac{y_1 + y_2}{2}\right)$

* Slope: $\frac{y_2 - y_1}{x_2 - x_1}$

* Slope-intercept form of a line: $y = mx + b$

* Point-slope form of a line: $y - y_1 = m(x - x_1)$

* Standard form of a line: $Ax + By = C$

* Slope of a line in standard form: $-\dfrac{A}{B}$

* Equation of a circle in standard form: $(x - h)^2 + (y - k)^2 = r^2$

* Magnitude of vector $v = \langle a, b \rangle$: $\|v\| = \sqrt{a^2 + b^2}$

---

de Morgan's Laws

* $\neg(p \wedge q) \equiv (\neg p) \vee (\neg q)$

* $\neg(p \vee q) \equiv (\neg p) \wedge (\neg q)$

Index

ALPHABETICAL LIST OF CONCEPTS WITH PROBLEM NUMBERS

This comprehensive index organizes the concepts and skills discussed within the book alphabetically. Each entry is accompanied by one or more problem numbers, in which the topics are most prominently featured.

All these numbers refer to problems, not pages, in the book. For example, 8.2 is the second problem in Chapter 8.

Numbers & Symbols

30°-60°-90° triangle: 12.29–12.33, 12.35–12.36, 12.41, 17.17, 17.26–17.27

45°-45°-90° triangle: 12.25–12.28, 12.34, 12.36, 13.15

A

AA similarity postulate: 8.23, 11.31–11.32, 11.34, 11.36, 11.39, 11.42, 13.30, 14.28

AAS theorem: 8.11, 8.25, 8.45

absolute values: 3.16–3.17

acute angle: 4.26–4.27, 4.42, 12.20

addition property of equality: 5.24, 5.36, 5.44

addition property of inequality: 10.4–10.5, 10.18

adjacent angles: 4.28, 4.30, 4.32–4.33, 4.37–4.38, 4.40, 4.44–4.45, 4.47

algebraic properties: see *properties of equality* and *properties of inequality*

alternate exterior angles: 6.7–6.8, 6.14, 6.22

alternate interior angles: 6.3–6.4, 6.13, 6.16, 6.28–6.29, 6.31, 6.33, 7.38, 8.10, 8.20, 8.40, 8.45, 11.36, 14.33, 15.40–15.41

altitude: 7.17–7.18, 11.39, 12.4–12.12, 15.33, 16.13–16.16, 16.46, 17.9, 18.2, 18.4, 18.11, 19.38, 20.18

and: see *conjunction*

angle: 4.24–4.47

angle addition postulate: 4.28, 4.30–4.31, 4.35, 4.37–4.40, 4.47, 5.44

angle bisector: 4.31, 4.34–4.36, 6.41, 7.19–7.21, 8.18, 8.28–8.30, 8.44, 15.16–15.20, 15.24, 15.32, 16.5–16.8, 16.18

antecedent: see *hypothesis*

apothem: 17.25–17.27, 17.29, 17.31, 17.45–17.46, 18.17

arc: 13.32–13.35, 13.43–13.44, 13.46, 14.1–14.11, 14.14, 14.16–14.25, 14.32–14.39

arc addition postulate: 13.44, 14.9, 14.16, 14.24, 14.33, 14.39

arc bisector: 14.4, 14.7–14.8

arc length: 17.37, 17.40

area

    (of a) circle: 17.32, 17.34–17.35, 17.38, 18.21, 18.23, 18.27

    (of a) parallelogram: 17.14, 17.17

    (of a) rectangle: 17.1, 17.3–17.5, 18.5, 18.9–18.10

    (of a) regular polygon: 17.29, 17.31, 18.17

    (of a) rhombus: 17.18

    (of a) sector: 17.38–17.39, 17.41

D

E

F

S

same-side exterior angles: 6.9–6.10, 6.22, 6.33

same-side interior angles: 6.5–6.6, 6.15, 6.17, 6.29, 10.21, 17.24

SAS inequality theorem: 10.40, 10.42, 10.45

SAS postulate: 8.12, 8.15–8.16, 8.22, 8.36, 8.38, 8.42, 8.44

SAS similarity theorem: 11.31, 11.33, 11.37–11.38

SSS similarity theorem: 21.37

scalar: 20.44–20.46

scale factor: 11.22–11.29, 11.35, 17.42–17.46, 18.38–18.42, 21.31–21.38

scalene: 7.6, 10.22

secant: 13.4, 13.6, 14.35–14.37, 14.40–14.46

sector: 17.38–17.39, 17.41

segment addition postulate: 4.18, 4.23, 5.28, 8.13, 10.1

segment proportionality theorem: 11.40–11.48

segments: 4.8–4.10, 4.12–4.16, 4.19–4.21

semicircle: 13.32–13.33, 13.43–13.44, 13.46, 14.16, 14.39, 15.5, 16.27, 16.44–16.46

similar figures: 8.23, 11.21–11.39, 11.42, 12.4–12.12, 17.42–17.46, 21.37–21.38

similar solids: 18.37–18.42

simple statement: 22.1, 22.25–22.28

sine: 12.37, 12.40–12.41, 13.29, 14.10, 17.24

skew: 4.7

slant height: 18.13, 18.15, 18.18, 18.24

slope: 19.27–19.38, 19.40–19.41, 19.43, 20.1–20.3, 20.8–20.12, 20.14–20.23, 21.15, 21.19

slope-intercept form of a line: 20.1, 20.3–20.5, 20.8, 20.15, 20.17, 20.19–20.20, 21.19

solving equations

for a specific variable: 2.17–2.23

linear: 2.1–2.12, 5.35, 5.37

quadratic: 3.38–3.39, 3.41–3.47

rational: 2.13–2.16

sphere

definition of: 13.1

surface area of: 18.28, 18.31–18.33

volume of: 18.28, 18.30, 18.32, 18.36

square: 9.23, 11.30, 17.3, 17.6, 19.16

square root: 3.11–3.12

SSS inequality theorem: 10.41, 10.43–10.44

SSS postulate: 8.4–8.5, 8.41, 15.10

SSS similarity theorem: 11.31, 11.35

standard form

(of a) circle: 20.25–20.29, 20.32–20.33

(of a) line: 20.4, 20.6–20.7, 20.9–20.12, 20.16, 20.18, 20.21–20.23, 20.34

standard position (of a vector): 20.37

straight angle: 4.26, 4.38, 4.45, 14.25

straightedge: 15.1

substitution method (for solving systems of equations): 2.26, 2.28, 2.30, 2.37–2.39

substitution property of equality: 5.21, 5.30, 5.40–5.41, 5.44

substitution property of inequality: 10.3, 10.12

subtraction property of equality: 5.29, 5.37–5.39, 5.41, 8.13

subtraction property of inequality: 10.6–10.7, 10.13–10.15, 10.18–10.19

sufficient: 5.6–5.8, 5.11

sum of perfect cubes: 3.31

supplementary angles: 4.38, 4.40, 4.43, 4.45, 4.47, 5.38–5.39, 5.42, 6.21, 6.27, 6.33, 8.14, 9.38, 10.21, 11.7

surface area

(of a) cone: 18.24, 18.27, 18.39

(of a) cylinder: 18.19, 18.22–18.23, 18.35

(of a) prism: 18.5–18.7, 18.10, 18.19, 18.42

(of a) pyramid: 18.14–18.15, 18.18, 18.24

(of) similar solids: 18.39, 18.42

(of a) sphere: 18.29, 18.31–18.33

symmetric property: 5.16

symmetry

line: 21.39–21.40

point: 21.42

rotational: 21.41–21.43

synthetic division: 1.42

system of equations: 2.24–2.41, 6.17, 6.29–6.30, 20.13, 20.24, 21.15

(of a) prism: 18.8–18.9, 18.19, 18.41
(of a) pyramid: 18.16–18.17, 18.24
(of) similar solids: 18.38, 18.40, 18.41
(of a) sphere: 18.28, 18.30, 18.32, 18.36, 18.38

W–X–Y–Z

width: 17.1–17.2

x-intercept: 20.7

y-intercept: 20.1, 20.3, 20.7

zero-product property: 3.37–3.38, 3.40, 3.48